Biomphalaria Snails and Larval Trematodes

Rafael Toledo • Bernard Fried
Editors

Biomphalaria Snails and Larval Trematodes

Springer

Editors
Rafael Toledo
Departamento de Biología Celular y
Parasitología
Facultad de Farmacia
Universidad de Valencia
Av. Vicent Andrés Estellés s/n
46100 Burjassot-Valencia
Spain
rafael.toledo@uv.es

Bernard Fried
Department of Biology
Lafayette College
Easton, PA 18042, USA
friedb@lafayette.edu

ISBN 978-1-4419-7027-5 e-ISBN 978-1-4419-7028-2
DOI 10.1007/978-1-4419-7028-2
Springer New York Dordrecht Heidelberg London

Library of Congress Control Number: 2010936302

Printed on acid-free paper

Springer is part of Springer Science+Business Media (www.springer.com)

Preface

The history of research on the relationships between *Biomphalaria* spp. and larval trematodes, particularly *Schistosoma mansoni*, is long, active, and ongoing today. Snails of the genus *Biomphalaria* are of significant medical importance with many species living in freshwater habitats associated with human settlements; many of these snail species are obligate intermediate hosts of the human blood fluke *S. mansoni*, the causative agent of hepatosplenic schistosomiasis. This parasitic disease continues to disrupt the lives of about 200 million people in over 70 countries and prevents these individuals from otherwise reasonable expectations of healthy and productive lives. Furthermore, it is estimated that within the developing world, especially in sub-Saharan Africa, more than 700 million people are at risk of becoming infected, despite the efforts to control transmission in human and snail populations by mass chemotherapy and the use of molluscicides. Several factors, such as the absence of a schistosome vaccine, the recent appearance of resistance to antischistosome drugs, and human activities that expand snail habitats, have increased the need for a better understanding of schistosome-snail interactions. In recent years, the application of new technologies has contributed to the accumulation of considerable new information on this topic that may be of great use to biomedical scientists.

In addition to the impact of *Biomphalaria* spp. on public health, these snails are also interesting models for the study of other topics such as population biology, including genetics and demography, proteomics, invertebrate immunobiology, mating systems, and biogeography, among others. *Biomphalaria* spp. snails have been extensively used as experimental biological models contributing significantly to new developments in many areas studied by biomedical scientists. Extensive coverage of these topics is included in this book, also considering trematode species other than schistosomes.

The aim of the present book is to provide an overview of the recent advances in the *Biomphalaria* spp.-larval trematode interactions, especially in *Biomphalaria*-schistosome systems. Emphasis is placed on gaps in our knowledge that must be filled to gain a better understanding of the relationships in these host–parasite systems. This may be critical for a deeper knowledge of the transmission of schistosomiasis and other snail-borne parasitic diseases.

The list of chapters includes basic information as well as new topics. All chapters are covered from a modern point of view, considering the new information that has accumulated by the application of novel techniques and analyzed in a contemporary context. In summary, the main goal of this book is to present *Biomphalaria* spp. and their interactions with larval trematodes in the context of modern biology and to provide an update of the current status of knowledge on these host–parasite systems.

Rafael Toledo
Bernard Fried

Contents

Contributors

Joanna M. Bridger
Laboratory of Nuclear and Genomic Health, Cell and Developmental Biology, Centre for Cell and Chromosome Biology, Brunel University, Kingston Lane, Uxbridge, Middlesex UB8 3PH, UK

Christine Coustau
Institut Sophia Agrobiotech, 06903, Sophia Antipolis, France

Patrice David
Centre d'Ecologie Fonctionnelle et Evolutive, UMR 5175 CNRS 1919 route de Mende, 34293 Montpellier cedex 5, France

José-Guillermo Esteban
Departamento de Biología Celular y Parasitología, Facultad de Farmacia, Universidad de Valencia, Av. Vicent Andrés Estellés s/n, 46100 Burjassot-Valencia, Spain

L. K. Eveland
Department of Biological Sciences, California State University, Long Beach, CA 90840, USA

Bernard Fried
Department of Biology, Lafayette College, Easton, PA18042, USA

M. A. Haseeb
Departments of Cell Biology and Pathology, Downstate Medical Center, State University of New York, Brooklyn, NY 11203, USA

Judith Humphries
Biology Department, Lawrence University, 711 East Boldt Way, Appleton, WI 54911, USA

Wannaporn Ittiprasert
Biomedical Research Institute, Rockville, MD 20852, USA; Institute of Molecular Biosciences

Philippe Jarne
Centre d'Ecologie Fonctionnelle et Evolutive, UMR 5175 CNRS 1919 route de Mende, 34293 Montpellier cedex 5, France

Matty Knight
Biomedical Research Institute, Rockville, MD 20852, USA

Fred Lewis
Biomedical Research Institute, Rockville, MD 20852, USA

Carla Muñoz-Antoli
Departamento de Biología Celular y Parasitología, Facultad de Farmacia, Universidad de Valencia, Av. Vicent Andrés Estellés s/n, 46100 Burjassot-Valencia, Spain

Jocelyn Myers
Biomedical Research Institute, Rockville, MD 20852, USA

Edwin C. Odoemelam
Laboratory of Nuclear and Genomic Health, Cell and Developmental Biology, Centre for Cell and Chromosome Biology, Brunel University, Kingston Lane, Uxbridge, Middlesex UB8 3PH, UK

Jean-Pierre Pointier
USR 3278 CNRS-EPHE CRIOBE, Université de Perpignan, 52 Av. Paul Alduy, 66860 Perpignan, Cedex, France

Nithya Raghavan
Biomedical Research Institute, Rockville, MD 20852, USA

David Rollinson
Wolfson Wellcome Biomedical Laboratories, Department of Zoology, Natural History Museum, London, SW7 5BD, UK

Joseph Sherma
Department of Chemistry, 326 Hugel Science Center, Lafayette College, Easton, PA 18042, USA

Rafael Toledo
Departamento de Biología Celular y Parasitología, Facultad de Farmacia, Universidad de Valencia, Av. Vicent Andrés Estellés s/n, 46100 Burjassot-Valencia, Spain

Maria Trelis
Departamento de Biología Celular y Parasitología, Facultad de Farmacia, Universidad de Valencia, Av. Vicent Andrés Estellés s/n, 46100 Burjassot-Valencia, Spain

Timothy P. Yoshino
Department of Pathobiological Sciences, University of Wisconsin, Madison, WI 53706, USA

Chapter 1
Biosystematics of *Biomphalaria* spp. with an Emphasis on *Biomphalaria glabrata*

Philippe Jarne, Jena-Pierre Pointier, and Patrice David

Abstract We review the characters and approaches that have been used to build the systematics of the genus *Biomphalaria* with special emphasis on *Biomphalaria glabrata*. Shell characters can be used to distinguish groups of species, but are of limited use for separating closely allied species because of wide ecophenotypic variation. The anatomy of reproductive tracts (e.g., shape and size of penial complex), reviewed here for most extant species, is more informative, but still not fully diagnostic. Recent molecular phylogenies clarified the relationships among most extant species. African species were shown to be of recent origin, and closely allied to *B. glabrata* and several taxa were synonymized (e.g., several African species) and/or grouped into species complexes (e.g., the *Biomphalaria straminea* group). Reproductive isolation has unfortunately been little used for clarifying species boundaries, as well as to detect potential areas of hybridization, in *Biomphalaria*. That some species of *Biomphalaria* mainly reproduce through self-fertilization certainly sets a practical limit to such investigation. Molecular studies have also indicated that *B. glabrata* is structured into several, rather deeply separated, main clades. As these clades are partially reproductively isolated from one another, *B. glabrata* might well be a species complex. *Biomphalaria* is the main intermediate host of the agent of intestinal schistosomiasis *Schistosoma mansoni*. Contrary to previous claims, snail susceptibility to this parasite cannot be used as a tool for building *Biomphalaria* systematics. We conclude that about 26 species of *Biomphalaria* can be recognized today, four of which are found in Africa only. We also plead for further work based on an integrated approach including morphology, molecular markers and experimental analysis of reproductive isolation, as well as ecology and biogeography.

J.-P. Pointier (✉)
USR 3278 CNRS-EPHE CRIOBE, Université de Perpignan,
52 Av. Paul Alduy, 66860, Perpignan Cedex, France
e-mail: pointier@univ-perp.fr

R. Toledo and B. Fried (eds.), Biomphalaria *Snails and Larval Trematodes*,
DOI 10.1007/978-1-4419-7028-2_1, © Springer Science+Business Media, LLC 2011

1.1 Introduction

By mid-2003, about 780 million people were at risk of being infected with schistosomiasis at a worldwide scale, and about 210 million people were infected (Bruun and Aagaard-Hansen 2008). Schistosomiasis also widely affects cattle in tropical countries, especially in Africa (Basch 1991). Although this tropical parasitosis has been largely neglected by all policy actors as a "poverty disease" (Watts 2008), these values have been, and still are, strong incentives to conduct research on various aspects of the epidemiology and ecology of schistosomiasis (Rollinson and Simpson 1987; Rollinson 2001). This certainly puts the focus on planorbids, a family of Pulmonate snails within the Basommatophorans, since several genera, including *Biomphalaria*, are involved in the transmission of schistosomiasis. *Biomphalaria* species have been the focus of hundreds of studies, mainly focusing on its relationship with the trematode *Schistosoma mansoni*, the agent of intestinal schistosomiasis in several African and American countries of the tropical area (Malek 1985; Brown 1994). However, less effort has been devoted to the systematics of planorbids in general, and particularly of *Biomphalaria* (see Mandahl-Barth 1957; Brown 1994 for an overview of African species; Pan American Health Organization 1968 for American species), although it is assumed to be a key aspect for understanding snail–parasite relationships, as well as for building control programs of schistosomiasis.

In addition, beyond medical and veterinary perspectives, *Biomphalaria* snails can also be considered as interesting biological models for studies on population biology, including genetics and demography, mating systems, and biogeography (Brown 1994; Dillon 2000). These snails indeed inhabit often-ephemeral habitats with fast turnover and potentially metapopulation dynamics (see Jarne and Delay 1991; Jarne and Théron 2001). With regard to mating systems, *Biomphalaria* species are self-fertile hermaphrodites that constitute appropriate models to address the evolution of the selfing rate and sexual selection (see reviews in Jordaens et al. 2007; Jarne et al. 2010). The mating system is also an issue when it comes to systematics, because self-fertilization may interfere when estimating the magnitude of reproductive isolation within and among species (Paraense 1959; Coyne and Orr 2004). *Biomphalaria* species can also be used as model systems in studies of biogeography, as they display a wide range of distribution areas, from very restricted in the Neotropical *Biomphalaria subprona* and *Biomphalaria nicaraguana* (Paraense 1996, 2003) to extremely large in the Neotropical *B. glabrata* (Paraense 2001) or the African *Biomphalaria pfeifferi* (Brown 1994).

Our aim here is to provide an overview on the systematics of *Biomphalaria* species in general, as well as to detail some aspects of the phylogeography of *Biomphalaria glabrata*, the main vector of schistosomes in the Neotropics. The Planorbidae belong to the Basommatophoran (freshwater Pulmonates), a group of about 200–300 species (Jarne et al. 2010) of which about 10% belongs to the *Biomphalaria* genus. A total of 19 extant species have classically been recognized in the Neotropics (Pan American Health Organization 1968) and 11 in Africa (Mandahl-Barth 1957). For a long time, *Biomphalaria* systematics has essentially been drawn from the analysis of morphologic traits including shell form and the anatomy of soft parts (reproductive tracts) (Mandahl-Barth 1957; Pan American Health Organization 1968), as in most molluscan species. More recently molecular markers, including allozymes and DNA sequences, provided new insights into the origin and evolution of *Biomphalaria*, particularly of *B. glabrata* (Woodruff and Mulvey 1997; DeJong et al. 2001). Here, we will review results derived from both morphologic and molecular markers in order to provide a general view on the systematics of the *Biomphalaria* genus, as well as on the phylogeography of *B. glabrata*. We will also devote some space to studies on mating systems, and their implications on systematics, and evaluate whether the host–parasite relationship between *Biomphalaria* and *S. mansoni* is of any use here.

1.2 The Shell of *Biomphalaria*, a Tricky Tool for Identification

The first binominal description of molluscs dates back to Linnaeus (1758)'s *Systema Naturae*, which included the description of 674 shells. Conchology, i.e., the study of shells, subsequently became the standard method for describing new species, especially during the nineteenth century with the blossoming of monographs and journals on conchology (e.g., *Conchologia Systematica*), and also well into the twentieth century. Molluscan shells indeed often display a stunning variety of forms and ornamentations (e.g., background colors, colored patterns, and sculptures), especially in marine species. Moreover, considerable variation arises in relation to developmental and environmental conditions, not necessarily to fixed differences among species. This becomes particularly critical in species with extremely wide distribution and/or occupying different types of habitats. Interpreting intraspecific variation as among-species diversity has resulted in taxonomic oversplitting and an avalanche of names for some species. At the same time, some species limits have probably been overlooked because of the presence of seemingly identical morphs in different species. Although shell characteristics are less diversified in freshwater molluscs (especially basommatophoran Pulmonates), this group displays significant ecophenotypic variation in shell form. As a consequence, species descriptions based on shell morphology, especially those published during the nineteenth century and well into the twentieth century, should be considered with caution.

Shells in basommatophorans have a general shape going from flat (discoid) to turriculate, and coiling may be sinistral (physids and planorbids, including *Biomphalaria*) or dextral (lymnaeids). In *Biomphalaria* species, the shell is discoid, sinistral with a rounded to oval aperture (i.e., opening profile; see Fig. 1.1). As it does not display any peculiar ornamentation, the taxonomy has been based on several quantitative parameters including (Fig. 1.1) (a) shell height and diameter taken at their largest size (Mandahl-Barth 1957); (b) whorl number; and (c) shape of whorls and aperture, which goes from rounded to angular. These characteristics are useful for gross determination of species. Adult size, for example, goes from about 10 mm in most species to 20 mm (*Biomphalaria tenagophila*, *Biomphalaria peregrina*, *Biomphalaria alexandrina*, and *Biomphalaria camerunensis*) and even to 30–40 mm (*B. glabrata*). The use of whorl number is more tricky because

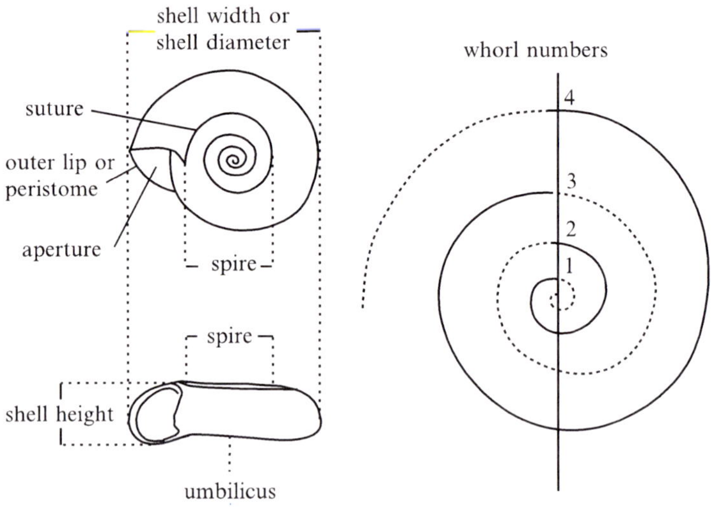

Fig. 1.1 Schematic representation of *Biomphalaria* shell with the main characteristics or parameters used to characterize variation within and among species. The figure on the right indicates how whorl numbers can be counted. The dorsal side of the living animal is oriented upward (*bottom left figure*) or toward the reader (*top left and right figure*)

precisely characterizing the actual original point of shell coiling is uneasy. Variation among species exists though as is obvious from the comparison of *Biomphalaria prona* and *B. glabrata* (Figs. 1.2a–c vs. 1.2d and e). In fact, most of the differences in whorl number between these groups could be more adequately described in terms of expansion rate (slowly and rapidly expanding spirals instead of many and few whorls, respectively, see Fig. 1.2).

Most *Biomphalaria* species were discovered during the nineteenth century and their description is often based on a single shell. The actual status of several taxa is questionable for reasons already mentioned above, including ecophenotypic variation. For example, the lacustrine forms of several African and American species generally have a smaller shell and the aperture widens rapidly during growth (whorl increment), producing a smaller height over diameter ratio than that in non-lacustrine forms. This can be visualized in Fig. 1.2 when comparing lacustrine (Fig. 1.2a) and non-lacustrine (Fig. 1.2b and c) forms of *B. prona* from Venezuela (Paraense et al. 1992). Another example may be given with *Biomphalaria pallida* from Cuba in which huge shell diversity may be observed within and among local populations at the scale of the whole island (Yong et al. 2001; Fig. 1.3). This variation cannot be attributed to habitat type (e.g., rivers vs. ponds). It is also noticeable that this is the only example of an analysis of shell variation at such a geographic scale in any *Biomphalaria* species. A further potential pitfall with shell morphology is related to developmental stage. It is not always obvious to distinguish species based on juvenile shells, as it is possible to confound juveniles of some species (e.g., *B. glabrata*) with adults of others (e.g., *Biomphalaria straminea* complex). However, juvenile individuals display distinctive characteristics such as much slender translucid shells, an aspect that cannot be missed by professional malacologists. As a result of these

Fig. 1.2 Shells of *Biomphalaria*. (**a**) *Biomphalaria prona*, Lake Valencia, Venezuela (6.3 mm). (**b**) *B. prona*, Maracay, Venezuela (8.4 mm); (**c**) *B. prona*, Acarigua, Venezuela (8.9 mm); (**d**) *Biomphalaria glabrata*, Lake Zuata, Venezuela (35 mm); (**e**) *B. glabrata*, Caripe, Venezuela (15 mm); (**f**) *B. glabrata*, Guadeloupe (brown form, 26 mm); (**g**) *B. glabrata*, Guadeloupe (white form, 27 mm); (**h**) *Helisoma duryi*, Venezuela (26 mm)

various sources of intraspecific variation, the literature on planorbid snails has been overloaded with a large number of names, the majority of which is not valid. Harry (1962) made an important work of collapsing hundreds of species names to a few tens (see also Paraense 1961a, b, 1963; Pan American Health Organization Guide 1968).

Biomphalaria glabrata is the largest *Biomphalaria* species with a shell that may reach 40 mm in diameter, although adult shells generally are in the range of 15–30 mm. A detailed description of its shell is provided in the Pan American Health Organization Guide (1968), and is reproduced in Box 1. It is based on the parameters mentioned above, as well as on some other characters (suture and occurrence of apertural lamellae). This description makes it clear that variation can be found in all these parameters and characters. Variation in whorl shape and aperture, which goes from rounded to ovate or angular, is obvious in Fig. 1.2d–g. Shell color also displays some variation, going from almost white (Fig. 1.2g) to brown (Fig. 1.2f). Variability comes as no surprise as the current distribution area of

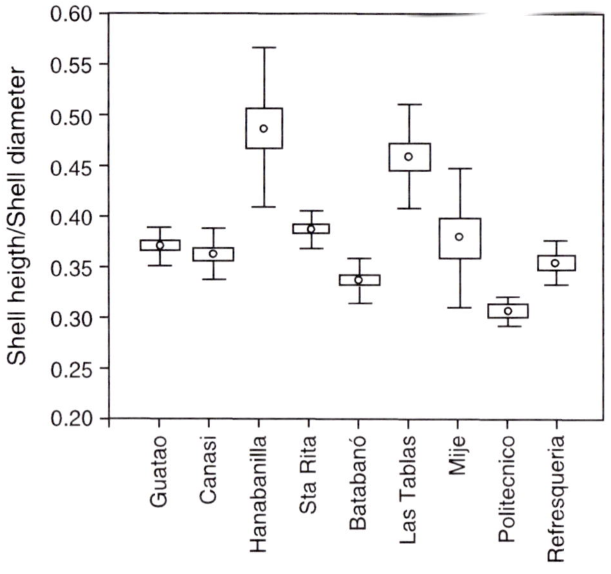

Fig. 1.3 Shell height to shell diameter ratio variation in nine populations of *Biomphalaria pallida* from Cuba. These populations are distributed all over Cuba. White circles = means; rectangles = confidence intervals; bars = minimum and maximum observed values. Fifteen individuals were analyzed per population

B. glabrata extends from the Greater Antilles to the south of Brazil, including the Lesser Antilles, Venezuela and Suriname (Paraense 2001). The question is whether it may result in species misidentification. The answer is probably no for adult individuals (or even shells) with some experience. We already mentioned the difficulty of distinguishing juvenile *B. glabrata* from adults of other species (*B. straminea* complex). It is perhaps more disturbing that *B. glabrata* can sometimes be confounded with *Helisoma duryi* (compare Fig. 1.2h with Fig. 1.2d–g; J. Mavárez, pers. commun.), a species from the sister genus of *Biomphalaria* (Morgan et al. 2002). *Helisoma duryi* indeed shows ample shell variation and can be found syntopically with *B. glabrata*. It is also not surprising that *B. glabrata* has been described under a variety of names, which should therefore be considered as synonyms (Paraense 1961a; Harry 1962; Pan American Health Organization 1968). The most common synonyms are listed in Box 2.

A major conclusion that can be drawn is that as a consequence of its large variability presumably resulting from both genetic and environmental factors, shell characters should be used with caution to distinguish *B. glabrata* from other species. At this point, there are two ways out of this deadend. (a) The description of shell variation has relied on classical morphology. However, more modern approaches, including geometric morphometrics (Zelditch et al. 2004) and multivariate analyses, might be used to evaluate how intraspecific variation overlaps with variation among species. In this context, it is noticeable that *Biomphalaria* shells could be described using geometric models, such as a logarithmic spiral (Raup 1966) which have proven valuable to describe the variation of gastropod shells

Box 1 Description of the shell of *Biomphalaria glabrata* by the Pan American Health Organization Guide (1968)

"The adult shell has about 5–6.5 whorls, increasing slowly or sometimes more rapidly in diameter. The whorls are normally rounded on the sides, though sometimes flattened, angular, or even carinate on the left and less frequently on the right. The suture ranges from shallow to deep between flattened and rounded or angular whorls, respectively. Each side varies from broadly and shallowly to deeply concave, generally in inverse proportion to the opposite side. The right side may be flat or even a little convex in individuals of some populations. The periphery varies from rounded to bluntly angular and is frequently shifted to the right. The aperture may be narrow, egg-shaped, semicircular, rounded, or transverse; it is usually subangular on the lower left, directed forward or bent more or less leftward and generally oblique to the right or to the left. The peristome is thin and continuous. Specimens of about 2–9 mm from habitats subject to seasonal drought may develop one or more sets of apertural lamellae (usually six per set), which later are wholly or partly resorbed. In some cases the whole set persists inside fully-grown shells."

Box 2 Some synonyms of *Biomphalaria glabrata* (from Harry 1962; Paraense 1961a, b, 1963). Although the genus name *Planorbis* has often been used, other genus names such as *Taphius* or *Australorbis* can also been found

Planorbis albescens "Spix" Wagner, 1827; *Planorbis antiguensis* "Guilding" Sowerby, 1877; *Planorbis antillarum* Beck, 1838; *Planorbis bahiensis* Dunker, 1886; *Planorbis becki* Dunker, 1850; *Planorbis blauneri* "Shuttleworth" Germain, 1921; *Planorbis bolivianus* "Philippi" Dunker, 1850; *Australorbis glabratus christophorensis* Pilsbry 1934; *Planorbis concavospira* Anton, 1839; *Planorbis confusus* Lutz, 1918; *Planorbis cumingianus* Dunker, 1848; *Planorbis dentifer* Moricand, 1853; *Planorbis ferrugineus* "Spix" Wagner, 1827; *Planorbis guadeloupensis* Sowerby, 1822; *Planorbis immunis* Lutz, 1923; *Planorbis lugubris* Wagner, 1827; *Planorbis lundii* Beck, 1838; *Planorbis lutescens* Lamarck, 1822; *Planorbis nigricans*, "Spix" Wagner, 1827; *Planorbis olivaceus* "Spix" Wagner, 1827; *Planorbis refulgens* Dunker, 1823; *Planorbis striatulus* "Richard" Beck, 1837; *Planorbis viridis* "Spix" Wagner, 1827; *Planorbis xerampelinus* Drouët, 1859.

(Samadi et al. 2000). (b) It would probably be more productive to find other types of biosystematic markers, such as characteristics of the reproductive tracts. This was the basis for the long-standing effort of taxonomic cleaning-up initiated in the 1950s by W.L. Paraense. This is described in more detail in the next section.

1.3 The Anatomy of Reproductive Tracts, a Reliable Tool for Identifying *Biomphalaria* Species

It became clear from the mid-twentieth century on that shell description suffers from several weaknesses (see above) as a taxonomical tool in molluscs, including *Biomphalaria*, providing the impetus for looking for other morphological characters. The first anatomical descriptions of the soft parts of *Biomphalaria* have been reported, sometimes incompletely, by Pilsbry (1934), Baker (1945), Ranson and Cherbonnier (1953), Ranson (1953), Malek (1954, 1955), Hubendick (1955), Mandahl-Barth (1957), and Schutte and Van Eeden (1960). This sets the basis for proposing several anatomical characteristics as diagnostic among species. For example, the presence of an elongated renal ridge distinguishes *B. glabrata* from all other *Biomphalaria* species, including the African ones (Paraense and Deslandes 1959; see Fig. 1.4). Note that this ridge is a simple black line in juvenile individuals (Fig. 1.4).

However, the most reliable characters for taxonomic purposes are those of the reproductive tracts. These tracts indeed have a complex morphology in snails (Geraerts and Joosse 1984; Jarne et al. 1993, 2009), especially in *Biomphalaria*. All basommatophorans are hermaphroditic and their reproductive tracts are subdivided into three parts (hermaphroditic, male and female), each including several structures (gland, pouch, duct, etc.) that can serve as taxonomic characters because they exhibit variation in presence, size, and form. These structures include the ovotestis (form and number of diverticulae), sperm duct (shape), penial complex (relative size of the preputium and penis sheath), *vas deferens* (shape and size relative to penis sheath), prostate (form and number of diverticulae), and vagina (size, presence/absence of a pouch and corrugations) (Fig. 1.5). Other characters show variability among species

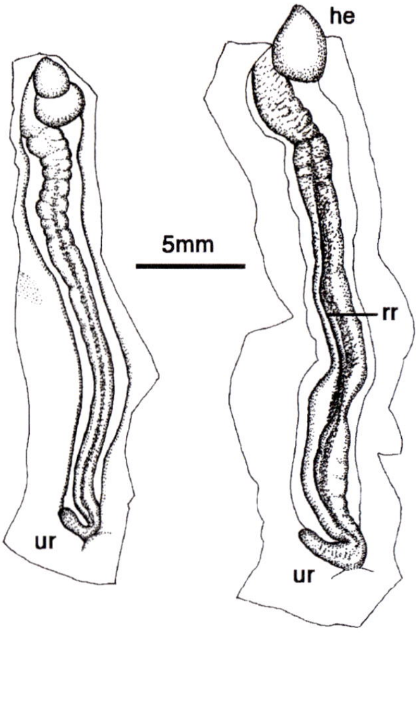

Fig. 1.4 Anatomy of the renal tube in *Biomphalaria glabrata*. Left: juvenile individual with a pigmented line along the tube. Right: renal tube in an adult showing a renal ridge. he = heart; rr = renal ridge; ur = ureter

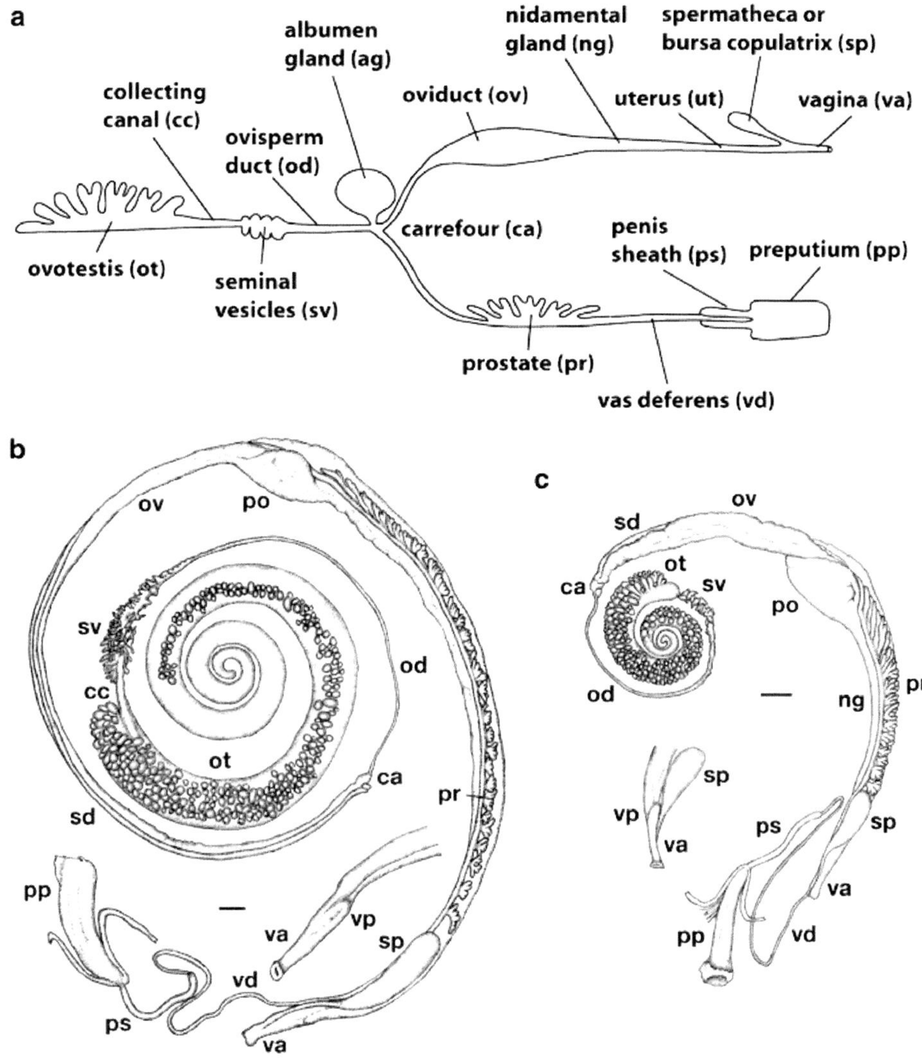

Fig. 1.5 Schematic anatomy of the reproductive tracts in *Biomphalaria* (**a**), and drawings of anatomy in *Biomphalaria glabrata* individuals from Lake Zuata, Aragua State (**b**) and Caripe, Monagas State (**c**) – both from Venezuela. In **b** and **c**, the vagina and spermatheca are represented twice: the view separated from the reproducted tract allows visualizing the vaginal pouch. ag = albumen gland; ca = carrefour; cc = collecting canal; ng = nidamental gland; od = ovispermiduct; ot = ovotestis; pe = penis; pg = preputial gland; pm = retractor muscle; po = oviduct pouch; pp = preputium; pr = prostate; ps = penis sheath; sd = spermiduct; sp = spermatheca; sv = seminal vesicles; va = vagina; vd = vas deferens; vp = vaginal pouch; bar = 1 mm

although they are more difficult to characterize clearly (e.g., form of diverticula of seminal vesicles) and will not be considered here. The interspecific variability of these characters for most *Biomphalaria* species is presented in Table 1.1. As for shells, it should be borne in mind that these structures display some intraspecific variability that must be taken into account for species diagnostic.

Table 1.1 Main characteristics of the reproductive tracts of the currently valid *Biomphalaria* species from the Neotropics and Africa

Biomphalaria species	Main species complex	Ovotestis diverticulae	Sperm duct	Penial complex	Vas deferens	Prostate diverticulae	Vagina	Maximum shell size of adult snails	References
schrammi		50, Simple	Straight	L: *ps*>>*pp* (4–7 times) W: *ps* ≈ *pp*	DSW < W*ps*	8–17, Short, branched	Regular, no VP	7 mm	Paraense et al. (1964) Paraense and Deslandes (1956)
peregrina		100, Mostly simple	Straight	L: *ps* ≈ *pp* W: *ps*<<*pp*	DSW < W*ps*	8–21, Long, branched or arborescent	Short, VP	16 mm	Paraense (1966)
helophila		20–30, Mostly simple	Straight	L: *ps*>*pp* (1–2 times) W: *ps* ≈ *pp*	DSW < W*ps*	5–16, Short, mainly bifurcate or trifurcate	Regular, no VP	7 mm	Paraense and Ibañez (1964) Paraense (1996)
havanensis		40–80, Mostly simple	Straight	L: *ps*<*pp* W: *ps*<*pp*	DSW < W*ps*	7–20, Long, branched	Regular, no VP	10 mm	Paraense and Deslandes (1958a) Paraense (1990) Yong et al. (1997, 2001)
pallida		40–60, Mostly simple	Straight	L: *ps*<*pp* (slightly) W: *ps* < *pp*	DSW < W*ps*	8–18, Short, branched	Regular, VP	10 mm	Paraense and Deslandes (1958a) Yong et al. (2001)
andecola	*andecola* complex	30–60, Mostly simple	?	L: *ps*<*pp* (slightly) W: *ps*<*pp*	MSW>DSW and < W*ps*	6–12, Very long, branched	Short, VP	22 mm	Paraense and Deslandes (1957) J.-P. Pointier (unpublished)
prona	*andecola* complex	70, Mostly simple	Flexuous	L: *ps*>*pp* (slightly) W: *ps*=*pp* Strong diaphragm	MSW > DSW and < W*ps*	7–12, Long, arborescent	Short, VP	11 mm	Paraense and Deslandes (1958b) Paraense et al. (1992)

edisoni	*andecola* complex	30–50 Mostly branched	Flexuous	L: $ps > pp$ W: $ps = pp$ Strong diaphragm	DSW < W*ps*	?, Arborescent	Short, VP	11 mm	Estrada et al. (2006)
occidentalis	*tenagophila* complex	>100 Mostly simple	Straight	L: $ps < pp$ W: $ps < pp$	MSW > DSW and DSW ≈ W*ps*	10–27, Long, branched	Regular, no VP	35 mm	Paraense (1981)
tenagophila	*tenagophila* complex	>100 Mostly branched	Straight	L: $ps \leq pp$ W: $ps < pp$	DSW ≈ W*ps*	10–30, Short, branched	Regular, VP	35 mm	Paraense (1961a) Pointier et al. (2005)
amazonica		30–50 Mostly simple	Slightly flexuous	L: $ps < pp$ W: $ps < pp$	DSW ≈ W*ps*	8–12, Short, branched	Short, VP	9 mm	Paraense (1966) Pointier et al. (2002)
intermedia	*straminea* complex	60, Mostly simple	Straight	L: $ps = pp$ W: $ps < pp$	DSW ≈ W*ps*	7–15, Mainly branched	Regular, no VP + swellings	12 mm	Paraense and Deslandes (1962)
straminea	*straminea* complex	60, Mostly simple	Flexuous	L: $ps > pp$ W: $ps < pp$	DSW ≈ W*ps*	10–20, Mainly branched	Regular, no VP + marked corrug	15 mm	Paraense and Deslandes (1955a)
kuhniana	*straminea* complex	60, Mostly simple	Flexuous	L: $ps \geq pp$ W: $ps < pp$	DSW ≈ W*ps*	4–7, Mainly branched	Regular, no VP + corrug	10 mm	Paraense (1988)
glabrata		>200, mostly branched	Straight	L: $ps = pp$ W: $ps < pp$	DSW ≈ W*ps*	15–30, Long, mainly branched	Regular, VP	40 mm	Paraense and Deslandes (1955b)
alexandrina		>150, Mostly branched	Straight	L: $ps < pp$ W: $ps < pp$	DSW ≈ W*ps*	18–24, Branched	Regular, VP	15 mm	J.-P. Pointier (unpublished)

(continued)

Table 1.1 (continued)

Biomphalaria species	Main species complex	Ovotestis diverticulae	Sperm duct	Penial complex	Vas deferens	Prostate diverticulae	Vagina	Maximum shell size of adult snails	References
camerunensis		200–300, Mostly branched	Flexuous	L: $ps = pp$ W: $ps < pp$	DSW ≈ W*ps*	20–25, Branched	Regular, VP	20 mm	Vollmer (1991) J.-P. Pointier (unpublished)
pfeifferi		60, Mostly branched	Flexuous	L: $ps < pp$ W: $ps < pp$	DSW ≈ W*ps*	7–12, Branched	Regular, VP	17 mm	Schutte and Van Eeden (1960) J.-P. Pointier (unpublished)
Other species									
cousini		15–25, Mostly simple	Straight	L: $ps < pp$ W: $ps \approx pp$	DSW < W*ps*	5–10, Branched	Regular, VP	8 mm	Paraense (1966)
trigyra		100, Mostly simple	Straight	L: $ps > pp$ W: $ps < pp$	DSW < W*ps*	5–10, Long, thin and branched	Regular, VP	14 mm	Pan American Health organization (1968)
orbignyi		100–180, Mostly simple	Straight	L: $ps = pp$ W: $ps < pp$	DSW ≈ W*ps*	8–20, Branched	Regular, VP	16 mm	Paraense (1975)
subprona	*andecola* complex ?	30, Mostly simple	Flexuous	L: $ps = pp$ W: $ps < pp$	DSW < W*ps*	6–11, Long, branched	Short, VP	5 mm	Paraense (1996)
nicaraguana	*andecola* complex ?	60, Mostly simple	Straight	L: $ps > pp$ W: $ps = pp$	DSW ≈ W*ps*	16–33, Short, branched	Regular, no VP	10 mm	Paraense (2003)
oligoza		20–30, Mostly simple	Straight	L: $ps = pp$ W: $ps \approx pp$	DSW < W*ps*	1–7, Long, mostly branched	Regular no VP	11 mm	Paraense (1974)

sericea	Numerous, mostly bifurcate or trifurcate	Straight	L: *ps* < *pp* W: *ps* < *pp*	DSW < W*ps*	12, Mostly branched	Regular, no VP	16 mm	Pan American Health organization (1968)
equatoria	80 Mostly simple	Straight	L *ps* > *pp* W: *ps* < *pp*	DSW < W*ps*	10 Branched	Short, VP	18 mm	Paraense (2004)

Species are listed in the same order than in the cladogram of Fig. 1.6 (bottom to top) to the exception of the last eight (Neotropical) species whose phylogenetic position is unknown. *Biomphalaria temescalensis* and *B. obstructa* are not included as they are synonyms of *B. havanensis* (Yong et al. 1997, 2001). *Biomphalaria choanomphala*, *B. smithi*, and *B. sudanica* belongs to the *B. alexandrina* complex (DeJong et al. 2001; Jørgensen et al. 2007). No data are available for *B. angulosa*, as well as for the dubious African species *B. salinarum* and *B. tchadiensis* (Brown 1994). *Biomphalaria oligoza* and *B. orbignyi* are closely related to *B. peregrina*. See Fig. 1.5 for a description of reproductive tracts. The "ovotestis diverticulae" and "prostate diverticulae" columns give the number and shape of diverticulae for these two organs. For the penial complex, we report the relative length (L) and width (W) of the penis sheath (*ps*) relative to the preputium (*pp*). For the vas deferens, relative widths of section (DSW = distal section, MSW = middle section) are provided, sometimes by comparison to the width of the penis sheath (W*ps*). Vagina size (relative to the distance separating the bursa copulatrix canal and the female pore) is given together with the presence/absence of a vaginal pouch. Corrug refers to corrugations. All values should be considered as indications showing some variation

This sets the basis for the first exhaustive description of 19 *Biomphalaria* species from the New World (Pan American Health Organization 1968). This guide provides a careful description of anatomical features of their reproductive tracts – note that a significant fraction of these species were described based on the work by W.L. Paraense. Relying on the same characters, several species names have subsequently been synonymyzed in that list as a result of taxonomical revisions (Malek 1985; Yong et al. 1997, 2001) and description of new species has been carried out (Paraense 1975, 1981, 2004; Estrada et al. 2006). Table 1.1 presents the main anatomical characteristics of all Neotropical and African *Biomphalaria* that are considered as valid species. Although several drawings of anatomical characters of reproductive tracts have already been published in African species (Mandahl-Barth 1957, Schutte and Van Eeden 1960; Vollmer 1991; Pointier et al. 2005), they are sometimes incomplete and lack the details characterizing work done in American species (Table 1.1). These characters are useful for distinguishing species since the interspecific variation is relatively limited. This does not come as a surprise because this species group is relatively young (see below).

Biomphalaria glabrata can be described using the same set of characters, and we report the description of the Pan American Health Organization Guide (1968) in Box 3. Drawings of *B. glabrata* reproductive tracts are presented in Fig. 1.5 (bottom), highlighting intraspecific variability. They represent individuals from two sites in Venezuela, Caripe and Zuata, separated by 350 km. Adult snails from Caripe have a shell diameter of 16 mm, whereas those of Zuata reach a size of 35 mm. These latters have a larger number of whorls and consequently have more elongated organs. For example, the prostate of Zuata individuals is about twice as long as that of Caripe snails and has 30 prostate diverticulae against 15 in Caripe (Fig. 1.5). However, the form of these diverticulae and the morphology of other organs are specific to *B. glabrata*, allowing a clear identification of this species.

As a conclusion, the anatomy of the reproductive tracts has been shown to be an interesting tool for identifying *Biomphalaria* species. However, a resolution limit is

Box 3 Description of the reproductive tracts of *Biomphalaria glabrata* by the Pan American Health Organization Guide (1968)

"There are usually over 200 ovotestis diverticula, which are predominantly trifurcate but may be divided into from two to five or more branches and in exceptional cases are even unbranched. The seminal vesicle has conspicuous diverticula. The vagina is cylindrical with a well-developed pouch to the right of the spermathecal duct. The spermatheca may be rounded, ovate, pear-shaped, or club-shaped, with a slender duct usually as long as the body. The prostate ususally has 15–30 long, slender, mainly arborescent, diverticula. The distal half of the vas deferens is nearly as wide as the penis sac. The penis sac is about as long as the preputium and distinctly narrower."

reached for groups of closely related species (e.g., the *B. straminea* or *B. prona* complex) as a consequence of intraspecific variation or incomplete variation sorting among species. This calls for more resolutive tools, and molecular markers are of marked interest here.

1.4 Molecular Phylogenetic and Phylogeographic Approaches

As for taxonomic studies in general (Hillis et al. 1996), molecular markers have substantially improved our knowledge of *Biomphalaria* systematics. The position of the *Biomphalaria* genus among planorbids has been clarified by Morgan et al. (2002) and Albrecht and Glaubrecht (2006). Molecular studies of the *Biomphalaria* group began with allozymic markers more than a decade ago (Bandoni et al. 1995; Woodruff and Mulvey 1997) and were enriched more recently by the analysis of DNA sequences (Campbell et al. 2000; DeJong et al. 2001; Jørgensen et al. 2007). Beyond clarifying the phylogenetic relationships among species, these studies set the basis for new biogeographic hypotheses. A summary of our current knowledge can be derived from the largest (in terms of number of species) of these studies (DeJong et al. 2001), with some refinements on the relationships among African species coming from Jørgensen et al. (2007). The work of DeJong et al. (2001), based on both mitochondrial (16S) and nuclear (ribosomal ITS1 and ITS2) nucleotidic sequences, included 23 *Biomphalaria* species, 16 from the Neotropics and 7 from Africa. Their results are summarized as a schematized cladogram in Fig. 1.6, which indicates that (a) the topology is consistent with a Neotropical origin of the group – the oldest fossils have been dated from ca. 60 millions years. (b) Several main species complex (groups of closely allied species) can be recognized, including the *B. havanensis*, *B. prona*, *B. straminea*, and *B. alexandrina* (Nilotic) complexes. This strongly suggests that some species names are not valid (e.g., *Biomphalaria temascalensis* and *Biomphalaria obstructa* in the *B. havanensis* complex), and the initial number of 16 and 7 species considered in this study boiled down to 11 and 3, respectively. (c) *Biomphalaria schrammi* is very distant from all other *Biomphalaria* species. (d) Perhaps the most important result is the close relationship between *B. glabrata* and the (monophyletic) African *Biomphalaria* species. This was already found in previous molecular studies (Bandoni et al. 1995; Woodruff and Mulvey 1997; Campbell et al. 2000), but is at variance with the previous state of knowledge based on morphological markers only and certainly came as a surprise. It is strongly suggestive of a recent origin of African species from Neotropical founders (i.e., proto-*B. glabrata*), which might have rafted west to east across the Atlantic. Dating based on molecular clocks suggests a transatlantic move 1.1–4.5 Mya ago. It remains to be explained why the earliest African fossil shells of *Biomphalaria* are much younger (0.25 Mya; B. van Boxclaer, pers. commun.). It is possible that the molecular clocks tick faster in freshwater snails than in the groups in which they have been calibrated. However, the lower time value (1.1 Mya) is based on a pretty generous clock (e.g., 1.2%/Mya for ITS sequences),

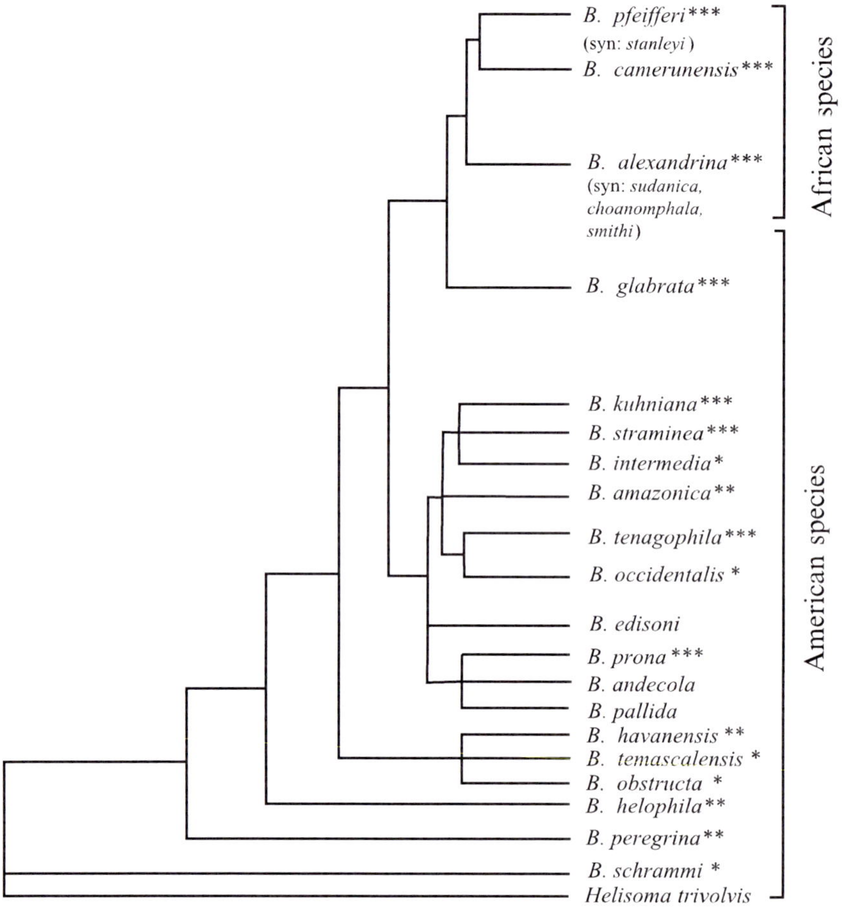

Fig. 1.6 Schematic representation of the phylogenetic relationships among 20 African and American species of *Biomphalaria* redrawn from the maximum parsimony tree of DeJong et al. (2001). This tree was built on combined mitochondrial (16S) and nuclear (ribosomal ITS1 and ITS2) sequences. This study initially included three other species which can be collapsed with species indicated in the figure as follows: *B. stanleyi* with *B. pfeifferi*; *B. sudanica*, *B. choanomphala*; and *B. smithi* with *B. alexandrina*. *Biomphalaria edisoni* has been added based on the more recent work of Estrada et al. (2006). *Helisoma trivolvis* served as outgroup in the phylogenetic analysis. Branch length has no evolutionary meaning, but the cladogram topology is indicative of phylogenetically supported branches in DeJong et al. (2001). Asterisks next to species names indicate species which have experimentally been found to be unsusceptible (*) or susceptible (**) to some strains of *Schistosoma mansoni*, and naturally infected in the field by *S. mansoni* (***). No star indicates no data

and reconciling molecular and fossil data would require a clock ticking five times faster. An alternative explanation is that fossilization conditions were not adequate for Basommatophorans for the period extending between 0.5 and 1 Mya ago. Very few fossils of other planorbids and lymnaeids have indeed been found during the same period over the whole African continent (B. van Boxclaer, pers. commun.). This result is also of major importance for the transmission of schistosomiasis (see below). (e) Another important result from the molecular studies is that the number of African *Biomphalaria* species is closer to four than to the 11 species proposed in Brown (1994), i.e., *B. camerunensis*, *B. pfeifferi*, *B. angulosa*, and a Nilotic complex including *B. sudanica* and *B. alexandrina*.

Molecular markers also helped to shed some light on the intraspecific phylogeny of *B. glabrata*. It is certainly natural to ask whether a species with such a wide distribution does not uncover cryptic species. Two phylogeographic analyses, both using nuclear and mitochondrial markers, have addressed this question (Mavárez et al. 2002; DeJong et al. 2003), the first one focusing more on the Northern part of the distribution (Venezuela and the Lesser Antilles) and the second more intensely considering samples from Brazil. A cladogram summarizing their main results is reported in Fig. 1.7. Let us retain that (a) *B. glabrata* is a monophyletic group including essentially five major clades. The time to the most recent common ancestor is of the order of two-third of the time separating *B. glabrata* from African *Biomphalaria* species. (b) Individuals from the Lesser Antilles (one clade) cluster together with those from Venezuela (one clade), and the two main, geographically segregated Brazilian clades form paraphyletic clades. (c) Snails from the Greater Antilles form another clade, perhaps basal to other clades (associated statistics are equivocal). These results, both in terms of time frame and cladogram topology, can be interpreted by taking into account the climatic fluctuations of the Quaternary and the refugia theory (Haffer 1969). The distribution of the rainforest associated to the Amazon basin indeed widely varied over the last million years, and periods of

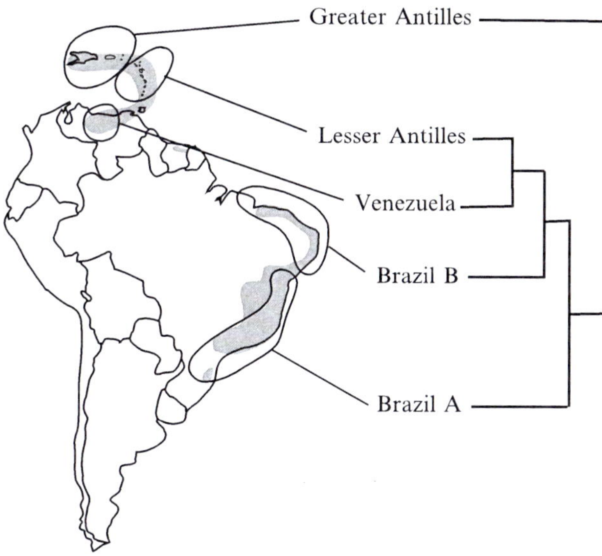

Fig. 1.7 Schematic representation of the phylogenetic relationships among the five main clades of *Biomphalaria glabrata* redrawn from the trees of Mavárez et al. (2002) and DeJong et al. (2003). These trees were built on both mitochondrial and nuclear sequences. The geographic distribution of *B. glabrata* is given in gray shade, and the main clades are connected to the area where they were sampled

maximum occupation were less favorable to *B. glabrata* (and to freshwater pulmonates in general). For example, *B. glabrata* does not occupy the rainforest area currently located between Eastern Venezuela and North-Eastern Brazil. Less rainy area with more open space (and ponds) can, therefore, be considered as refugia during hot periods. The rainforest is also certainly a strong barrier to gene flow (except when aided by human activities; see DeJong et al. 2003). This might explain why clades from the Lesser Antilles and Venezuela are separated from those from Brazil. The occurrence of several clades within Brazil might result from more local refugia as has been proposed for frog species (Carnaval et al. 2009). That the clade from the Greater Antilles is both separated from and (perhaps) basal to other clades suggests a more ancient history, and no gene flow over several hundreds of thousands of years. This is not completely surprising given the large geographic distance between the Lesser and the Greater Antilles.

1.5 Reproductive Isolation

Tests of reproductive isolation remain the ultimate way to recognize species limits (see Coyne and Orr 2004). Reproductive isolation can conveniently be subdivided into pre- and post-zygotic processes that might, therefore, involve various traits (e.g., mating behavior and gamete compatibility if we are to cite characteristic pre- and post-zygotic processes). Incomplete reproductive isolation can lead to the creation of hybrid zones (review in Coyne and Orr 2004). From a practical point of view, testing for reproductive isolation is relatively straightforward in animal species with internal fertilization and copulation, especially in freshwater Pulmonates. An important aspect that should be taken into account is whether reproductive barriers between species are bidirectional or asymmetrical, i.e., stronger in one direction (e.g., male A/female B) than in the other (male B/female A). Asymmetry could occur in freshwater snails because matings are unidirectional, with one individual playing the male role and the other individual the female role, even if the two individuals are hermaphrodites. A further complication is self-fertilization: in predominantly outcrossing species, loss of fertility in some crosses can be compensated by selfing, yielding apparently fertile crosses even when reproductive isolation exists. In this case, molecular or morphological markers are necessary to confirm the hybrid status of progenies. Paraense (1959) noted that hybridization may occur in one direction, while selfing occurs when attempting to perform the reciprocal cross; he referred to this situation as one-sided reproductive isolation. In addition, many species regularly practice selfing, even in the presence of mates of their own species; in this case, reproductive isolation is all the more difficult to assess because it will affect only the small fraction of outcrossed eggs produced. Note that reproductive isolation should not in principle be detected within species. However, various levels of compatibility are generally observed, very often as a function of geographic distance or ecological isolation (see examples in Coyne and Orr 2004).

On the whole, there have been few attempts to test for reproductive isolation both within and among species of *Biomphalaria* (summarized in Tables 1.2 and 1.3), and a good part dates back to the 1950s. Here again, W.L. Paraense's work stands out. Researchers took advantage of the common occurrence of albino phenotypes with monofactorial recessive inheritance in natural populations of *Biomphalaria* to check for the occurrence of hybrids: when crossing pigmented and albino individuals, maternal hybrid offspring of albinos should be pigmented and albino offspring are produced by self-fertilization. However, as mentioned above, reproductive isolation may begin with the courting/mating behavior and extends to post-zygotic processes (see Coyne and Orr 2004). It is, therefore, worth analyzing and quantifying each of these steps, and following individuals over several generations. A weakness of most studies in *Biomphalaria* is that results are not quantified in an appropriate way, and reported in a more qualitative than quantitative fashion. There have been attempts to cross *B. glabrata* with at least three species, namely, *B. alexandrina*, *B. tenagophila*, and *B. straminea*. Unsurprisingly, these crosses failed to produce any hybrids at all in most cases (Table 1.2). However, hybrids have been derived from crosses between *B. glabrata* and *B. alexandrina* (*boissyi*) (Barbosa 1956; Barbosa et al. 1958a; Table 1.2) which is not surprising given that they belong to the same phylogenetic clade. This is consistent with the occurrence of hybrids between these two species in the Nile delta during the last two decades of the twentieth century following accidental introduction (Kristensen et al. 1999). Neither hybrids, nor *B. glabrata* have been detected in a recent survey (Lofty et al. 2005).

The report of hybrids between *B. glabrata* and *B. tenagophila* (Mello-Silva et al. 1998) is much more surprising, all the more that hybrid survival was pretty high and seemingly fit F2 individuals were obtained in the same experiment. The two species indeed belong to two well-differentiated phylogenetic clades. Barbosa (1964) suggested, based on the observation of intermediate morphological forms, that such hybrids occur in natural populations from Brazil. He later claimed based on similar arguments (Barbosa 1973) that hybrids between *B. glabrata* and *B. straminea* also occur in the wild as a result of the invasion of *B. straminea* populations in areas formerly occupied by *B. glabrata*. Quite clearly, such claims would today have to be substantiated through molecular analyses. Even more surprising is the report that *B. peregrina* can hybridize with both *B. straminea* and *B. tenagophila* (Barbosa et al. 1958a) as the first species is separated from the other two by presumably several tens of millions of years – *B. peregrina* is the second most basal species in the cladogram of Fig. 1.6. It is possible that this is due to an error in taxonomical identification as species characteristics were not fully stabilized at the time of this study. *B. straminea* and *B. tenagophila* might produce non-viable hybrids (Mello-Silva et al. 1998), although Paraense and Deslandes (1955a) failed to detect such hybrids. Part of the reason for such a discrepancy may be related to the mating system of *B. straminea*: this species is likely to be a selfer, judging from results obtained in the closely related *Biomphalaria kuhniana* (Dupuy et al. 2009). Copulations are pretty rare in such species (as in *B. pfeifferi*, another predominantly selfing species, Tian-Bi et al. 2008) and this certainly impairs our capacity to implement the biological species concept. If there is variation in the selfing rate, it remains possible that the second

Table 1.2 Results of crosses in the F1 generation among *Biomphalaria* species

		F1 hybrids		
Species 1 (pop)	Species 2 (pop)	Occurrence	Survival	Reference
B. glabrata	*B. alexandrina*	Yes	Yes	Barbosa (1956)
Recife, PE	Cairo, Egypt			
B. glabrata	*B. tenagophila*	No	–	Paraense and Deslandes (1955a)
Santa Luzia, MG	Itajubá, MG	Yes	Non viable	
Puerto Rico ♀	Paulista, PE ♂	Yes	> 79%	Mello-Silva et al. (1998)
Paulista, PE ♂	Joinville, SC ♀			
B. glabrata	*B. straminea*	No	–	Paraense and Deslandes (1955c)
?	?	No	–	
Paulista, PE ♂	Sete Lagoas, MG ♀			Mello-Silva et al. (1998)
B. peregrina	*B. straminea*	Yes	Yes	Barbosa et al. (1958a)
Ecuador	?, PE			
B. peregrina	*B. tenagophila*	Yes	Yes	Barbosa et al. (1958a)
Ecuador	?, SP			
B. straminea	*B. tenagophila*	No	–	Paraense and Deslandes (1955c)
?	?	Yes	Non-viable	
Belém, PA ♂	Joinville, SC ♀			Mello-Silva et al. (1998)

Biomphalaria straminea is referred to as *B. centimetralis* in Paraense and Deslandes (1955b). Brazilian localities are followed by an acronym indicating the state to which they belong: MG = Minas Gerais, PA = Pará, PE = Pernambuco, SC = Santa Catarina, SP = São Paulo. ♀ and ♂ indicate which population served as female and male respectively in crosses (when the information is available). Mello-Silva et al. (1998) also reported on the occurrence of F2 hybrids in the *B. glabrata* (Joinville) × *B. tenagophila* (Paulista) cross. – = no data. ? = Data not available

Table 1.3 Results of crosses among populations of *Biomphalaria glabrata*

		F1 hybrids			
Pop 1 (female)	Pop 2 (male)	Occurrence	Survival	F2	Reference
Santa Luzia MG	Recife PE	Yes	100%	Yes	Paraense (1956)
Santa Luzia MG	Estremoz Lake RGN	Yes Hy = not all offspring	100%	Yes	Paraense (1956)
Santa Luzia MG	João Pessoa PA	2/15 Fertile crosses Hy = 1% offspring	Not reported	Not studied	Paraense (1956)
Santa Luzia MG	Capanema PA	1/3 Offspring Decrease in egg production when paired	Not studied	Not studied	Paraense (1959)
Capanema PA	Santa Luzia MG	All offspring	100%	Used to infer hybrid occurrence in F1	Paraense (1959)
Santa Luzia MG	Caracas	3 hyb from 3 ind /10 Decrease in egg production when paired	Not studied	Not studied	Paraense (1959)
Caracas Venezuela	Santa Luzia MG	All offspring	Not studied	Used to infer hybrid occurrence in F1	Paraense (1959)
Puerto Rico	Paulista PE	No	–	–	Mello-Silva et al. (1998)

Pop 1 acted as female and pop 2 as male. Santa Luzia and Puerto Rico = albino lines. Brazilian localities are followed by an acronym indicating the state to which they belong: MG = Minas Gerais, PA = Pará, PE = Pernambuco, RGN = Rio Grande do Norte, SC = Santa Catarina. The clades from Fig. 1.5 to which these populations putatively belong (authors' interpretation) are Paulista and Santa Luzia = Brazil 2; Recife, João Pessoa, Estremoz Lake, Capanema = Brazil 1; Caracas = Venezuela; Puerto Rico = Greater Antilles. – = Irrelevant study was conducted in a highly selfing population of *B. straminea*, while the first was conducted in a more outcrossing population.

Attempts have also been made to cross populations of *B. glabrata* separated by various geographic distances using an albinism marker to check for crossing between populations and the occurrence of self-fertilization (Paraense 1956, 1959; Mello-Silva et al. 1998). Results of these studies are summarized in Table 1.3. Two interesting results can be derived from this series of analyses. (a) There is a clear trend of decreased hybrid fitness, going down to an absence of hybrids, when crosses involve populations separated by increasing geographic distance. Using the cladogram from Fig. 1.7, populations can be attributed to phylogenetic clades based on their geographic location. Crosses from the same Brazilian clade lead to full fertility of hybrids, and there is a decrease in hybrid survival when crossing different Brazilian clades, a clade from Brazil and one from Venezuela, and finally a clade from Brazil and one from Puerto Rico (Greater Antilles). The latter result, if it came to be replicated, would suggest that *B. glabrata* might better be considered as a species complex than as a single species. (b) The results of Paraense (1959) on hybrid production (no data on survival) indicate that reproductive isolation may be "one-sided," i.e., that hybrid production depends on crosses direction (i.e., on which population is playing "female"). The difference between such crosses resides in what is uniparentally transmitted over the whole reproductive process. It might bear on substances transmitted with the semen or fertilization success – this is quite possible if populations have diverged in this respect because of genetic drift, natural selection, or sexual selection (Jarne et al. 2010). Alternatively, nucleo-cytoplasmic incompatibilities might well affect egg survival or early juvenile survival asymmetrically (see Levin 2003; Coyne and Orr 2004). An argument in favor of the first hypothesis is that decreased hybrid production is associated with a general decrease in the production of eggs and a deterioration in egg capsule structure when partners were maintained together (selfing did not reimplace outcrossing as observed in some other crosses), though this was reversible since reasonably fit offspring were produced once partners were separated (Paraense 1959).

Incompletely isolated genetic entities form hybrid zones when they occur in sympatry (Coyne and Orr 2004; Arnold 2006). They may take various forms, such as the elongated line running in Europe between the two subspecies of the house mouse (Boursot et al. 1993), a mosaic of areas with hybrids interspersed with "pure" (single species) areas as observed along the Western Atlantic coast for the mussels *Mytilus edulis* and *Mytilus galloprovincialis* (Bierne et al. 2003) or in oak trees (Mir et al. 2006). Studying hybrid zones is of wide interest in evolutionary biology because it allows dissecting selective tensions between incompletely separated genomes (Arnold 2006). No such hybrid zones have been reported in freshwater pulmonates. We mentioned above that *B. glabrata* and *B. alexandrina* hybridized over a few years in the Nile delta (Kristensen et al. 1999; Lofty et al. 2005) but the situation has not been well characterized and appears to have been very ephemeral whatsoever. A similar comment can be made on the possible hybridization between *B. glabrata* and *B. straminea* in North-Eastern Brazil (Barbosa 1973). It is likely that such zones would not be detected as long as the population genetic structure of *Biomphalaria* species has not been studied at wide geographic scales, perhaps focusing on areas where species with "not too large" phylogenetic distance co-occur (see Fig. 1.6). Within *B. glabrata*, it might be worth following situations of introduction of new clades in a given area. DeJong et al. (2003) suggested, for example, that individuals from the Southern Brazil clade have been introduced into Northern Brazil, and that individuals from the Lesser Antilles clade have been introduced into Southern Brazil. Given the results reported in Table 1.3 on incompatibility within *B. glabrata*, and the availability of microsatellite markers, there is certainly here a potential for studying the competitive interactions and/or hybridization between differentiated genomes.

1.6 *Biomphalaria* Systematics and the Interaction with *Schistosoma mansoni*

Close specificity between species pairs of hosts and parasites (mutualists) might sometimes be used to build a "cosystematics" of hosts and parasites (Nieberding and Olivieri 2006). In other words, hosts (parasites, respectively) could be recognized based on the systematics of their parasites (hosts resp.). Such an approach could be extended within species: variation among populations in host specificity might in principle reveal that the host species is indeed a species complex. Whether implemented within or among species, this approach should be manipulated with caution since strict specificity is usually not the rule. This has been a creeping issue in *Biomphalaria* systematics with regard to its relationship with the trematode *S. mansoni*, the agent of intestinal schistosomiasis in the Neotropics (see Rollinson 2001; several chapters in this book). More specifically, the question is whether susceptibility to *S. mansoni* tells us anything about the specific status of the infected host. The answer is negative as we will show below before proposing an explanation for why it is so.

Five Neotropical species of *Biomphalaria* have been found to act as intermediate hosts of *S. mansoni* in the field (Pan American Health Organization 1968; Malek 1985; W.L. Paraense and C. Balzán, pers. commun.). This includes *B. glabrata,* the main snail host in both the Lesser and Greater Antilles, Venezuela, Suriname, and Brazil; *B. tenagophila* in Southern Brazil; *B. straminea* in North-Eastern Brazil, *B. kuhniana* in Venezuela and Martinique (Lesser Antilles) and *B. prona* in Venezuela. Other Neotropical species have been shown to be susceptible in experimental infection, though not found infected in natural populations: *B. amazonica* from Brazil (Correa and Paraense 1971), *B. havanensis* from USA and Puerto Rico (Cram et al. 1945; Richards 1963), *B. helophila* from Puerto Rico (Richards 1963), *B. peregrina* from Brazil and Ecuador (Barbosa et al. 1958b; Paraense and Correa 1973), and *B. sericea* from Brazil (Barbosa et al. 1963). The susceptibility of the other species has not been evaluated. Moreover, most African species are known as active intermediate hosts of schistosomes in the field (Brown 1994).

These results can be reset in the phylogenetic perspective proposed in Fig. 1.6. It becomes clear that both the "*B. glabrata* + African species" clade, and the clades going (in the figure) from *B. kuhniana* down to *B. pallida* include species that can serve as intermediate hosts for *S. mansoni*, even if transmission seems to be less efficient in this second clade (Fig. 1.6). More basal species (*B. havanensis, B. helophila*, and *B. peregrina*) can be experimentally infected, though they are not found to be infected in the wild. Incidentally, we can certainly hypothesize from these results that other species such as *B. nicaraguana, B. trygira, B. subprona,* or *B. cousini* (see Table 1.1) are potential hosts for *S. mansoni*. The only exception is *B. schrammi*, the most basal species in the cladogram, which is not susceptible to *S. mansoni* (Paraense et al. 1964). However, this species is phylogenetically and morphologically very distant from other *Biomphalaria* species and might even be considered as a distinct genus-level lineage (DeJong et al. 2001; see Fig. 1.5).

All these results show that the susceptibility/unsusceptibility of *Biomphalaria* species to schistosomes is irrelevant to the biosystematics of this snail group. This is not extremely surprising given what we know of the history of the *Biomphalaria*/*Schistosoma* interaction (see DeJong et al. 2003; Morgan et al. 2005). The genus *Schistosoma* was not present in America prior to its introduction via the slave trade, and *S. mansoni* in Africa is exclusively transmitted by *Biomphalaria* species, unlike its most closely related species *Schistosoma rodhaini*. Other schistosome species are transmitted by snails from the genus *Bulinus*, which is widely separated from *Biomphalaria* (by perhaps 150–200 million years; Morgan et al. 2002; Albrecht and Glaubrecht 2006). As it is known that the African *Biomphalaria* clade is not older than three to four millions years (Campbell et al. 2000; DeJong et al. 2003), the ancestor of *S. mansoni* and *S. rodhaini* presumably captured an African *Biomphalaria* species quite recently. Going from *Bulinus* to *Biomphalaria* constitutes a large phylogenetic jump, certainly requiring a broad capacity to infect Basommatophoran snails. Most other African bulinids (e.g., *Segmentorbis*) that are as closely related to *Bulinus* as to *Biomphalaria* do not transmit schistosomes (Morgan et al. 2002). A phylogeographic analysis has shown that Neotropical *S. mansoni* individuals belong to a single (out of five) clade of this species which is mainly found in the most Western part of Subsaharan Africa (Morgan et al. 2005). Even if *S. mansoni* specialized and co-evolved with African *Biomphalaria* species for several hundreds of thousands of years, *S. mansoni* retained its broad capacity to infect most *Biomphalaria* species from the New world. However, it was not broad enough to infect species beyond the *Biomphalaria* genus.

Up to now, we considered the susceptibility of *Biomphalaria* spp. to *S. mansoni* as a general 0/1 process, and our view should be clarified by more quantitative experimental approaches incorporating various aspects of snail and schistosomes fitness traits in a geographic perspective (Thompson 2005). This would allow considering local coadaptation. Variation in the capacity of schistosome strains to infect the same snail strain has indeed been demonstrated (Théron and Coustau 2005; Webster and Woolhouse 1998). Moreover, Théron et al. (2008) demonstrated strong differences in compatibility depending on whether one uses large field isolates of snails and miracidia or their corresponding laboratory established strains. Genetic change along laboratory culture (e.g., loss of diversity through repeated genetic bottlenecks), mainly in parasite strains, may explain such a discrepancy. This idea could explain much of the highly variable strain-by-strain compatibility variations that are so apparent in the literature on schistosomes, such as, for example, the variation in susceptibility of *B. glabrata* and *B. tenagophila* populations observed in Brazil by Paraense and Correa (1963a, b). An aspect that has not been considered is whether any pattern in schistosome susceptibility emerges from a phylogeographical perspective in *B. glabrata* (Fig. 1.7). In other words, is there variation in susceptibility among snail populations that is structured by their geography? A tentative answer is that it is unlikely given the limited variation in *S. mansoni* populations in the New World (Morgan et al. 2005).

As a summary, all Neotropical *Biomphalaria* species (except *B. schrammi*) might be susceptible to infection by *S. mansoni* in field conditions, provided they are in contact with the panel of variability (at genes involved in the interaction with this trematode) displayed by this trematode in the New World. Moreover, it is unlikely that populations of *B. glabrata* may be distinguished, especially geographically, through their susceptibility to *S. mansoni*. This suggests that little can be inferred on both the phylogeny of *Biomphalaria* species and the phylogeography of *B. glabrata* from patterns of infection by the much-studied *S. mansoni*, and that susceptibility/unsusceptibility to a particular strain of *S. mansoni* is not a strong argument for considering two population samples as distinct species. One can, for example, doubt that susceptibility in *B. straminea* and unsusceptibility in *B. kuhniana* (Floch and Fauran 1954; Paraense 1988; Caldeira et al. 2000) may be used as a character distinguishing these two snail species. Moreover, *B. kuhniana* has been found to be naturally infected in populations from Martinique in 1967 (W.L. Paraense, pers. commun.).

1.7 Conclusions and Perspectives

For the sake of clarity, we will first conclude on the *Biomphalaria* genus before focusing on *B. glabrata*. The general conclusion we can draw from the studies reviewed here is that the systematics of the *Biomphalaria* genus has been fairly well sorted out. We began this chapter mentioning that 30 species of *Biomphalaria* have been recognized in classical works, 19 in the Neotropics (Pan American Health Organization 1968) and 11 in Africa (Brown 1994). Subsequent work based on the anatomy of the reproductive tracts and molecular phylogeny led to both merge and synonymize species and created a few new ones as well (e.g., *B. edisoni*). The current picture might, therefore, be nearer 25 species, with 21 in the Neotropics and four in Africa (see Table 1.1). Of course, this count is open to discussion because one might ask why *B. glabrata* is considered as a single species with wide variation, while *B. straminea*, *B. kuhniana*, and *B. intermedia* should be *bona fide* species. There are indeed no more phylogenetic differences between these species than among clades of *B. glabrata* (DeJong et al. 2001). However, even if new species are described, it is unlikely that the species number will increase much in the Neotropics. It is also clear that more species have been described in Africa (Brown 1994) than the actual number (ca. four): it would indeed be surprising that Africa harbors half the number of species reported in the Neotropics for a much shorter evolutionary history of *Biomphalaria*.

The systematics of *Biomphalaria* has been tackled with various tools and approaches, including shell morphology, reproductive anatomy, molecular phylogeny, crosses among species, and schistosome susceptibility. We have shown unambiguously that both shell morphology and schistosome susceptibility are of very limited value. It is also clear that the other approaches used singly, although much more powerful, have their own limitations. We therefore plead for an integrated approach to the systematics of *Biomphalaria* based on all these approaches used as once. This will certainly be more costly, but more rewarding too. Such an approach might be implemented keeping the following ideas in mind:

- Although shell morphology is of tricky use for distinguishing species, it should not completely be discarded. Future studies should aim at including both variation among populations, eventually at species scale, and ecological information on the sampling sites, since this can affect variation. The use of modern morphometrical tools is also to be advised. Such studies might worthfully be complemented by laboratory analysis of shell variation across generations using quantitative genetics (Lynch and Walsh 1998).
- The genital anatomy has been described in most *Biomphalaria* species (see Table 1.1). Work by Paraense (Paraense 1974, 1975, 1981, 1988, 1996) has set a standard for reporting drawings that should be followed. As for shells, it would also be worth to have a clearer idea about geographic variation.
- There is room for completing the phylogeny of *Biomphalaria* species, both in terms of species that still have to be positioned (compare Table 1.1 and Fig. 1.6) and in the relationships among species in certain groups (e.g., *B. straminea, B. havanensis*, or *B. prona*). Studying at least one nuclear and one mitochondrial gene is certainly to be advised (see DeJong et al. 2001).
- Perhaps the most critical aspect is the necessity to test for reproductive isolation among species, once again taking potential within-species variation into account. Appropriate genetic markers should be used. Albino markers are certainly of practical interest, but are likely to be influenced by selection (Jarne et al. 1993). Microsatellite markers are available in some species (e.g., *B. glabrata*, Mavárez et al. 2002; *B. kuhniana*, Dupuy et al. 2009) and might be developed in others. Claims of successful crosses that are surprising in the light of molecular phylogenies have to be substantiated (e.g., between *B. straminea* and *B. tenagophila*).
- Susceptibility to *S. mansoni* is simply useless for clarifying the biosystematics of *Biomphalaria*.
- The ecology and biogeography of species (including potential zones of sympatry or hybridization) have been briefly touched on in this review. However, more information would be useful, especially when comparing closely related species.

The same conclusions and suggestions can be made for *B. glabrata*. A good point in this species is that we have a pretty good idea of the phylogeographic structure (Mavárez et al. 2002; DeJong et al. 2003). It should be used for further studies on shell morphology, genital anatomy, and reproductive isolation. On the latter point, *B. glabrata* is certainly a good model for studying how crossing between various genetic entities affects fitness over the full continuum going from within populations (including self-fertilization) to among clades (e.g., Puerto Rico and Brazil; see Fig. 1.7). Such studies should clarify the strength of genetic cohesion in this species (Escobar et al. 2008). A final point on the susceptibility to *S. mansoni*: the *B. glabrata* / *S. mansoni* is one of the most studied pairs of host/parasite. However, the variability exhibited in the Neotropics by these two species has hardly been considered in host/parasite studies. Moreover, the relationship between these species has a very young history and *S. mansoni* seems to have a broad capacity to infect *Biomphalaria*. These facts should also be considered when interpreting results on host/parasite interactions. They even suggest that the *B. glabrata*/*S. mansoni* pair is not a good model for studying coevolutionary dynamics. A more optimistic view would be to propose to consider results on the interaction between *B. glabrata* and *S. mansoni* in the light of those between *S. mansoni* and African *Biomphalaria* species, since they have coevolved for a much longer time.

Acknowledgments PD and PJ's work is supported by CNRS and the French Ministry for research. The authors thank A. Théron for discussions and comments on the manuscript.

References

Albrecht C, Glaubrecht M (2006) Brood care among Basommatophorans: a unique reproductive strategy in the freshwater limpet snail *Protancylus* (Heterobranchia: Protancylidae), endemic to ancient lakes on Sulawesi, Indonesia. Acta Zool 87:49–58

Arnold ML (2006) Evolution through genetic exchange. Oxford University Press, Oxford

Baker FC (1945) The Molluscan family Planorbidae. University of Ilinois Press, Urbana

Bandoni SM, Mulvey M, Loker ES (1995) Phylogenetic analysis of eleven species of *Biomphalaria* Preston 1910 (Gastropoda: Planorbidae) based on comparisons of allozymes. Biol J Linn Soc 54:1–27

Barbosa FS (1956) Cross-breeding of *Australorbis glabratus* and *Biomphalaria boissyi*. Trans Roy Soc Trop Med Hyg 50:296–297

Barbosa FS (1964) The renal ridge: a disputed feature of anatomy of planorbid snail *Australorbis tenagophilus*. Rev Inst Med Trop São Paulo 6:64–70

Barbosa FS (1973) Possible competitive displacement and evidence of hybridization between Brazilian species of planorbid snails. Malacologia 14:401–408

Barbosa FS, Barbosa I, Carneiro E (1958a) The anatomy of *Tropicorbis philippianus* (Dunker) and its relationships to the Brazilian Planorbidae (Mollusca, Pulmonata). J Conchyliologie 98:180–185

Barbosa FS, Barbosa I, Rodríguez JD (1958b) *Tropicorbis philippianus* (Dunker), a potential intermediate host of *Schistosoma mansoni* in Ecuador. J Parasitol 44:622

Barbosa FS, Barbosa I, Carneiro E (1963) Description of *Australorbis sericeus* (Dunker), a possible intermediate host of *Schistosoma mansoni* in Ecuador. Ann Trop Med Parasitol 57:52–58

Basch PF (1991) Schistosomes. Development, reproduction and host relations. New York, Oxford: Oxford University press.
Bierne N, Bonhomme F, David P (2003) Habitat preference and the marine-speciation paradox. Proc R Soc B 270:1399–1406
Boursot P, Auffray JC, Britton-Davidian J, Bonhomme F (1993) The evolution of house mice. Annu Rev Ecol Syst 24:119–152
Brown DS (1994) Freshwater snails of Africa and their medical importance. Taylor & Francis, London
Bruun B, Aagaard-Hansen J (2008) The social context of schistosomiais and its control. An introduction and annotated bibliography. World Health Organization, Genève, Special Programme for Research and Training in Tropical Disease
Caldeira RL, Vidigal THDA, Matinella L, Simpson AJG, Varvalho OS (2000) Identification of planorbids from Venezuela by Polymerase chain reaction amplification and restriction fragment length polymorphism of internal transcriber spacer of the RNA ribosomal gene. Mem Inst Oswaldo Cruz 95:171–177
Campbell G, Jones CS, Lockyer AE, Hughes S, Brown DS, Noble LR, Rollinson D (2000) Molecular evidence supports an African affinity of the Neotropical freshwater gastropod, *Biomphalaria glabrata*, Say 1818, an intermediate host for *Schistosoma mansoni*. Proc R Soc Lond 267:2351–2355
Carnaval AC, Hickerson MJ, Haddad CFB, Rodrigues MT, Moritz C (2009) Stability predicts genetic diversity in the Brazilian Atlantic forest hotspot. Science 323:785–789
Correa LR, Paraense WL (1971) Susceptibility of *Biomphalaria amazonica* to infection with two strains of *Schistosoma mansoni*. Rev Inst Med Trop São Paulo 13:387–390
Coyne JA, Orr HA (2004) Speciation. Sinauer Associates, Sunderland
Cram EB, Jones MF, Wright WH (1945) A potential intermediate hosts of *Schistosoma mansoni*. Science 101:302
DeJong RJ, Morgan JAT, Paracnse WL, Pointier JP, Amarista M, Ayeh-Kumi PFK, Babiker A, Barbosa CS, Brémond P, Canese AP, Pereira de Souza C, Domínguez C, File S, Gutiérrez A, Incani RN, Kawano T, Kazibwe F, Kpikpi J, Lwambo NJS, Mimpfoundi R, Njiokou F, Poda JN, Sene M, Velásquez LE, Yong M, Adema CM, Hofkin BV, Mkoji GM, Loker ES (2001) Evolutionary relationships and biogeography of *Biomphalaria* (Gastropoda: Planorbidae) with implications regarding its role as host of the human bloodfluke, *Schistosoma mansoni*. Mol Biol Evol 1812:2225–2239
DeJong RJ, Morgan JAT, Wilson WD, Al-Jaser MH, Appleton CC, Coulibaly G, D'Andrea PS, Doenhoff MJ, Haas W, Idris MA, Magalhães LA, Moné H, Mouahid G, Mubila L, Pointier JP, Webster JP, Zanotti-Magalhães EM, Paraense WL, Mkoji GM, Loker ES (2003) Phylogeography of *Biomphalaria glabrata* and *B. pfeifferi*, important intermediate hosts of *Schistosoma mansoni* in the New and Old World tropics. Mol Ecol 12:3041–3056
Dillon RJ (2000) The ecology of freshwater molluscs. Cambridge: Cambridge: University press.
Dupuy V, Nicot A, Jarne P, David P (2009) Development of ten microsatellite loci in the pulmonate snail *Biomphalaria kuhniana* (Mollusca, Gastropoda). Mol Ecol Resour 9:255–257
Escobar JS, Nicot A, David P (2008) The different sources of inbreeding depression, heterosis and outbreeding depression in a metapopulation of *Physa acuta*. Genetics 180:1593–1608
Estrada VE, Velásquez LE, Caldeira RL, Bejarano EE, Rojas W, Carvalho OS (2006) Phylogenetics of South American *Biomphalaria* and description of e new species (Gastropoda: Planorbidae). J Molluscan Stud 72:221–228
Floch H, Fauran P (1954) Essais infructueux d'infection expérimentale de *Tropicorbis kuhnianus* (Clessin) par *Schistosoma mansoni*. Bull Soc Pathol Exot 47:452–459
Geraerts WPM and Joosse J (1984) Freshwater snails (Basommatophora). In The Mollusca. Reproduction, A.S. Tompa, N.H. Verdonk, and J.A.M. van den Biggelaar JAM (eds), pp. 142-208, London: Academic Press.
Haffer J (1969) Speciation in Amazonian forest birds. Science 165:131–137

Harry HW (1962) A critical catalogue of the nominal genera and species of Neotropical Planorbidae. Malacologia 1:33–53

Hillis DM, Moritz G, Mable BK (1996) Molecular systematics. Sinauer Associates, Sunderland

Hubendick B (1955) Phylogeny in the Planorbidae. Trans Zool Soc Lond 28:453–542

Jarne P, Delay B (1991) Population genetics of freshwater snails. Trends Ecol Evol 6:383–386

Jarne P, Théron A (2001) Genetic structure in natural populations of flukes and snails: a practical approach and review. Parasitology 123:S27–S40

Jarne P, Vianey-Liaud M, Delay B (1993) Selfing and outcrossing in hermaphrodite freshwater gastropods (Basommatophora): where, when and why. Biol J Linn Soc 49:99–125

Jarne P, Pointier JP, David P, Koene J (2010) Basommatophoran Gastropods. In: The Evolution of Primary Sexual Characters in Animals, Leonard JL, Córdoba-Aguilar (eds), pp. 173–196, London Academic Press

Jordaens K, Dillen L, Backeljau T (2007) Effects of mating, breeding system and parasites on reproduction in hermaphrodites: pulmonate gastropods (Mollusca). Anim Biol 57:137–195

Jørgensen A, Kristensen TK, Stothard JR (2007) Phylogeny and biogeography of African *Biomphalaria* (Gastropoda: Planorbidae), with emphasis on endemic species of the great East African lakes. Zool J Linn Soc 151:337–349

Kristensen TK, Yousif F, Raahauge P (1999) Molecular characterization of *Biomphalaria* spp. in Egypt. J Molluscan Stud 65:133–136

Levin DA (2003) The cytoplasmic factor in plant speciation. Syst Bot 28:5–11

Lofty WM, DeJong RJ, Abdel-Kader A, Loker ES (2005) A molecular survey of *Biomphalaria* in Egypt: is *B. glabrata* present? Am J Trop Med Hyg 73:131–139

Lynch M, Walsh B (1998) Genetics and analysis of quantitative traits. Sinauer Associates, Sunderland

Malek EA (1954) Morphological studies on the family Planorbidae (Mollusca: Pulmonata). II. The genital organs of *Biomphalaria boissyi* (Subfamily Planorbinae H.A. Pilsbry 1934). Trans Am Microsc Soc 73:285–296

Malek EA (1955) Anatomy of *Biomphalaria boissyi* as related to its infection with *Schistosoma mansoni*. Am Midl Nat 54:394–404

Malek EA (1985) Snail hosts of schistosomiasis and other snail-transmitted diseases in tropical America: a manual. Washington: Pan American Health Organization, Scientific Publication No. 478.

Mandahl-Barth G (1957) Intermediate hosts of *Schistosoma*. African Biomphalaria and Bulinus: I. Bull World Health Organ 16:1103–1163

Mavárez J, Steiner C, Pointier JP, Jarne P (2002) Evolutionary history and phylogeography of the schistosome-vector freshwater snail *Biomphalaria glabrata* based on nuclear and mitochondrial DNA sequences. Heredity 89:266–272

Mello-Silva CC, Grault CE, Almeida da Costa V, Barbosa FS (1998) Possible hybridization of Brazilian planorbid snails and its importance in population dynamics. Mem Inst Oswaldo Cruz 93:S227–S232

Mir C, Toumi L, Jarne P, Sarda V, Di Giusto F, Lumaret R (2006) Endemic North African *Quercus afares* Pomel originates from hybridization between two genetically very distant oak species (*Q. suber* L. and *Q. canariensis* Willd.). I. Evidence from allozyme analysis. Heredity 96:175–184

Morgan JAT, DeJong RJ, Jung YH, Khallaayoune K, Kock S, Mkoji GM, Loker ES (2002) A phylogeny of planorbid snails, with implications for the evolution of *Schistosoma* parasites. Mol Phylogenet Evol 25:477–488

Morgan JAT, Dejong RJ, Adeoye GO, Ansa EDO, Barbosa CS, Bremond P, Cesari IM, Charbonnel N, Correa LR, Coulibaly G, D'Andrea PS, De Souza CP, Doenhoff MJ, File S, Idris MA, Incani RN, Jarne P, Karanja DMS, Kazibwe F, Kpikpi J, Lwambo NJS, Mabaye A, Magalhaes LA, Makundi A, Moné H, Mouahid G, Muchemi GM, Mungai BN, Sene M, Southgate V, Tchuem-Tchuenté LA, Théron A, Yousif F, Magalhaes EMZ, Mkoji GM, Loker ES (2005)

Origin and diversification of the human parasite *Schistosoma mansoni*. Mol Ecol 14:3889–3902
Nieberding CM, Olivieri I (2006) Parasites: proxies for host genealogy and ecology? Trends Ecol Evol 22:156–165
Pan American Health Organization 1968. A guide for the identification of the snail intermediate hosts of schistosomiasis in the Americas. Washington: Pan American Health Organization, Scientific Publication No. 168.
Paraense WL (1956) A genetic approach to the systematics of planorbid mollusks. Evolution 10:403–407
Paraense WL (1959) One-sided reproductive isolation between geographically remote populations of a planorbid snail. Am Nat 93:93–101
Paraense WL (1961a) The nomenclature of Brazilian planorbids. I. "*Australorbis glabratus*" (Say 1818). Rev Bras Biol 21:287–296
Paraense WL (1961b) The nomenclature of Brazilian planorbids. II. "Australorbis tenagophilus" (Orbigny 1835). Rev Bras Biol 21:343–349
Paraense WL (1963) The nomenclature of Brazilian planorbids. III "*Australorbis stramineus* (Dunker 1848)". Rev Bras Biol 23:1–7
Paraense WL (1966) The synonymy and distribution of "*Biomphalaria peregrina*" in the Neotropical region. Rev Bras Biol 26:269–296
Paraense WL (1974) *Biomphalaria oligoza* n.n. for *Tropicorbis philippianus* (Dunker) sensu Lucena. Rev Bras Biol 34:379–386
Paraense WL (1975) *Biomphalaria orbignyi* sp.n. from Argentina (Gastropoda: Basommatophora: Planorbidae). Rev Bras Biol 35:211–222
Paraense WL (1981) *Biomphalaria occidentalis* sp. n. from South America (Mollusca, Basommatophora, Planorbidae). Mem Inst Oswaldo Cruz 76:199–211
Paraense WL (1988) *Biomphalaria kuhniana* (Clessin 1883), planorbid mollusc from South America. Mem Inst Oswaldo Cruz 83:1–12
Paraense WL (1990) *Biomphalaria obstructa* (Morelet 1849): a study of topotypic specimens (Mollusca: Pulmonata: Planorbidae). Mem Inst Oswaldo Cruz 85:391–399
Paraense WL (1996) *Biomphalaria subprona* (Martens 1899) (Gastropoda: Planorbidae). Mem Inst Oswaldo Cruz 91:187–190
Paraense WL (2001) The schistosome vectors in the Americas. Mem Inst Oswaldo Cruz 96:7–16
Paraense WL (2003) A bird's eye survey of Central American planorbid molluscs. Mem Inst Oswaldo Cruz 98:51–67
Paraense WL (2004) Planorbidae, Lymnaeidae and Physidae of Ecuador (Mollusca: Basommatophora). Mem Inst Oswaldo Cruz 99:357–362
Paraense WL, Correa LR (1963a) Variation in susceptibility of populations of *Australorbis glabratus*, infected to a strain of *Schistosoma mansoni*. Rev Inst Med Trop São Paulo 5:14–22
Paraense WL, Correa LR (1963b) Susceptibility of *Australorbis tenagophilus* to infection with *Schistosoma mansoni*. Rev Inst Med Trop São Paulo 5:23–29
Paraense WL, Correa LR (1973) Susceptinility of *Biomphalaria peregrina* from Brazil and Ecuador to two strains of *Schistosoma mansoni*. Rev Inst Med Trop São Paulo 15:127–130
Paraense WL, Deslandes N (1955a) Isolamento reproductivo entre *Australorbis glabratus* e *A. nigricans*. Mem Inst Oswaldo Cruz 53:321–327
Paraense WL, Deslandes N (1955b) Observations on the morphology of *Australorbis glabratus*. Mem Inst Oswaldo Cruz 53:87–103
Paraense WL, Deslandes N (1955c) Studies on *Australorbis centimetralis*. II. Biospecific characterization. Rev Bras Biol 14:341–343
Paraense WL, Deslandes N (1956) Diagnostic characters of the Brazilian species of *Australorbis*. Rev Bras Biol 16:281–286
Paraense WL and Deslandes, N. 1957. A redescription of *Taphius andecolus* (Orbigny 1835) Revista Brasileira de Biologia 17:235–243.

Paraense WL, Deslandes N (1958a) Observations on *Taphius havanensis*. Rev Bras Biol 18:87–91

Paraense WL and Deslandes N (1958b) *Taphius pronus* (Martens 1873) Revista brasileira de Biologia 18: 367-373.

Paraense WL, Deslandes N (1959) The renal ridge as a reliable character for separating *Taphius glabratus* from *T. tenagophilus*. Am J Trop Med Hyg 8:456–472

Paraense WL, Deslandes N (1962) *Australorbis intermedius* sp. n. from Brazil. Rev Bras Biol 22:343–350

Paraense WL, Ibañez N (1964) *Australorbis helophilus*. Rev Bras Biol 24:249–258

Paraense WL, Fauran P, Courmes E (1964) Observation sur la morphologie, la taxonomie, la répartition géographique et les gîtes *d'Australorbis schrammi*. Bull Soc Pathol Exot 57:236–254

Paraense WL, Pointier JP, Delay B, Pernot AF, Incani RN, Balzan C, Chrosciechowski P (1992) *Biomphalaria prona* (Gastropoda: Planorbidae): morphological and biochemical study. Mem Inst Oswaldo Cruz 87(2):171–179

Pilsbry HA (1934) Zoological results of the Dolan West China expedition of 1931. Part II. Mollusks. Proc Acad Nat Sci Philadelphia 86:5–28

Pointier JP, Paraense WL, DeJong R, Loker ES, Bargues MD, Mas-Coma S (2002) A potential snail hosts of schistosomiasis in Bolivia: *Biomphalaria amazonica* Paraense 1966. Mem Inst Oswaldo Cruz 97:793–796

Pointier JP, DeJong RJ, Tchuem Chuenté LA, Kristensen TK, Loker ES (2005) A neotropical snail hosts of *Schistosoma mansoni* introduced into Africa and consequences for the schistosomiasis transmission: *Biomphalaria tenagophila* in Kinshasa (Democratic Republic of Congo). Acta Trop 93:191–199

Ranson G (1953) Observations sur les Planorbidae africains. Bull Soc Pathol Exot 46:783–786

Ranson G, Cherbonnier G (1953) Appareil genital et radules de trios planorbes africains. Bull Museum Natl Hist Nat, Paris 25:176–180

Raup DM (1966) Geometric analysis of shell coiling: general problems. J Paleontol 40:1178–1190

Richards CS (1963) Infectivity of *Schistosoma mansoni* for Puerto Rican mollusks, including a new potential intermediate host. Am J Trop Med Hyg 12:26–33

Rollinson, D. (ed) (2001) Flukes and snails re-visited. Parasitology 123:S1–S292

Rollinson D, Simpson AJG (eds) (1987) The biology of schistosomes – from genes to latrines. Academic Press, London

Samadi S, Roumégoux A, Bargues MD, Mas-Coma S, Pointier JP (2000) Morphological studies of lymnaeid snails from the human fascioliasis endemic zone of Bolivia. J Molluscan Stud 66:31–44

Schutte CHJ, Van Eeden JA (1960) Contributions to the morphology of *Biomphalaria pfeifferi* (Krauss). 2. Internal anatomy. Ann Mag Nat Hist 2:136–156

Théron A, Coustau C (2005) Are *Biomphalaria* snails resistant to *Schistosoma mansoni?* J Helminthol 79:187–191

Théron A, Coustau C, Rognon A, Gourbière S, Blouin MS (2008) Effects of laboratory culture on compatibility between snails and schistosomes. Parasitology 135:1179–1188

Thompson JN (2005) The geographic mosaic of coevolution. University of Chicago Press, Chicago

Tian-Bi TN, N'Goran KE, N'Guetta SP, Matthys B, Sangare A, Jarne P (2008) Prior selfing and the selfing syndrome in animals: an experimental approach in the freshwater snail *Biomphalaria pfeifferi*. Genet Res 90:61–72

Vollmer W (1991) Anatomical features and their variation in the African schistosome hosts, *Biomphalaria sudanica* and *B. camerunensis* (Pulmonata; Planorbidae). Proceedings of the Second International Workshop on Malacology, Hong Kong, pp 393–408.

Watts S (2008) Foreword. In: Bruun B, Aagaard-Hansen J (eds) The social context of schistosomiais and its control. An introduction and annotated bibliography, pp. VII–IX. World Health Organization, Gèneve, Special Programme for Research and Training in Tropical Diseases.

Webster JP, Woolhouse MEJ (1998) Selection and strain specificity of compatibility in the snail-schistosome system. Parasitology 123:S41–S56

Woodruff DS, Mulvey M (1997) Neotropical schistosomiasis: African effinities of the host snail *Biomphalaria glabrata* (Gastropoda: Planorbidae). Biol J Linn Soc 60:505–516

Yong M, Pointier JP, Perera G (1997) The type locality of *Biomphalaria havanensis* (Pfeiffer 1839). Mal Rev 30:115–117

Yong M, Gutiérrez A, Perera G, Durand P, Pointier JP (2001) The *Biomphalaria havanensis* complex (Gastropoda: Planorbidae) in Cuba: a morphological and genetic study. J Mol Stud 67:103–111

Zelditch ML, Swiderski DL, Sheets HD, Fink WL (2004) Geometric morphometrics for biologists. A primer. Elsevier Academic Press, San Diego

Chapter 2
Laboratory Rearing of *Biomphalaria glabrata* Snails and Maintenance of Larval Schistosomes In Vivo and In Vitro

L.K. Eveland and M.A. Haseeb

Abstract This chapter describes the methods used by the current authors and numerous other investigators to optimize conditions for successfully growing and maintaining both uninfected and infected *Biomphalaria glabrata* in the laboratory, with particular emphasis on the optimal physical, chemical, and biological factors that affect successful snail rearing and/or breeding, and the ways that they can be measured. Consideration is also given to incidental organisms that can alter aquarium ecology and affect both the snails and their intended trematode parasites. Various methods of infecting snails and maintaining snail tissues and cells in vitro are also described. The goal of this chapter is to aid researchers who have the need or desire to maintain the life cycle of this pulmonate and its trematode parasites in the laboratory.

2.1 Introduction

Schistosomiasis and other snail-transmitted trematode infections continue to be parasitic scourges of humanity; difficulties studying them in the laboratory include the proper rearing and maintenance of their snail intermediate hosts. Snails thrive in nature under diverse and changing environmental conditions, but the exact physicochemical factors for their survival are difficult to duplicate in the laboratory, where diurnal and seasonal fluctuations in light, temperature, salinity, and other unknown ecological conditions may not be fully understood or cannot be replicated (Berrie 1970; Webbe and James 1971).

Because the maintenance of the schistosome life cycle can be quite involved in terms of manpower, space, and other resources, the National Institutes of Health – National Institute of Allergy and Infectious Diseases (NIH-NIAID) provides snails

L.K. Eveland (✉)
Department of Biological Sciences, California State University, Long Beach, CA 90840, USA
e-mail: eveland@csulb.edu

R. Toledo and B. Fried (eds.), Biomphalaria *Snails and Larval Trematodes*,
DOI 10.1007/978-1-4419-7028-2_2, © Springer Science+Business Media, LLC 2011

and developmental stages of schistosomes. Depending upon the laboratory's research focus and requirements for snail-derived schistosome stages (i.e., sporocysts, cercariae), it may sometimes be preferable to maintain the life cycle in the laboratory, which requires sustaining both the intermediate and an appropriate definitive host. When resources cannot be committed, and experiments can be timed and planned in advance, both snails and definitive hosts with prepatent infections can be procured free of charge. Both schistosome and snail cDNA libraries and a *Biomphalaria glabrata* (M-line) embryonic (Bge) cell line are also available (Lewis et al. 2008).

This chapter is intended for researchers who have the need and resources to maintain the entire life cycle of schistosomes or other trematodes in their laboratory. It attempts to describe the optimal parameters and methodologies for rearing and maintaining the snail intermediate hosts, and both classical and surgical techniques for infecting the snails with schistosomes. It describes the methods for maintaining snail tissues and cells to support in vitro cultivation and the growth of intramolluscan developmental forms of schistosomes and other trematodes. It also attempts to highlight some of the experimental studies on rearing and maintaining *Biomphalaria* snails that have dealt with the challenges of maintaining and studying trematodes in the laboratory.

2.2 Snail Rearing and Maintenance

B. glabrata snails are bred and maintained in the laboratory for several purposes. They are most often used to support the life cycles of schistosomes and other trematodes, and to generate a supply of schistosome sporocysts or cercariae (Lee and Lewert 1956; Lewis et al. 1986). They are also used in the laboratory as first and second intermediate hosts, particularly for echinostomatid trematodes whose cercariae encyst in snails (Huffman and Fried 1990; Fried and Huffman 1996), and as a source of organs and cells for in vitro culture (see Sect. 2.4). In addition, *Biomphalaria* snails are used in undergraduate research and laboratory teaching exercises.

2.2.1 Measures of Successful Snail Rearing

Successful snail rearing is the first requirement for maintaining the schistosome life cycle. Growth rate, fecundity, shell size, survival, and death rates are important indicators of the physiological state of the snail host. Large snails produce greater numbers of cercariae and lay more and large egg masses (or clutches), even though snail size does not appear to influence cercarial infectivity (Eveland and Ritchie 1972).

2.2.1.1 Survival and Longevity

In nature *B. glabrata* can live for 12–18 months, but their survival in the laboratory under closely monitored conditions may or may not exceed 12 months. Since the generation time is approximately 5 weeks, this snail can undergo several generations over a 12-month period (Ritchie et al. 1963; Loker 2006).

2.2.1.2 Growth and Maturation

Although it is impractical to determine the exact growth and maturation potentials of *Biomphalaria* snails in nature, theoretical growth curves have been developed for *B. glabrata* and several other species of snails in their natural habitats (Leveque and Pointier 1976). An approximation of those potentials has also been reported under laboratory conditions (Ritchie et al. 1963, 1966). When three different strains of *Biomphalaria* snails were cultivated individually in 200 ml of water, they grew more slowly and began egg-laying later than when cultured in a circulating water system. The best mean growth in the stagnant system was 19.7 mm in 180 days, and in the circulating system snails reached 27 mm in 150 days. Also, in the circulating system egg-laying began sooner than in the stagnant system (Ritchie et al. 1963). Another study reported better growth and fecundity of *Biomphalaria pfeifferi* in nonaerated water that was renewed weekly, and better snail growth but less fecundity in nonaerated water that was changed monthly (Frank 1963). *B. glabrata* growth parallels that of another planorbid snail (*Helisoma trivolvis*) maintained under identical laboratory conditions (Schneck and Fried 2005).

2.2.1.3 Fecundity

B. glabrata is hermaphroditic and is capable of both self- and cross-fertilization, although the latter is the preferred mechanism. Under ideal conditions, the egg to egg cycle may be as short as 1 month (Ritchie et al. 1963). A mature snail may lay 10,000 or more eggs/year. The eggs are laid in gelatinous masses of several eggs each that are found on the wall of the laboratory aquaria or other, e.g., filtration devices or the glass-covered heaters. We have used small 2–3 × 4–5″ cellophane sheets (rafts) that float in aquaria and snails deposit egg masses on their under surface; the sheets are easily transferred to other aquaria. We have observed that snails laid more eggs when maintained in plastic trays with shallow water.

2.2.2 Physical and Biological Factors that Affect Rearing

Although snails are prolific in nature, a number of important physical and biological factors must be taken into consideration for their successful cultivation in the laboratory, including but not limited to temperature, diet, crowding, and parasitism.

2.2.2.1 Temperature

Numerous studies have been conducted to determine optimal temperatures for *Biomphalaria* growth and development (Webbe and James 1971). Although minor differences in both growth and fecundity have been reported in studies on different *Biomphalaria* spp., uninfected snails grow best over a temperature range of 24–28°C, with 25°C being optimal for *B. glabrata* (Chernin and Schork 1959). *B. glabrata* grows more slowly in the laboratory at 20°C (Sturrock and Sturrock 1972). However, infected snails can be kept at a somewhat higher temperature (26–28°C) to allow for the rapid development of the trematode stages. Noninfected snails can be maintained in rooms with a relatively constant temperature, and submerged aquarium heaters can be used to achieve higher temperatures in aquaria containing infected snails. In rooms where constant temperature cannot be easily maintained, thermostatically controlled heaters should be used in aquaria, and the room temperature should be kept slightly lower than that of the aquaria.

2.2.2.2 Diet

As stated above, shell size is an important determinant of snail fecundity and cercarial production, although it does not influence cercarial infectivity in mice. The snails can live solely on a lettuce diet, but they grow much larger and lay more egg masses when their diet is supplemented. Romaine lettuce is preferable to iceberg, and maximum growth was obtained when lettuce was supplemented with a formula consisting of young grass shoots from wheat, barley, and rye crops, mixed together with an extract of thyroid and endocrine glands, wheat germ and powdered milk in a ratio of 4:2:2:1. This formula results in snails nearly 3 times larger than the lettuce-fed snails, and the larger snails lay 7 times more egg masses. Snails fed on lettuce supplemented with Gaines Meal dog food or commercial chicken food grow nearly twice as large as those fed on lettuce alone, and lay 6 and 4 times as many egg masses, respectively. The egg masses from snails on the supplemented diets are also larger and contain more embryos (Eveland and Ritchie 1972). Infected *B. glabrata* grows twice as large when fed on Purina Rat Chow for 7 weeks than when fed on boiled lettuce (Mecham and Holliman 1972). Cumulative cercarial outputs have also been shown to be higher from snails fed Romaine lettuce than from those fed iceberg lettuce, but cercarial output was not affected by adding the formula mentioned above to the diet (Cohen 1984). We and others (Boston et al. 1994) have also supplemented lettuce diet with Tetramin fish food.

It has been documented that some snail fauna are scarce in calcium-poor British freshwaters (Boycott 1936). *B. glabrata* growth rates, relative shell weights, fecundity correlate with increasing calcium concentrations (Thomas et al. 1974), and snails maintained in water with <2 mg/L Ca^{++} have smaller and extremely fragile shells (Mishkin and Jokinen 1986). Snail mortality is significantly greater when calcium levels are very low (1.5 mg/L) or very high (75 mg/L), with 30 mg/L being the optimal calcium concentration for the greatest snail fecundity (Mishkin and Jokinen 1986). Deleterious effects of high aquarium calcium concentrations on *B. glabrata* fecundity, growth, and survival have been reported (Frank 1963). Small pieces of chalk (calcium carbonate, $CaCO_3$) have been added to aquaria as a calcium source (MacInnis 1970; Ulmer 1970) with the aim of promoting snail growth and shell hardening. In fact, calcium reserves in *B. glabrata* increase when snails are infected with *Schistosoma mansoni* (Shaw and Erasmus 1987; Ong et al. 2004). Hemolymph calcium levels of snails kept in aquaria with chalk sticks for 3 days were higher than those of snails maintained without chalk (Chernin 1963). Another study that compared calcium content of *B. glabrata* maintained with and without chalk documented higher calcium levels in both soft tissues and shells of snails raised without chalk (Boston et al. 1994). Also, *B. alexandrina* and *Bulinus truncatus* with patent *S. mansoni* and *Schistosoma haematobium* infections were found to have relatively lower calcium content in shells and higher calcium content in soft tissues than the uninfected snails (Mostafa 2007). It seems that the substantial amount of calcium in both artificial spring water (Ulmer 1970) and lettuce (Boston et al. 1994) is apparently enough to meet the calcium requirement of most snails.

2.2.2.3 Crowding

Extreme crowding may influence both snail growth and fecundity. However, uninfected *Biomphalaria* grow and reproduce well over a wide population density range from 2 to 50 snails per unit volume of water when adequate nutrition is provided and the water is properly conditioned, either by the use of an activated charcoal and/or a water circulation system such as that mentioned in Sect. 3.2.1.2 on *Growth and Maturation* (Wright 1960; Ritchie et al. 1963). Due to their more fragile condition, infected snails should be kept at between 15 and 20 snails per liter. When 20 uninfected *B. glabrata* were maintained for as few as 4 days in one liter of stagnant water, the average O_2 uptake was considerably lower than for uncrowded snails (Coelho et al. 1977).

2.2.2.4 Parasitism

Many bacteria (Michelson 1961; Richards 1978; Ducklow et al. 1979, 1981; Bean-Knudsen et al. 1988), ichthyosporea (Hertel et al. 2002, 2004), microsporidia (Bayne et al. 1975; Lai and Canning 1980), fungi (Bagy et al. 1992), protozoa (Richards 1968), and numerous metazoa have been described living in or on *B. glabrata* collected from the field; also, many larval trematodes use *B. glabrata* as their first intermediate host. Occurrence of many diverse organisms in snails is not surprising since their natural habitats are shared by a variety of flora and fauna. Their relationship with *B. glabrata* remains imprecisely defined; most are probably symbionts and commensals although some may be pathogenic. Some may even exclusively parasitize trematode parasites of snails without causing apparent injury to the latter and may qualify as hyperparasites. Properly maintained snail populations seldom show signs of distress (declining snail number or reduced breeding potential) even when multiple species are present, but poorly maintained aquaria may permit unchecked growth of microorganisms that harm young and adult snails. Because

B. glabrata are maintained in the laboratory to support life cycles of schistosomes and other trematodes, the effect of unintended trematode infections on the growth, survival, and fecundity of snails is of wide interest.

Trematodes (Schistosomes and Other Trematodes)

Growth: Influence of infection on snail growth is obviously complex and varies with host and parasite species and strains, nutrition, and a number of other unknown factors (Thompson 1983). For example, one study using *S. mansoni*-infected *B. pfeifferi* provided evidence of "parasitic gigantism," or accelerated shell growth during the prepatent period of infection (Sturrock 1966), while another reported that *S. mansoni* infection of *B. glabrata* retards growth (Meier and Meier-Brook 1981).

Survival: Schistosome infections adversely affect snail survival, and the average life span of *B. glabrata* infected with *S. mansoni* is 39 days (Barbosa 1962). In one study, 77% of *S. mansoni*-infected *B. glabrata* died within 18 weeks postinfection, compared with only 11% mortality in noninfected controls (Pan 1965). Another study with these species demonstrated that infected snail deaths rose at a steady rate following infection until it reached 30% per week by 12 weeks following infection (Schwanbek et al. 1986). While infections cause profound mortality in *Biomphalaria* spp. regardless of their age, some studies report more deleterious effects in young snails, while others indicate that *S. mansoni* causes more damage to older snails (Sturrock 1966; Meier and Meier-Brook 1981). Other studies suggest that schistosome-infected snails live as long or longer than noninfected snails (Short 1952; McClelland and Bourns 1969; Fryer 1986). The number of miracidia used for infection can also profoundly affect snail survival. In a study with *S. haematobium* in *Bulinus truncatus*, 53% of uninfected snails survived to week 22, two miracidia per snail resulted in 10% survival and none of the snails exposed to 5 or 20 miracidia survived (Chu et al. 1966).

Fecundity: Schistosomes and other trematodes adversely affect the reproductive system of their snail hosts. *S. mansoni*-infected *B. glabrata* lays fewer eggs and their viability is greatly reduced, although complete cessation of egg-laying does not occur (Barbosa 1962). Reduction in or cessation of egg-laying has been observed in *B. glabrata* and *B. pfeifferi* infected with *S. mansoni* around the time infection becomes patent, and its extent may depend upon factors, such as the age of the snails at the time of infection or strains of parasites and snails (Pan 1965; Etges and Gresso 1965; Sturrock 1966). This effect has often been referred to as *parasitic castration*, even though it may or may not imply gonadal invasion and physical damage. Also, despite experiencing a dramatic decline in egg production, *S. mansoni*-infected *B. glabrata* continues to produce and transfer sperm to uninfected snails (Cooper et al. 1996). Other studies have shown increased egg-laying in *B. glabrata* infected with *S. mansoni* during the first 5 weeks of infection (Rodgers et al. 2005).

Other possibilities for the reduction in host fecundity include interference with host development or sexual maturation by nutrient depletion, or deleterious effects on the host accessory glands or endocrine function (Hurd 1990). The larval schistosomes may compete for scarce resources such as glucose required by the albumin gland of the snail host (von Brand 1979). They may also cause physiological stress, preventing the host from feeding and/or utilizing its nutritional intake (Becker 1980; Thompson 1983). Schistosomes produce substances that cause endocrine dysfunction and thus interfere with the host gonad development or oogenesis (Meulman 1972; Meier and Meier-Brook 1981). For example, substances present in hemolymph as well as excretory–secretory products from secondary sporocysts of *S. mansoni* suppress both polysaccharide and galactogen synthesis in *B. glabrata* (Crews and Yoshino 1990). The schistosome-induced host-derived factor, schistosomin, interferes with glucose and galactogen incorporation by the snail albumin gland (Dictus et al. 1987). This heat stable, pronase-labile factor also suppresses both male and female reproduction in *Lymnaea stagnalis* infected with the avian schistosome *Trichobilharzia ocellata* (de Jong-Brink et al. 1988, 1992). In this system, schistosomin appears and fecundity decreases at the same time that cercariae begin to differentiate in daughter sporocysts (Shallig et al. 1991). Schistosomin is believed to be a gonadotropin-antagonist that interferes with the neuroendocrine mechanisms in the host (Hurd 1990; de Jong-Brink 1995). However, in *B. glabrata* snails infected with *S. mansoni* or *Echinostoma paraensei,* a schistosomin-like substance was detectable during early snail development but did not become elevated when mature snails were infected, suggesting that schistosomin may not be responsible for castration in *B. glabrata* (Zhang et al. 2009) and that some other factor(s) may be involved. Effects of larval schistosomes on the snail are more completely covered in Chap. 5.

Organisms Other than Trematodes

Ichthyosporea and Microsporidia: One of the first microsporidia found in *B. glabrata* epithelial cells, *Coccospora brachynema* (Richards and Sheffield 1970) was renamed *Steinhausia brachynema* (Sprague et al. 1972). Transovarial transmission of a related species *Steinhausia mytilovum* in bivalves without apparent pathogenesis has been described (Sagrista et al. 1998; Matos et al. 2005). Microsporidia use both horizontal and vertical transmission (Dunn et al. 2001), making it more likely for these organisms to have survived in laboratory-reared *B. glabrata* that were isolated from the field many years ago. *Pleistophora* sp. found in cultured cells from the heart and gonads of juvenile *B. glabrata* do not impair host reproduction (Bayne et al. 1975; McClymont et al. 2005). *Nosema algerae*, a parasite of mosquitoes, was found to experimentally infect *S. mansoni* sporocysts in *B. glabrata* but not snail cells (Lai and Canning 1980).

Originally described as an ameba, *Nuclearia* sp. isolated from *B. glabrata* hearts attaches to and kills both *S. mansoni* sporocysts and *B. glabrata* embryonic cells (Stibbs et al. 1979; Owczarzak et al. 1980). This organism has been redescribed as a unicellular eukaryotic symbiont *Capsaspora owczarzaki* (Hertel et al. 2002). It is more prevalent in laboratory-maintained snails that are resistant to *S. mansoni*, but may not be the sole reason for the resistance (Hertel et al. 2004). Another

Table 2.1 Directly observed metazoa found associated with *Biomphalaria glabrata* or in aquaria with snails[a]

Taxonomic group	Organism	Comment
Ostracoda	*Cypridopsis vidua*	Detritus feeder, robs snail nourishment
Turbellaria	*Macrostomum* sp.	Consumes young snails
Rotifera	*Rotaria rotatoria*	May not affect snails directly but indicate poor aquarium conditions; reduce cercarial output, motility and infectivity
	Philodina acuticornis	
Copepoda	*Cyclops* sp.	May not affect snails directly but indicate poor aquarium conditions
Oribatida (water mite)	*Hydrozetes lemnae*	
Bryozoa	*Plumatella* sp.	
Oligochaeta	*Chaetogaster l. limnaei*	Ingests miracidia and cercariae
	Tubifex sp.	Consumes dead snails; keeps aquaria clean
Hirudinida	*Helobdella* sp.	Feed on snails; reduces snail fecundity

[a] Adapted from Webbe and James (1971) and Lewis et al. (1986)

unicellular eukaryotic symbiont (*Anurofeca* sp.) was originally described as an alga that inhibits anural tadpole growth (Richards 1949; Hertel et al. 2004). Based on DNA sequence homology, two ciliates (*Paruroleptus* sp. and *Trichodina* sp.) were also identified in field-collected *B. glabrata* in Brazil (Hertel et al. 2004).

Eukaryotic Metazoa: Of the many metazoa associated with *B. glabrata*, the few properly identified and documented are listed in Table 2.1. Perhaps the most closely examined is *Chaetogaster limnaei limnaei*. It has been found in association with *B. glabrata* and many other snail species, and has been regarded as a commensal or a symbiont. A large number of studies have demonstrated that *C. l. limnaei* protects *B. glabrata* against infection with *S. mansoni* either by ingesting miracidia or preventing their penetration into the snail (Khalil 1961; Michelson 1964; Wajdi 1964), but does not interfere with *B. glabrata* growth or fecundity (Rodgers et al. 2005). It also ingests cercariae of *Fasciola hepatica* when living in *Lymnaea tomentosa* (Rajasekariah 1978) and those of *Echinostoma trivolvis* and *Zygocotyle lunata* when living in *Helisoma trivolvis* (Fried et al. 2008). Immersion of infested snails in 1% urethane for 10–20 min has been reported helpful in partial elimination of *C. l. limnaei* (Michelson 1964), however, complete elimination of this and other infestations can only be accomplished by thorough cleaning of aquaria, including an ethanol rinse, and using new water, gravel, and known clean snails (see Sect. 2.2.3).

2.2.3 Methods of Snail Rearing

Snails reared for experimental purposes are typically maintained in monocultures (i.e., one snail species in one container) to prevent the introduction of competitor snails, commensals and pathogens, including trematode cercariae that encyst in these molluscs. They are most often kept in a room where temperature, humidity, and light can be regulated. Temperature and humidity are usually kept constant and the room with uninfected breeding snails is illuminated 12–14 h daily with fluorescent light

(Lee and Lewert 1956). Snails require 8 h of light for breeding (Webbe and James 1971). Schistosome-infected snails are maintained either in a separate room which is kept dark or in tanks that have been covered with dark or black paper or other material that prevents light from reaching snails and thus cercarial shedding.

Containers: The choice of containers depends on the number of snails to be maintained and the laboratory space available for this purpose. In developing countries, where ambient summer temperatures are high and the cost of temperature regulation is prohibitive, both field isolates and experimental snails are maintained in large earthen pots that may hold 2–10 Ga of water. Earthen pots are porous and have a large surface area that allows much greater evaporation, thus lowering the water temperature. These vessels require frequent replenishment of water.

In laboratories where temperature regulation is feasible, snails are maintained in glass or plexiglass tanks of various capacities depending on the shelving or storage arrangement. 10-Ga tanks are used more often because they are easy to clean and allow access to snails or debris at the tank bottom. The 10-Ga aquaria are also more suitable because of the ease of availability of various devices such as aerators, filters, and immersed heaters that fit and operate with these tanks.

Water: Snails do not usually survive for more than a day or two in freshly collected chlorinated water from municipal supplies. Tap water can be made usable for snail rearing by storage in aquaria with large surface area or repeatedly filtered through activated charcoal, which eliminates harmful substances such as chloramines. Charcoal-filtered water from aquaria with healthy snails (snail-conditioned water) can also be added. Where available, investigators have reliably used well water. For maintaining small number of snails or for experimental studies, deionized or distilled water can be reconstituted with different combinations of salts to prepare artificial spring water or "snail water" (Hopf and Muller 1962; Ulmer 1970; Cohen et al. 1980; Thomas 1986). Poorly conditioned new aquaria adversely affect snail survival, growth, and reproduction, but these effects may not be noticed over several days. Introduction of guppies (*Lebistes reticulatus*) to new aquaria conditions water and their survival indicates that snails would thrive unaffected in these aquaria. The fish also eat *Chaetogaster* sp. that may contaminate the tanks. Guppies should be removed once the snails have been established because they will also eat snail egg masses. Growth of snails is also believed to be enhanced because snails will feed on the organic waste of the fish (Ulmer 1970).

pH: Studies have shown that pH in the range of 4.9–8.9 does not adversely affect snail rearing and breeding (Chernin and Schork 1959; Webbe and James 1971). Optimal pH has been reported to be 7.0 ± 0.2 (Etges and Ritchie 1966), and we maintain aquaria over a narrow pH range of 6.9–7.1.

Aeration and Filtration: Aquaria with standing water and those with water circulating through a series of aquaria are in use in different laboratories. In either case, the water is filtered through filter wool and activated charcoal or beds of sand which also aerates water; direct aeration of aquaria is often unnecessary. A variety of aeration and filtration devices that run on electricity or pressurized air are available from biological supply houses. When pressurized air is used, it is prudent to use an in-line air filter to trap oil and debris. Besides aeration, properly operational filters remove much of the snail feces, thus facilitating the cleaning of the aquaria; the settled snail feces and decaying lettuce remains still require removal by nets or siphoning off using a suction device. Care must be exercised not to remove the juvenile snails. Aquatic plants may also be added to aquaria; some provide favorite sites for snails to deposit their egg masses (Standen 1951).

2.2.3.1 Axenic Snail Rearing

Axenically grown snails may be used in studies of their nutritional requirements and for the screening of microbial agents which may prove useful in snail control. *B. glabrata* have been successfully grown axenically from individual eggs separated from egg masses (Chernin 1957). The snails were maintained in either aquarium water or a "handling solution", and fed autoclaved brewer's yeast and formalin-killed *Escherichia coli* (Chernin 1957; Chernin and Schork 1959). Some of the snails maintained in aquarium water reached adult size, their growth was suboptimal, and they never laid eggs. This retarded growth and development was thought to be due to the limited food supply (Chernin and Schork 1959). Also, the snails did not grow when either brewer's yeast or *E. coli* was missing, and the growth was dependent upon the concentration of *E. coli*. Snails did not grow when *E. coli* were heat-killed, suggesting the presence of a heat-labile essential growth factor(s) in bacteria. Although penicillin had no adverse effect on the snails, streptomycin severely inhibited their growth (Chernin and Schork 1959).

A subsequent study using a medium containing both a solid and a liquid portion demonstrated optimal *B. glabrata* growth in the absence of egg-laying, but when vitamin E (α-Tocopherol) was added to the medium all of the seven noncontaminated snails laid eggs (Vieira 1967). Another study using an even more complex medium with both solid and liquid portions containing multiple vitamins, including vitamin E, also demonstrated optimal snail growth and egg-laying. This study used two slightly different solid media: medium one was used for growing snails for up to 8 weeks, and medium two was used thereafter. The solid media were sterilized either by autoclaving or by irradiation, and the liquid medium was autoclaved. The snails grew larger and laid many more eggs when the solid portion of the medium was sterilized by irradiation (de Souza et al. 1977).

2.3 Methods of Infecting Snails

2.3.1 Miracidial Infection

Ideally, one should infect snails using as few miracidia as necessary because polymiracidial infections may result in shorter snail longevity (see Sect. 2.2.1.1).

However, high infection rates are often difficult to obtain with monomiracidial infections. Using increasing numbers of miracidia generally results in increased parasite load within the surviving snails, although this increase is not necessarily linear. One study reported a mean of 34 secondary sporocysts resulting in *B. glabrata* monomiracidially infected with *S. mansoni*, but when the miracidia number was increased to two and four per snail the numbers of secondary sporocysts were 54 and 52, respectively (Christie and Prentice 1978). Snail mortality was not reported in this study.

Preparation of Miracidia: The organs with the largest number of schistosome eggs (e.g., liver) are surgically recovered (or harvested) from animals 7–10 weeks after exposure to 100–300 cercariae. The animal tissue is homogenized in a glass homogenizer or in a blender for 20–30 s in 0.85% NaCl (salt prevents hatching of miracidia). The homogenate is poured into a flask and allowed to settle for 20–30 min by which time eggs sink to the bottom of the flask. The supernatant is poured off. This washing step is repeated for two or more times as required by adding 0.85% NaCl and pouring off the supernatant until it becomes clear. In the final step, spring or freshwater is added instead of saline, and the entire flask contents are transferred into a side-arm flask which is carefully filled to the rim with water. The flask is placed in bright light for 5 min to stimulate miracidial hatching and then covered with foil or other suitable material, leaving the side-arm uncovered. A bright light is directed at the side-arm to attract miracidia which can then be transferred using a Pasteur pipet (MacInnis 1970).

Infection: For monomiracidial infections, individual snails should be incubated in small clean glass containers with a single miracidium in just enough water to cover the snail for at least 1 h. For polymiracidial infections, 20–25 snails can be put together for 1 h in a large Petri dish containing 100–200 miracidia and enough water to cover the snails. To achieve accurate counting, heavy suspensions of miracidia may first need to be serially diluted before precise numbers can be retrieved using a pipet. From 50 to 100 infected snails should then be maintained in well-aerated 10-Ga aquaria covered with brown paper or contact paper throughout the prepatent period.

2.3.2 *Transplantation of Parthenitae*

Trematode parthenitae recovered from infected snails and transplanted into non-infected snails thrive and undergo usual/normal/natural asexual multiplication. The first successful transplantations included those of *S. mansoni* sporocysts in *B. glabrata* (Chernin 1966) and *Echinoparyphium dunni* rediae in *Lymnaea rubiginosa* (Heyneman 1966). It should be noted that echinostome rediae produced patent infections following their introduction into body cavities of recipient snails, whereas schistosome sporocysts required implantation into tissues of recipient snails via the cephalopedal sinus. Other studies on transplantation of echinostome rediae have also been reported (Dönges 1963, 1968, 1971; Dönges and Götzelmann 1988).

The original transplantation technique (Chernin 1966) has undergone minor modifications and has helped elucidate the aspects of intramolluscan development of schistosomes. For example, daughter sporocysts of *S. mansoni* when transplanted into noninfected snails produce an additional generation of sporocysts (DiConza and Hansen 1972), while normal intramolluscan development of *S. mansoni* involves continuous multiplication of sporocysts (Jourdane et al. 1980).

Perhaps the technique's greatest benefit has been the perpetual maintenance of schistosome clones (Jourdane and Théron 1980; Nojima et al. 1980; Cohen and Eveland 1984; Jourdane 1984, 1990). In addition to circumventing the requirement for a vertebrate-host (Jourdane 1990), schistosome clones obtained by this technique can be maintained and biologically characterized (Cohen and Eveland 1988). Schistosome clones are biologically different from one another and from respective developmental stages resulting from polymiracidial infections (Rowntree and James 1977; Smith and Clegg 1979; Cohen 1984; Cohen and Eveland 1988; Al-Adhami et al. 2001). One study has revealed genetic heterogeneity among individuals within clones (Bayne and Grevelding 2003).

Sporocysts of other schistosome species have also been transplanted into their respective snail hosts: *S. haematobium* in *Bulinus truncatus* (Jourdane et al. 1981); *S. bovis* in *Bulinus truncatus* (Jourdane et al. 1984); and *S. japonicum* in *Oncomelania hupensis* (Jourdane et al. 1985; Jourdane and Xia 1986; Xia et al. 1998).

Although these techniques offer obvious advantages for the maintenance of clones of various schistosome species and strains, they also present some challenges. For example, they involve excision of schistosome sporocysts which normally are entangled in snail tissues, preparation of sporocysts for implantation, anesthetization of recipient snails to prevent them from retracting into their shells, surgical implantation of sporocysts, and postsurgical care of recipient snails. Some of these procedures used in earlier studies are summarized in Table 2.2.

The donor monomiracidially-infected snails are swabbed with 70% ethanol prior to the removal of soft tissues into a balanced salt solution that approximates the snail hemolymph composition (Chernin 1963; Cohen 1984; Cohen and Eveland 1984; Jourdane et al. 1985). This balanced salt solution has been used with various combinations of antibiotics (Chernin 1963; Cohen and Eveland 1984; Jourdane et al. 1985). Diluted Schneider's Drosophila medium (110 mOsm/Kg H_2O) was used in one study (Nojima et al. 1980). Parasitized tissues (hepatopancreas and ovotestis) are excised and cut into 1–2 mm^3 fragments using iris scissors (Chernin 1966; Cohen 1984).

Early trials of anesthetizing molluscs with chloral hydrate, cocaine, carbon dioxide, chlorotone, nembutal, and aqueous and alcoholic solutions of menthol were unsatisfactory. However, urethane (ethyl carbamate) was reported to be a successful snail relaxant (see Michelson 1958). Aqueous solutions of urethane (0.75–1.5%) administered for 2–5 h have been used to anesthetize 10–20 mm snails (Chernin 1966; Cohen and Eveland 1984). A 5-h treatment in urethane was required for adequate anesthetization and treatments longer than 8 h resulted in snail mortality (Cohen and Eveland 1984). Other studies have used 0.8% nembutal

Table 2.2 Summary of procedures used in the transplantation of larval schistosomes in snails

Schistosome species	*S. haematobium*	*S. japonicum*	*S. mansoni*
Miracidial infection of donor snail	Polymiracidial	Polymiracidial	Polymiracidial
Recipient snail species	*Bulinus truncatus truncatus*	*Oncomelania hupensis hupensis*	*Australorbis glabratus*[a]
Anesthetic	Nembutal[b] (0.15%)	Nembutal[b] (0.08%) for 7 h	Urethane (1.25–1.50%) for 2–5 h
Medium in which sporocysts excised	CBSS	CBSS	CBSS
Route/site of implantation	Cephalopedal sinus	Cephalopedal sinus	Cephalopedal sinus
Postoperative snail care	Aerated water containing nitrofurantoin for 48 h; then in aerated water at 28°C	Tap water containing nitrofurantoin for 24 h; then in water at 26°C with 12 h light and dark cycles	Aerated water in beakers and petri dishes
Outcome measures	84.6 and 33.3% infected on 34th day	80% survival at 48 h; 75% infected	77% infected
Comments	Higher infection rate for the sympatric combination and lower for the allopatric		Other strains and species of snails transplanted
Reference	Jourdane et al. 1981	Jourdane et al. 1985	Chernin 1966

CBSS Chernin's balanced salt solution
[a]*Australorbis glabratus* (=*Biomphalaria glabrata*)
[b]Nembutal = sodium pentobarbital

(sodium pentobarbital) for 5–8 h (Jourdane and Théron 1980) or 0.08% nembutal for 7 h (Jourdane et al. 1985). Exposure to urethane for 2 h affects acid phosphatase in hemocytes and numbers of circulating hemocytes but has little or no effect on phagocytosis of mouse erythrocytes or snail susceptibility to *S. mansoni* infection (Granath and Yoshino 1985).

While most investigators have implanted donor snail fragments into the exposed cephalopedal sinus of anesthetized recipient snails, sporocysts have also been implanted into the cephalopedal sinuses of nonanesthetized snails via a 1.5-mm hole in the shell (Nojima et al. 1980). A later study found success of transplantation (determined by resultant patent infections) to be greater in anesthetized snails with exposed cephalopedal sinuses than in nonanesthetized snails implanted through a hole in the shell (Cohen and Eveland 1984). Holes in snail shells were sealed with "Dental Cyanon" (α-cyanoacrylate) (Nojima et al. 1980) or with plasticene (Cohen and Eveland 1984). In addition, implantation of sporocysts developed in vitro in the presence of *B. glabrata* embryonic (Bge) cells or in Bge cell-conditioned medium resulted in patent infections (Ivanchenko et al. 1999; Kapp et al. 2003).

2.4 In Vitro Culture of Snail Tissues and Cells

The intramolluscan development of trematodes with multiple larval generations is far too complex to permit the analysis of the host-parasite interaction in vivo. In vitro cultivation is essential for understanding nutritional requirements and metabolic pathways of both the snails and trematodes and the factors that affect parasite development and snail susceptibility. Development of schistosomes in *B. glabrata* has been studied most intensively and our present knowledge is the result of incremental gains from studies of each of the successive larval generations from miracidium to cercaria (Basch and DiConza 1973; Ivanchenko et al. 1999; Coustau and Yoshino 2000).

Prior to 1949, maintenance of molluscan explants (organ culture) succeeded in maintaining healthy explants for only a few days (Bayne 1976). Early attempts to culture *B. glabrata* tissues used tentacles (Benex 1961, 1967), the usual sites of miracidial penetration, and hearts (Chernin 1963). Tentacle ciliary and muscular activity continued for 3 weeks of culture in a complex medium (Benex 1961). *S. mansoni* miracidia penetrated the cultured tentacles and transformed into mother sporocysts that differentiated into daughter sporocysts (Benex 1967). The snail hearts were cut into fragments, trypsinized, and cultured in a balanced salt solution (Chernin's balanced salt solution based on hemolymph analysis) containing bovine amniotic fluid, lactalbumin hydrolysate, yeast extract, beef embryo extract, horse serum, penicillin, and streptomycin. The hearts continued to beat up to 47 days in this solution and released amebocytes and "epithelial" cells, which did not divide (Chernin 1963). Also, digestive gland and ovotestis explants derived from *B. glabrata* infected with *S. mansoni* 40 days previously yielded infective cercariae for 1–2 weeks. When snail tissues were explanted 19 days after infection, cercariae emerged after a lag of 8–13 days demonstrating cercarial development in vitro (Chernin 1964). Thus, snail organ culture has permitted development in vitro of vboth early (Benex 1967) and late (Chernin 1964) schistosome larval stages (Bayne 1976).

The complex Medium 199 supplemented with salts, *Helix pomatia* extract (prepared by homogenizing the snail tissues in distilled water), fetal calf serum, streptomycin, and other antibiotics sustained explanted *B. glabrata* gonads at pH 7.0 and muscle tissue at pH 8.5 for 60 days and the cells that dissociated from the sheared tissue were subcultured (Burch and Cuadros 1965). However, another laboratory could not replicate these results using the same medium (Basch and DiConza 1973).

A study to establish primary cultures of *B. glabrata* cells used snail embryos at the trochophore stage with rudimentary digestive tracts and circulatory systems whose hearts had just begun to beat. Embryos of an albino strain of *B. glabrata* (Newton 1955) were freed from their capsules into a saline solution of six inorganic salts, and glucose, and then teased into fragments. The salt solution was replaced with one of 120 culture media tested, four of which were found to be superior. One of these media was based on an analysis of adult *B. glabrata* hemolymph. It consisted of salts, amino acids, coenzymes, vitamins, nucleic acid precursors, glucose, and antibiotics, including streptomycin, inactivated fetal calf serum, and whole egg ultrafiltrate. The medium had an osmolality of 155 mOsm/Kg H_2O which was higher than that of the *B. glabrata* hemolymph (112 [range: 108–115] mOsm/Kg H_2O]. The cultures were healthy during 6 weeks of observation, had numerous mitotic figures, and cells differentiated into five cell types: fibroblast-like; epithelial-like; contractile muscle cells; macrophage-like; and hemocytes (Basch and DiConza 1973). Another study reported values of 102–119 mOsm/Kg H_2O for the hemolymph of normal adult *B. glabrata* (Lee and Cheng 1972).

Primary cultures of *B. glabrata* (albino strain) trochophore embryos were more extensively examined in another study in which individual eggs were removed from masses, and incubated at 23°C in autoclaved aquarium water containing antibiotics, including streptomycin, until embryos reached the trochophore stage. They were then teased out of capsules into buffered saline with antibiotics (no streptomycin), trypsinized and transferred to a medium based on Schneider's *Drosophila* medium supplemented with fetal calf serum and several antibiotics, including streptomycin. The cultures were incubated at 25°C, 27°C, or 30°C. The cultures grew dense fibroblast-like cells, small irregular cells, and epithelioid patches. Several subcultures were made following trypsinization in two slightly different media consisting of buffered salts, amino acids, sugars, vitamins, nucleic acid precursors and fetal calf serum, peptone, and lactalbumin hydrolysate. At or about 3 weeks small colonies of cells were observed and subcultured. These cells grew well and were designated Bge (*B. glabrata* embryo) cells. By the seventh passage, Bge cells were cultured in another medium which lacked vitamins, nucleic acid precursors, and peptone but was supplemented with yeast hydrolysate. This medium was also based on diluted Schneider's medium, and was used for all subsequent cultures (Hansen 1976).

Bge cells are clear and approximately 20 μm in diameter when rounded, and have relatively large nuclei measuring about 9 μm. When attached, they are diploid (2n = 36), fibroblast-like or irregular in shape, with two to five protoplasmic extensions; the population doubles in 18–20 h at 27°C. Other subcultures with slightly different characteristics were also described (Hansen 1976).

The previously mentioned penetration of *S. mansoni* miracidia, their transformation into mother sporocysts, and differentiation of daughter sporocysts in *B. glabrata* tentacles maintained in vitro (Benex 1967) has been replicated for *S. japonicum* in cocultures of Bge cells (Coustau et al. 1997). *S. mansoni* daughter sporocysts have also been developed from mother sporocysts in Bge cocultures (Yoshino and Laursen 1995).

These and many other attempts employing modifications in procedures, medium composition, and the gaseous atmosphere have resulted in complete development and continuous propagation of snail-associated developmental stages of *S. mansoni*, from miracidium to cercariae, in Bge cell cocultures and in Bge cell – conditioned medium (Ivanchenko et al. 1999). Specifically, mercaptoethanol was used which

Table 2.3 Survival and development of various trematodes in vitro with and without *Biomphalaria glabrata* embryonic (Bge) cells[a]

Trematode	Natural snail host	Maximum survival (days) in medium only	Maximum survival (days) in medium + Bge cells	Development of daughter sporocysts/ rediae	References
Schistosoma mansoni	*Biomphalaria*	18	Continuous	+	Yoshino and Laursen 1995; Ivanchenko et al. 1999
S. japonicum	*Oncomelania*	28	210	+	Coustau et al. 1997
S. mattheei	*Bulinus*	7	25	–	Coustau and Yoshino 2000
S. intercalatum	*Bulinus*	21	63	–	Coustau and Yoshino 2000
Echinostoma caproni	*Biomphalaria*	14	119	–	Ataev et al. 1998
Fascioloides magna	*Lymnaea*	12	>90	+	Laursen and Yoshino 1999

[a] Adapted from Coustau and Yoshino 2000

possibly reduced oxidative stress or chelated metals ions, and nitrogen was used to modify the O_2 and CO_2 concentrations. The in vitro-derived cercariae were infective to hamsters and the sporocysts were infective to *B. glabrata* when implanted through a hole in the shell (Ivanchenko et al. 1999). Larval trematodes that develop in vitro in the presence of Bge cells are listed in Table 2.3.

It is noteworthy that Bge cells also support the in vitro development of snail-associated stages of trematodes that normally use a snail other than *B. glabrata* (Coustau et al. 1997; Laursen and Yoshino 1999).

In addition, primary cultures of *B. glabrata* ovotestis (Iwanaga 2002; Barbosa et al. 2006) and amebocyte-producing organ (APO) (Barbosa et al. 2006) have been carried out. The ovotestis cultures grew two cell types (Iwanaga 2002) and those of the APO grew three cell types (Barbosa et al. 2006).

2.5 Concluding Remarks

Because of its essential role as the intermediate host of *S. mansoni*, *B. glabrata* has been maintained and studied for almost a century in numerous laboratories throughout the world. Even though a great deal has been learned about its biology from both field and laboratory studies, laboratory rearing and maintenance of this species remain quite challenging and extremely labor intensive. However, numerous studies done in our laboratory and others have succeeded in defining the best possible conditions for breeding and maintaining the snails under artificial conditions. Of the many factors that determine success, diet, temperature, proper aeration, and water quality factors such as proper pH and salt content appear to be the most important. Also important in maintaining healthy snails is the early recognition and elimination of unintended organisms from the aquaria.

The more recent success in schistosome cloning and their maintenance by microsurgical transplantation of sporocysts has been highlighted in this chapter. Development of successive snail-associated schistosome stages, from miracidial transformation into sporocysts to the generation of infective cercariae in Bge cell culture is discussed as a major achievement. Schistosome clones maintained in snails by microsurgical transplantation of sporocysts should also make a major contribution to our efforts at dissecting and understanding the human immune response in schistosomiasis, which may ultimately help in designing a vaccine that will reduce the human misery caused by schistosomiasis in endemic areas of the globe.

Efforts at deciphering and elucidating the *B. glabrata* genome are progressing rapidly (see Chap. 9). This knowledge will further our understanding of the biology of the snail and its relationship with its trematode parasites. Another potential benefit of this understanding could be to control the natural populations of *B. glabrata*, thus reducing or eliminating transmission of schistosomiasis.

References

Al-Adhami BH, Doenhoff M, Thornhill J, Akhkha A, White E, Kusel JR (2001) A study of some characteristics of individual clones of *Schistosoma mansoni* with emphasis on the biological and metabolic activities. Parasitology 123:563–572

Ataev GL, Fournier A, Coustau C (1998) Comparison of *Echinostoma caproni* mother sporocyst development in vivo and in vitro using *Biomphalaria glabrata* snails and a *B. glabrata* embryonic cell line. J Parasitol 84:227–235

Bagy MMK, Khallil AM, Obuidallah AH (1992) Fungi inhabiting some aquatic macroinvertebrates and water plants of the Nile at Egypt. Zentralblatt für Mikrobiologie 147:459–475

Barbosa FS (1962) Aspects of the ecology of the intermediate hosts of *Schistosoma mansoni* interfering with the transmission of bilharziasis in north-eastern Brazil. In: Wolstenholme GEW, O'Connor M (eds) Bilharziasis. Ciba foundation symposium. Little, Brown and Co, Boston, pp 23–34

Barbosa L, Silva LM, Coelho PMZ, Santos SR, Fortes-Dias CL (2006) Primary culture of the region of the amebocyte-producing organ of the snail *Biomphalaria glabrata*, the intermediate host of *Schistosoma mansoni*. Memórias do Instituto Oswaldo Cruz 101:639–643

Basch PF, DiConza JJ (1973) Primary cultures of embryonic cells from the snail *Biomphalaria glabrata*. Am J Trop Med Hyg 22:805–813

Bayne CJ (1976) Culture of molluscan organs: a review. In: Maramorosch K (ed) Invertebrate tissue culture: research applications. Academic, New York, pp 61–74

Bayne CJ, Grevelding CG (2003) Cloning of *Schistosoma mansoni* sporocysts in vitro and detection of genetic heterogeneity among individuals within clones. J Parasitol 89:1056–1060

Bayne CJ, Owczarzak A, Noonan WE (1975) In vitro cultivation of cells and a microsporidian parasite of *Biomphalaria glabrata* (Pulmonata: Basommatophora). Conference on Pathobiology of Invertebrate Vectors of Disease. Ann N Y Acad Sci 266:513–527

Bean-Knudsen DE, Uhazy LS, Wagner JE, Young BM (1988) Systemic infection of laboratory-reared *Biomphalaria glabrata* (Mollusca, Gastropoda) with an acid-fast bacillus. J Invertebr Pathol 51:291–293
Becker W (1980) Metabolic interrelationship between parasitic trematodes and mollusks, especially *Schistosoma mansoni* in *Biomphalaria glabrata*. Zeitschrift für Parasitenkunde 63:101–111
Benex J (1961) La survie d'explants de Planorbes (*Australorbis glabratus*) en milieu synthétique, antiseptique et nutritif. Comptes Rendus de l'Académie des Sciences 253:734–736
Benex J (1967) Possibilities of organotypic culture in liquid medium in the study of parasitic problems. II. Comparative study of the evolution of *Schistosoma mansoni* in explants of various species of *Planorbis* kept in organotypic cultures. Analysis of parasitic specificity. Annales de Parasitologie Humaine et Comparee 42:493–524. (In French)
Berrie AD (1970) Snail problems in African schistosomiasis. Adv Parasitol 8:43–96
Boston CJ, Layman LR, Fried B, Sherma J (1994) Calcium content of planorbid snails maintained in artificial spring water on leaf lettuce with and without chalk. Veliger 37:121–123
Boycott AE (1936) The habitats of fresh-water Mollusca in Britain. J Anim Ecol 5:116–186
Burch JB, Cuadros C (1965) A culture medium for snails and tissues. Nature 206:637–638
Chernin E (1957) A method of securing bacteriologically sterile snails (*Australorbis glabrata*). Proc Soc Exp Biol Med 96:204–210
Chernin E (1963) Observations on hearts explanted in vitro [sic] from the snail *Australorbis glabratus*. J Parasitol 49:353–364
Chernin E (1964) Maintenance in vitro of larval *Schistosoma mansoni* in tissues from the snail, *Australorbis glabratus*. J Parasitol 50:531–545
Chernin E (1966) Transplantation of larval *Schistosoma mansoni* from infected to uninfected snails. J Parasitol 52:473–482
Chernin E, Schork AR (1959) Growth in axenic culture of the snail *Australorbis glabratus*. Am J Hyg 69:146–160
Christie JD, Prentice MA (1978) The relationship between numbers of *Schistosoma mansoni* daughter sporocysts and miracidia. Ann Trop Med Parasitol 72:197–198
Chu KY, Sabbaghian H, Massoud J (1966) Host-parasite relationship of *Bulinus truncatus* and *Schistosoma haematobium* in Iran. Bull World Health Organ 34:121–130
Coelho PMZ, Figueiredo YP, Pellegrini J, Melo GR (1977) *Biomphalaria glabrata*: the crowding effect related to aquarium water components. J Parasitol 63:284
Cohen LM (1984) *Schistosoma mansoni*: cloning and its applications. Ph.D. thesis. State University of New York, Downstate Medical Center
Cohen LM, Eveland LK (1984) *Schistosoma mansoni*: long-term maintenance of clones by microsurgical transplantation of sporocysts. Exp Parasitol 57:15–19
Cohen LM, Eveland LK (1988) *Schistosoma mansoni*: Characterization of clones maintained by microsurgical transplantation of sporocysts. J Parasitol 79:963–969
Cohen LM, Neimark H, Eveland LK (1980) *Schistosoma mansoni*: response of cercariae to a thermal gradient. J Parasitol 66:362–364
Cooper LA, Larson SE, Lewis FA (1996) Male reproductive success of *Schistosoma mansoni* – infected *Biomphalaria glabrata* snails. J Parasitol 82:428–431
Coustau C, Ataev G, Jourdane J, Yoshino TP (1997) *Schistosoma japonicum*: In vitro cultivation of miracidium to daughter sporocyst using a *Biomphalaria glabrata* embryonic cell line. Exp Parasitol 87:77–87
Coustau C, Yoshino TP (2000) Flukes without snails: advances in the in vitro cultivation of intramolluscan stages of trematodes. Exp Parasitol 94:62–66
Crews AE, Yoshino TP (1990) Influence of larval schistosomes on polysaccharide synthesis in albumin glands of *Biomphalaria glabrata*. Parasitology 101:351–359
De Jong-Brink M (1995) How schistosomes profit from the stress responses they elicit in their hosts. Adv Parasitol 35:177–256
De Jong-Brink M, Elsaadany MM, Boer HH (1988) *Trichobilharzia ocellata*: interference with endocrine control of female reproduction of *Lymnaea stagnalis*. Exp Parasitol 65: 91–100

De Jong-Brink M, Hordijk PL, Vergeest DP, Schallig HD, Kits KS, ter Maat A (1992) The anti-gonadotropic neuropeptide schistosomin interferes with peripheral and central neuroendocrine mechanisms involved in the regulation of reproduction and growth in the schistosome-infected snail *Lymnaea stagnalis*. Prog Brain Res 92:385–396

de Souza CP, das Gracas Vilela M, de Carvalho MA, Cardillo-Vieira E (1977) Rearing of the snail *Biomphalaria glabrata* under axenic conditions in isolators. Am J Trop Med Hyg 26:1013–1017

DiConza JJ, Hansen EL (1972) Multiplication of transplanted *Schistosoma mansoni* daughter sporocysts. J Parasitol 58:181–182

Dictus WJ, De Jong-Brink M, Boer HH (1987) A neuropeptide (calfluxin) is involved in the regulation of calcium into mitochondria of the albumen gland of the fresh-water snail *Lymnaea stagnalis*. Gen Comp Endocrinol 65:439–450

Dönges J (1963) Die experimentelle bestimung der anzahl der rediengenerationen bei Trematoden. Naturwissenschaften 50:103–104

Dönges J (1968) Der beweis potentiell unbeschränkter generansfolge bei redien von *Isthmiophora melis* (Trematoda, Echinostomatidae) durch das transplantations-experiment. Verhandlungen der Deutschen Zoologischen Gesellschaf 32:S550–S558

Dönges J (1971) The potential number of redial generations in echinostomatids (Trematoda). Int J Parasitol 1:51–59

Dönges J, Götzelmann M (1988) Digenetic trematodes: multiplication of the intramolluscan stages in some species is potentially unlimited. J Parasitol 74:884–885

Ducklow HW, Boyle PJ, Maugel PW, Strong C, Mitchell R (1979) Bacterial flora of the schistosome vector snail *Biomphalaria glabrata*. Appl Environ Microbiol 38:667–672

Ducklow HW, Clausen K, Mitchell R (1981) Ecology of bacterial communities in the schistosomiasis vector snail *Biomphalaria glabrata*. Microb Ecol 7:253–274

Dunn AM, Terry RS, Smith JE (2001) Transovarial transmission in the microsporidia. Adv Parasitol 48:57–100

Etges FJ, Gresso W (1965) Effect of *Schistosoma mansoni* infection upon fecundity in *Australorbis glabratus*. J Parasitol 51:757–760

Etges FG, Ritchie LS (1966) Comparative observations on growth rate and reproduction of *Australorbis glabratus* in field and laboratory conditions. Bull World Health Organ 34:963–966

Eveland LK, Ritchie LS (1972) Infectivity of cercariae of *Schistosoma mansoni* from snails on inadequate diets. Parasitology 64:441–444

Frank GH (1963) Some factors affecting the fecundity of *Biomphalaria pfeifferi* (Krauss) in glass aquaria. Bull World Health Organ 20:531–537

Fried B, Huffman JE (1996) The biology of the intestinal trematode *Echinostoma caproni*. Adv Parasitol 38:311–368

Fried B, Peoples RC, Saxton TM, Huffman JE (2008) The association of *Zygocotyle lunata* and *Echinostoma trivolvis* with *Chaetogaster limnaei*, an ectosymbiont of *Helisoma trivolvis*. J Parasitol 94:553–554

Fryer SE (1986) Studies on the epidemiology of a Nigerian strain of *Schistosoma haematobium* with particular reference to the molluscan hosts. Ph.D. thesis. University of Wales

Granath WO Jr, Yoshino TP (1985) *Biomphalaria glabrata* (Gastropoda): Effect of urethane on the morphology and function of hemocytes, and on susceptibility to *Schistosoma mansoni* (Trematoda). J Invertebr Pathol 45:324–330

Hansen EL (1976) A cell line from embryos of *Biomphalaria glabrata* (Pulmonata): establishment and characteristics. In: Maramorosch K (ed) Invertebrate tissue culture: research applications. Academic, New York, pp 77–99

Hertel LA, Bayne CJ, Loker ES (2002) The symbiont *Capsaspora owczarzaki*, nov. gen. nov. sp., isolated from three strains of the pulmonate snail *Biomphalaria glabrata* is related to members of the Mesomycetozoea. Int J Parasitol 32:1183–1191

Hertel LA, Barbosa CS, Santos RAAL, Loker ES (2004) Molecular identification of symbionts from the pulmonate snail *Biomphalaria glabrata* in Brazil. J Parasitol 90:759–763

Heyneman D (1966) Successful infection with larval echinostomes surgically implanted into body cavity of the normal snail host. Exp Parasitol 18:220–223

Hopf HS, Muller RL (1962) Laboratory breeding and testing of *Australorbis grabratus* for molluscicidal screening. Bull World Health Organ 27:783–789
Huffman JE, Fried B (1990) *Echinostoma* and echinostomiasis. Adv Parasitol 29:215–269
Hurd H (1990) Physiological and behavioural interactions between parasites and invertebrate hosts. Adv Parasitol 29:271–318
Ivanchenko MG, Lerner JP, McCormick RS, Toumadje A, Allen B, Fischer K, Hedstrom O, Helmrich A, Barnes DW, Bayne CJ (1999) Continuous in vitro propagation and differentiation of cultures of the intramolluscan stages of the human parasite *Schistosoma mansoni*. Proc Natl Acad Sci USA 96:4965–4970
Iwanaga Y (2002) In vitro cultivation of cells from ovotestis tissue of pigmented *Biomphalaria glabrata*. Zool Sci 19:271–274
Jourdane J (1984) Long term maintenance of a male and a female clone of *Schistosoma mansoni* by microsurgical transplantation of sporocysts. Reliability of the technique. Annales de Parasitologie Humaine et Comparee 59:361–367. (In French)
Jourdane J (1990) Cloning of schistosomes. Parasitol Today 6:363–365
Jourdane J, Kechemir N, Combes C (1981) Mise en évidence d'une réplication des sporocystes fils de *Schistosoma haematobium* aprés transplantation microchirurgicale chez *Bulinus truncatus*. Comptes Rendus de l'Académie des Sciences 293:531–533
Jourdane J, Liang Y, Bruce JI (1985) Transplantation of *Schistosoma japonicum* daughter sporocysts in *Oncomelania hupensis*. J Parasitol 71:244–247
Jourdane J, Mouahid A, Touassem R (1984) Evolution of the sporocysts of *Schistosoma bovis* after microsurgical transplantation into *Bulinus truncatus*. Annales de Parasitologie Humaine et Comparee 59:459–466. (In French)
Jourdane J, Théron A (1980) *Schistosoma mansoni*: cloning by microsurgical transplantation of sporocysts. Exp Parasitol 50:349–357
Jourdane J, Théron A, Combes C (1980) Demonstration of several sporocyst generations as a normal pattern of reproduction of *Schistosoma mansoni*. Acta Trop 37:177–182
Jourdane J, Xia MY (1986) Study of schistosome-mollusk vector compatibility by the technique of microsurgical transplantation of sporocysts in two *Schistosoma japonicum – Oncomelania hupensis* combinations originating in mainland China. Annales de Parasitologie Humaine et Comparee 61:559–564. (In French)
Kapp K, Coustau C, Wippersteg V, Jourdane J, Kunz W, Grevelding CG (2003) Transplantation of in vitro-generated *Schistosoma mansoni* mother sporocysts into *Biomphalaria glabrata*. Parasitol Res 91:482–485
Khalil LF (1961) On the capture and destruction of miracidia by *Chaetogaster limnaei*. J Helminthol 35:269–274
Lai PF, Canning EU (1980) Infectivity of a microsporidium of mosquitoes (*Nosema algerae*) to larval stages of *Schistosoma mansoni* in *Biomphalaria glabrata*. Int J Parasitol 10:293–301
Laursen JR, Yoshino TP (1999) *Biomphalaria glabrata* embryonic (Bge) cell line supports *in vitro* miracidial transformation and early larval development of the deer liver fluke, *Fascioloides magna*. Parasitology 118:187–194
Lee C-L, Lewert RM (1956) The maintenance of *Schistosoma mansoni* in the laboratory. J Infect Dis 99:15–20
Lee FO, Cheng TC (1972) *Schistosoma mansoni*: alterations in total protein and hemoglobin in the hemolymph of infected *Biomphalaria glabrata*. Exp Parasitol 31:203–216
Leveque C, Pointier JP (1976) Study of the growth of *Biomphalaria glabrata* (Say) and other Planorbidae in Guadeloupe (West Indies). Ann Trop Med Parasitol 70:199–204
Lewis FA, Liang Y-s, Raghavan N, Knight M (2008) The NIH-NIAID Schistosomiasis Resource Center. PloS Negl Trop Dis 2:e267
Lewis FA, Stirewalt MA, Souza CP, Gazzinelli G (1986) Large-scale laboratory maintenance of *Schistosoma mansoni*, with observations on three schistosome/snail host combinations. J Parasitol 72:813–829
Loker ES (2006) Research on the molluscan intermediate hosts for schistosomiasis: What are the priorities? Report of the Scientific Working Group meeting on Schistosomiasis. Nov 14–16 2005, Geneva, pp 95–104

McClelland G, Bourns TKR (1969) Effects of *Trichobilharzia ocellata* on growth reproduction and survival of *Lymnaea stagnalis*. Exp Parasitol 24:137–146
McClymont HE, Dunn AM, Terry RS, Rollinson D, Littlewood DTJ, Smith JE (2005) Molecular data suggest that microsporidian parasites in freshwater snails are diverse. Int J Parasitol 35:1071–1078
MacInnis AJ (1970) Maintenance of *Schistosoma mansoni* and *Schistosomatium douthitti*. In: MacInnis AJ, Voge M (eds) Experiments and techniques in parasitology. W.H. Freeman, San Francisco, pp 141–143
Matos E, Matos P, Azevedo C (2005) Observations on the intracytoplasmic microsporidian *Steinhausia mytilovum*, a parasite of mussel (*Mytella guyanensis*) oocytes from the Amazon river estuary. Brazilian. J Morphol Sci 22:183–186
Mecham JA, Holliman RB (1972) An improved feeding procedure for *Biomphalaria glabrata*. J Parasitol 58:835
Meier M, Meier-Brook C (1981) *Schistosoma mansoni*: effect on growth, fertility, and development of distal male organs in *Biomphalaria glabrata* exposed to miracidia at different ages. Zeitschrift für Parasitenkunde 66:121–131
Meulman EA (1972) Host-parasite interrelationships between the freshwater pulmonate *Biomphalaria pfeifferi* and the trematode *Schistosoma mansoni*. Netherlands J Zool 22:355–427
Michelson EH (1958) A method for the relaxation and immobilization of pulmonate snails. Trans Am Microsc Soc 77:316–319
Michelson EH (1961) An acid-fast pathogen of fresh-water snails. Am J Trop Med Hyg 10:423–433
Michelson EH (1964) The protective action of *Chaetogaster limnaei* on snails exposed to *Schistosoma mansoni*. J Parasitol 50:441–444
Mishkin EM, Jokinen EH (1986) Effects of environmental calcium on fecundity and cercarial production of *Biomphalaria glabrata* (Say) infected with *Schistosoma mansoni* Sambon. J Parasitol 72:885–890
Mostafa O (2007) Effects of *Schistosoma mansoni* and *Schistosoma haematobium* infections on calcium content in their intermediate hosts. Parasitol Res 101:963–966
Newton WL (1955) The establishment of a strain of *Australorbis glabratus* which combines albinism and high susceptibility to infection with *Schistosoma mansoni*. J Parasitol 41:526–528
Nojima H, Noda S, Sato A (1980) Serial implantations of larval *Schistosoma mansoni* from infected to uninfected snails. J Parasitol 66:478–482
Ong JHL, Chejlava M, Fried B, Koehnlein KM, Bosavage GL, Sherma J (2004) Effects of *Schistosoma mansoni* infection on inorganic elements in the snail *Biomphalaria glabrata*. J Helminthol 78:343–346
Owczarzak A, Stibbs HH, Bayne CJ (1980) The destruction of *Schistosoma mansoni* mother sporocysts *in vitro* by amoebae isolated from *Biomphalaria glabrata*: an ultrastructural study. J Invertebr Pathol 35:26–33
Pan CT (1965) Studies on the host-parasite relationship between *Schistosoma mansoni* and the snail *Australorbis glabratus*. Am J Trop Med Hyg 14:931–976
Rajasekariah GR (1978) *Chaetogaster limnaei* K von Baer 1872 on *Lymnaea tomentosa*: Ingestion of *Fasciola hepatica* cercariae. Experientia 34:1458–1459
Richards CM (1949) The inhibition of growth in crowded *Rana pipiens* tadpoles. Physiol Zool 16:138–151
Richards CS (1968) Two new species of *Hartmannella* amebae infecting freshwater mollusks. J Protozool 15:651–656
Richards CS (1978) Spirochetes in planorbid molluscs. Trans Am Microsc Soc 97:191–198
Richards CS, Sheffield HG (1970) Unique host relations and ultrastructure of a new microsporidian of the genus *Coccospora* infecting *Biomphalaria glabrata*. Proceedings of the IV Intrenational Colloqium of Insect Pathology. pp. 439–452
Ritchie LS, Berrios-Duran LA, DeWeese R (1963) Biological potentials of *Australorbis glabratus*: growth and maturation. Am J Trop Med Hyg 12:264–268
Ritchie LS, Hernandez A, Rosa-Amador R (1966) Biological potentials of *Australorbis glabratus*: life span and reproduction. Am J Trop Med Hyg 15:614–617

Rodgers JK, Sandland GJ, Joyce SR, Minchella DJ (2005) Multi-species interactions among a commensal (*Chaetogaster limnaei limnaei*), a parasite (*Schistosoma mansoni*), and an aquatic snail host (*Biomphalaria glabrata*). J Parasitol 91:709–712

Rowntree S, James C (1977) Single sex cercariae of *Schistosoma mansoni*, a comparison of male and female infectivity. J Helminthol 51:69–70

Sagrista E, Bozzo MG, Bigas M, Poquet M, Durfort M (1998) Developmental cycle and ultrastructure of *Stainhausia mytilovum*, a microsporidian parasite of oocytes of the mussel, *Mytilus galloprovincialis* (Mollusca, Bivalvia). Eur J Protistol 34:58–68

Schneck JL, Fried B (2005) Growth of *Biomphalaria glabrata* (NMRI strain) and *Helisoma trivolvis* (Colorado strain) under laboratory conditions. Am Malacol Bull 20:71–73

Schwanbek A, Becker W, Rupprecht H (1986) Quantification of parasite development in the host-parasite system *Biomphalaria glabrata* and *Schistosoma mansoni*. Zeitschrift für Parasitenkunde 72:365–373

Shallig HD, Sassen MJ, Hordijk PL, de Jong-Brink M (1991) *Trichobilharzia ocellata*: influence of infection on the fecundity of its intermediate snail host *Lymnaea stagnalis* and cercarial induction of the release of schistosomin, a snail neuropeptide antagonizing female gonadotropic hormones. Parasitology 101:351–359

Shaw MK, Erasmus DA (1987) *Biomphalaria glabrata*: changes in calcium reserves following parasitism by larval *Schistosoma mansoni*. Parasitology 95:267–276

Short RB (1952) Sex studies on *Schistosomatium douthitti* (Cort, 1914) Price, 1931 (Trematoda: Schistosomatidae). Am Midl Nat 47:1–54

Smith MA, Clegg JA (1979) Different levels of immunity to *Schistosoma mansoni* in the mouse: the role of variant cercariae. Parasitology 78:311–321

Sprague V, Ormieres R, Manier JF (1972) Creation of a new genus and a new family in the Microsporida. J Invertebr Pathol 20:228–231

Standen OD (1951) Some observations upon the maintenance of *Australorbis glabratus* in the laboratory. Ann Trop Med Parasitol 45:80–83

Stibbs HH, Owczarzak A, Bayne CJ, DeWan P (1979) Schistosome sporocyst-killing amoebae isolated from *Biomphalaria glabrata*. J Invertebr Pathol 33:159–170

Sturrock BM (1966) The influence of infection with *Schistosoma mansoni* on the growth rate and reproduction of *Biomphalaria pfeifferi*. Ann Trop Med Parasitol 60:187–197

Sturrock RF, Sturrock BM (1972) The influence of temperature on the biology of *Biomphalaria glabrata* (Say), intermediate host of *Schistosoma mansoni* on St. Lucia, West Indies. Ann Trop Med Parasitol 66:385–390

Thomas JD (1986) The chemical ecology of *Biomphalaria glabrata* (say): Sugars as attractants and arrestants. Comp Biochem Physiol A 83:457–460

Thomas JD, Lough A, Aram RH (1974) The effects of calcium in the external environment on the growth and natality rates of *Biomphalaria glabrata* (Say). J Anim Ecolo 43:839–860

Thompson SN (1983) Biochemical and physiological effects of metazoan endoparasites on their host species. Comp Biochem Physiol B 74:183–211

Ulmer MJ (1970) Notes on rearing of snails in the laboratory. In: MacInnis AJ, Voge M (eds) Experiments and techniques in parasitology. W.H. Freeman, San Francisco, pp 143–144

Vieira EC (1967) Influence of vitamin E on reproduction of *Biomphalaria glabrata* under axenic conditions. Am J Trop Med Hyg 16:792–796

von Brand T (1979) Pathophysiology of the host. In: Von Brand T (ed) Biochemistry and physiology of endoparasites. Elsevier/North Holland Biomedical Press, Amsterdam, pp 321–390

Wajdi N (1964) The predation of *Schistosoma mansoni* by the oligochaete annelid *Chaetogaster*. J Helminthol 38:391–392

Webbe G, James C (1971) The importation and maintenance of schistosomes of human and veterinary importance. In: Taylor AER, Muller R (eds) Isolation and maintenance of parasites in vivo. IX Symposium of the British Society of Parasitology, London, pp 77–107

Wright CA (1960) The crowding phenomenon in laboratory colonies of freshwater snails. Ann Trop Med Parasitol 54:224–232

Xia MY, Jourdane J, Combes C (1998) Local adaptation of *Schistosoma japonicum* in its snail host demonstrated by transplantation of sporocysts. Proceedings of the IX International Congress of Parasitology. pp. 573–576

Yoshino TP, Laursen JR (1995) Production of *Schistosoma mansoni* daughter sporocysts from mother sporocysts maintained in synxenic culture with *Biomphalaria glabrata* embryonic (Bge) cells. J Parasitol 81:714–722

Zhang S, Nian H, Wang B, Loker ES, Adema CM (2009) Schistosomin from the snail *Biomphalaria glabrata*: Expression studies suggest no involvement in trematode-mediated castration. Mol Biochem Parasitol 165:79–86

Chapter 3
Biomphalaria: Natural History, Ecology and Schistosome Transmission

David Rollinson

Abstract This chapter reviews the environmental and ecological factors that influence the local distribution of *Biomphalaria* spp. in Africa and South America, including temperature, light, water chemistry and conductivity, depth and types of water body and aquatic vegetation. The role of these intermediate snail hosts in the transmission of *Schistosoma mansoni* is reviewed, including aspects related to species specificity, the impact of seasonality and focality of transmission. Monitoring and surveillance techniques are also considered, in particular new molecular techniques for detecting prepatent and patent infections, as well as Geographical Information Systems and their application for risk mapping in-line with schistosomiasis control programmes. The need to document the ecological requirements of species in this medically important genus is emphasised due to ongoing environmental and predicted climatic changes.

3.1 Introduction

Biomphalaria Preston, 1910, is a genus of significant medical importance, as many species have a close association with freshwater habitats associated with human settlements and act as intermediate hosts for compatible strains of the helminth parasite *Schistosoma mansoni*, the cause of the intestinal schistosomiasis or bilharzia. These snails are found in varied aquatic habitats including small pools, lakes, streams and irrigation channels, generally in shallow water on the edge of the habitat bringing them into the vicinity of areas commonly frequented by people. They occur within a defined climatic temperature range, in South and Central America, Arabia, Africa and Madagascar. There are about 12 species in Africa (Mandahl-Barth 1958; Brown 1994) and 11 in South America (Caldeira et al. 2009). Detailed

D. Rollinson (✉)
Wolfson Wellcome Biomedical Laboratories, Department of Zoology,
Natural History Museum, London, SW7 5BD, UK
e-mail: d.rollinson@nhm.ac.uk

R. Toledo and B. Fried (eds.), Biomphalaria *Snails and Larval Trematodes*,
DOI 10.1007/978-1-4419-7028-2_3, © Springer Science+Business Media, LLC 2011

maps of the distribution of *Biomphalaria* spp. in Africa are provided by Brown (1994). The evolutionary history of the group suggests an origin in the Americas with a more recent spread and invasion of the African continent (Campbell et al. 2000; Dejong et al. 2003 and elsewhere in this volume). For a recent review on the genetic diversity of schistosomes and snails, see Rollinson et al. (2009).

Given the medical importance of the genus, many investigations have attempted to define the range of environmental conditions necessary for the success of different species. The logic is simple, in which, if the likely distribution of the snail host can be defined, this will provide an insight into the distribution of disease. Moreover, many studies have looked at seasonal patterns of snail abundance, growth and reproduction in an attempt primarily to determine the peak times of schistosome transmission. Such longitudinal studies are demanding in time and effort, and the available literature tends to reflect the earlier detailed efforts of the pioneers in this subject. With the advent of good chemotherapy for the treatment of schistosomiasis, ecological studies regarding the snail host have tended to decline. However, more recently, the growing realisation that chemotherapy alone may not be sufficient to control the disease has once again opened the doors for studies related to snails and transmission control (King 2009). Furthermore, it is well recognised that environmental and climatic changes are likely to impact on snail distributions and associated disease outbreaks. A sound knowledge of the many factors governing snail distributions is required to predict how such changes, for example, in rainfall or temperature, will impact on snail populations.

Of course there will be many inter-linked factors governing the suitability of a freshwater habitat for a snail. This chapter reviews the environmental and ecological factors that influence the local distribution of *Biomphalaria* spp. and considers the role of this intermediate snail host in the transmission of *S. mansoni*. Monitoring and surveillance techniques and their potential application in schistosomiasis control programmes are also considered, in particular the use of geographical information systems (GIS) for mapping snail habitats and molecular techniques for detecting prepatent and patent schistosome infections in snail populations.

3.2 Environmental Factors Affecting *Biomphalaria*

3.2.1 Temperature and Light

Diurnal fluctuations in air temperature can be extreme, but water temperatures, depending on the size of the water body, tend to be more stable. Determining the effects of temperature on snails in the field is a difficult task, even with modern environmental data capture techniques, because there are considerable temperature gradients, both spatial and temporal, within even small water bodies, and snails can seek the microhabitats where temperature is most favourable

(Shiff 1960). Notwithstanding this, water temperature does appear to be a major factor in the distribution of freshwater snails. Temperature will influence rates of evaporation and is therefore directly linked to many other environmental factors influencing snail abundance.

Reproduction and growth of juvenile snails as well as survival of adults may all be influenced by temperature. Above-optimal temperatures are believed to impair development of the gonad in *B. pfeifferi*, so reducing fecundity. In laboratory studies, Appleton and Eriksson (1984) showed that the fecundity of *B. pfeifferi* fell off rapidly following exposure to temperature regimes incorporating more than 46 deg. h. > 27°C/day before or during gametogenesis. In addition, the gross development of the ovotestis was impaired. In the field, spells of above-optimal temperatures tend to be irregular and often exceed 45 deg. h. > 27°C/day. Such isolated spells of very hot water temperatures may in fact be more damaging to the processes of gametogenesis and ovotestis development. Optimal temperatures for *B. pfeifferi* range from 20 to 29°C (De Kock and Van Eeden 1981).

Detailed laboratory studies on the effects of temperature on *B. pfeifferi* are reviewed by Brown (1994) and support the field observation that although tropical conditions are required, relatively high temperatures are an important factor in limiting the occurrence of this snail, with a sharp fall in egg production occurring above 27°C. Short periods of intensely hot weather that may be experienced in the field were considered to be more damaging to *B. pfeifferi* than longer exposure to lower levels of above-optimal heat in laboratory studies. Similarly for *B. glabrata*, egg production was highest in the range 20–27°C (Pimentel-Souza et al. 1990). Comprehensive laboratory experiments on the direct effect of temperature on the embryonic development, growth, sexual maturity, survival and reproduction of *B. alexandrina* were reported by El Hassan (1974). Ishii (1984) studied the effects of temperature on the larval development of *Angiostrongylus cantonensis* in *B. glabrata* and showed that the threshold of development was 15.8°C, while Yousif and Lämmler (1975) concluded that 26°C was the optimal temperature for infection in this combination. The influence of climatic (and thus water) temperature on snail distribution has been widely studied, as listed in Table 3.1.

Temperature may restrict snails at certain times of the year and give rise to cycles of growth and reproduction associated with seasons. For example, there is little or no breeding by *B. pfeifferi* during the winter in subtropical areas such as the plateau of Zimbabwe and the low-veld of south-eastern Transvaal (Brown 1994). *B. pfeifferi* is notably absent from the Western Cape area of South Africa due to the prevailing cool climatic temperatures present there (Brown 1994). Joubert et al. (1986) reported that *B. glabrata* withstood cold conditions (0–6°C) less well than *Bulinus tropicus* and *Lymnaea natalensis*. Sturrock (1966) suggested that the absence of *B. pfeifferi* from otherwise suitable habitats on the coastal plain of East Africa might be due to the high temperatures present there. This was corroborated by the studies of Appleton (1977a, b).

The discontinuous geographic distribution of *B. pfeifferi* and the observed seasonal changes in the occurrence of this snail in several rivers and farms in the Awash Valley in Ethiopia indicate that a combination of environmental factors and

Table 3.1 Studies on the influence of climatic temperature on snail distribution

Species	Territory	References
B. pfeifferi	South Africa	Appleton (1977b), Appleton and Bruton (1979) and De Kock et al. (2004)
	Zimbabwe	Woolhouse and Chandiwana (1989) and Woolhouse (1992)
	Madagascar	Pflueger (1978)
	Mali	Coulibaly and Madsen (1990)
	Nigeria	Olofintoye and Odaibo (1996)
B. pfeifferi and *B. sudanica*	Ethopia	Goll (1982), Kloos et al. (1978, 1988) and Erko et al. (2006)
B. sudanica and *B. stanleyi*	Uganda	Kazibwe et al. (2006)
B. alexandrina	Egypt	Yousif et al. (1993) and Kader (2001)
B. glabrata	Brazil	Barbosa et al. (1987)
	St. Lucia, West Indies	Sturrock and Sturrock (1972)
B. tenagophila	Brazil	Chieffi and Moretti (1979) and Santana Teles et al. (2002)
B. occidentalis	Argentina	Rumi and Hamann (1992)

agricultural practices affects the stability of snail habitats – high temperatures and silt content of water probably cause marked variations in the survival rates of *B. pfeifferi* (Kloos and Lemma 1974; Kloos 1985). The seasonal fluctuation of *B. sudanica* in Lake Ziway, Ethiopia (Erko et al. 2006), was found to be highly associated with rainfall, water level of the lake and vegetation availability and abundance.

The influence of experimental illumination and seasonal variation on cross-breeding mating in *B. glabrata* was investigated in Brazil (Barbosa et al. 1987; Pimentel-Souza et al. 1988), and it was found that mating increased significantly in autumn and winter and was dependent on both the intensity of illumination and the schedule of illumination exposure. *B. alexandrina* was unable to tolerate maintenance under darkness (El-Emam and Madsen 1982) and had a lower tolerance to starvation and temperature fluctuations than *Bulinus truncatus* or *Helisoma duryi*, the rams-horn snail, a proposed competitive biological control agent for *B. alexandrina*. Interactions between St. Lucian *B. glabrata* and *H. duryi* were reported by Christie et al. (1981), who found that the time required for elimination of the vector snail depended on environmental temperature and numbers of the competitor snail initially introduced into the drains.

B. pfeifferi and *B. sudanica* are considered to be moderately successful aestivators, in which both can survive through periods of seasonal desiccation in some localities especially those associated with aquatic or marginal vegetation (Brown 1994).

3.2.2 Water Chemistry and Conductivity

The concentration of dissolved salts will limit the distribution of pulmonate snails when it is particularly high or low, but water chemistry generally affects the abundance and life cycles of snails rather than their presence or absence (Williams 1970).

Seasonal changes may result in evaporation from water bodies leading to increases in salinity, causing snail populations to disappear, as was reported for *B. pfeifferi* in Niger (Gretillat and Gaston 1975). *B. glabrata* was reported to be more tolerant of salty environments in the coastal areas of Pernambuco, Brazil, and this constitutes a risk for the spread of schistosomiasis (Da Silva et al. 2006). In an experimental study, Martin et al. (2007) showed that *B. glabrata* survived surprisingly well in artificial ocean water diluted to 10% with deionised water, although was less tolerant at 20% and did not survive at 25%. Grisolia and de Freitas (1985) reported that *B. glabrata* colonised water with chloride concentrations ranging from 1.0 to 3,500 mg/L. High calcium levels in the predominantly limestone environment of Virgem das Graças appeared to promote large *B. glabrata* populations in wells and springs (Kloos et al. 2004).

The effects of water conductivity, as well as other factors, on *B. stanleyi* and *B. sudanica* in Lake Albert, Uganda, were reported by Kazibwe et al. (2006). The local water conductivity range across sampling sites (44–700 IS) at Lake Albert indicated that the total dissolved salts in the water were high. While there was a positive correlation between water conductivity and densities of infected *B. stanleyi*, the inverse was true for *B. sudanica*. Excess of magnesium seems unfavourable for *B. pfeifferi*, which is generally less tolerant of soft water than *Bulinus* spp., although locally adapted populations of *B. pfeifferi* may thrive in soft water (Brown 1994).

Other studies on the influence of conductivity include those for *B. pfeifferi* and *Bulinus globosus* in Ibadan, Nigeria (Olofintoye and Odaibo 1996), *B. tenagophila* from a dam in Brazil (Grisolia and de Freitas 1985), *B. peregrina* in Mendoza, Argentina (Ciocco and Scheibler 2008), *B. straminea*, *B. tenagophila*, *B. peregrina* and *B. orbignyi* in rice fields Corrients province, Argentina (Rumi and Hamann 1990), and *B. alexandrina* and *Bulinus truncatus* around Suez, Egypt (Yousif et al. 1998a, b).

That *B. pfeifferi* thrives in water contaminated with organic matter (60 mg/L ≤ COD ≤ 1,060 mg/L) was observed in a study of survival in a sewage/wastewater purified in stabilisation ponds in Burkina Faso (Klutse and Baleux 1996). The effects on *B. tenagophila* of the quality of the water entering the irrigation system of a watercress garden in Rio de Janeiro, Brazil, were investigated by Baptista and Jurberg (1993), who found significant differences between colonised and non-colonised areas in pH, chlorides, conductivity, organic nitrogen and nitrates. A decrease in salinity is a key factor in the spread of *Biomphalaria* in Senegal after construction of the Diama dam (Vercruysse et al. 1994). An inverse relationship between the number of *B. alexandrina* snails and salinity of the water around Lake Maryut in Egypt was noted as early as 1978 (Mohamed et al. 1978) – the decrease in salinity here was related to increased discharges of agricultural and domestic waste into the lake.

It is of interest to record the abundance of other freshwater snails in habitats frequented by *Biomphalaria* species. General studies on the ecology of freshwater snails (14 species, including nine pulmonates) include those of Owojori et al. (2006), Ndifon and Ukoli (1989) and Olofintoye and Odaibo (1996) in Nigeria, which considered rainfall, pH, oxygen concentration, conductivity and presence or absence of macrophytes. A study of the distribution of freshwater snails in irrigation schemes in the Sudan (Madsen et al. 1988) showed positive and negative associations between snail and plant species. In the Gezira area, *B. pfeifferi* was positively correlated with the presence of *Bulinus truncatus*, *Lymnaea natalensis*, *Cleopatra bulimoides* and *Lanistes carinatus*, while in the Rahad area, *Melanoides tuberculata* was positively correlated with *B. pfeifferi*.

The environmental variables measured by Rumi et al. (2002) in their study of *B. straminea* and *B. tenagophila*, in comparison with the non-vector snail *Drepanotrema* spp., in Chaco, Argentina, included substratum (macrophytes), water quality (pH, O_2, nutrients, etc.), as well as other gastropods (Ancylidae, Hydrobiidae, Ampullaridae). Seven of 26 variables explained 62% of the variation in planorbid association, including dissolved oxygen and ammonium, and the plants *Eichhornia crassipes*, *Pistia stratiotes*, *Panicum elephantipes*, *Hydrocotyle ranunculoides* and *Canna glauca*, the latter being the most significant. In an earlier study (Rumi and Hamann 1990) in rice-fields in Argentina, these authors found positive correlations between *Biomphalaria* spp. abundance and conductivity, hardness, calcium, nitrites plus nitrates, ammonium and bicarbonates, and a negative correlation with water temperature; no correlation was found for total iron, phosphates, pH or soil granulometry; *B. orbignyi* was the most abundant species in these rice-fields.

3.2.3 Depth and Type of Water Body

Generally, *Biomphalaria* spp. prefer shallow water (0–7 cm), close to the shoreline (0–40 cm) and low velocity (optimum 13.3 cm/s, range 12–21). These microhabitat preferences have been noted for *B. pfeifferi* in Tanzania (Utzinger et al. 1997; Utzinger and Tanner 2000), for *B. tenagophila* in the Pamulha Reservoir, Brazil (Freitas et al. 1987), for *B. glabrata* in Minas Gerais, Brazil (Guimarães et al. 2009), and for *B. sudanica* in Uganda, while *B. stanleyi* preferred deeper water habitats (Kazibwe et al. 2006). Interestingly, where high temperatures occur in spring in Natal, South Africa, *B. pfeifferi* tends to be found in deeper water (Appleton and Bruton 1979). In Nigeria, *B. pfeifferi* was reported to prefer permanent bodies of water such as dams (Ndifon and Ukoli 1989). In contrast, Kloos et al. (2004) reported that *B. glabrata* had successfully adapted to both standing and flowing waters in their study area in Minas Gerais, Brazil. In Tanzania, *B. pfeifferi* preferred plant detritus and bedrock as dominant substrates (Utzinger and Tanner 2000), which was in contrast to the findings of Utzinger et al. (1997), where no significant preferences were reported for substratum type.

In South Africa, *B. pfeifferi* occurs mainly on hard rock formations that are resistant to erosion and which thus favour formation of permanent pools in rivers, providing refuges for snails in times of flood (Brown 1994). Physical characteristics of the environment, together with water velocity and amount of organic matter, were the most important factors in habitat preference for *B. glabrata* and *B. tenagophila*, plus other freshwater snails such as *Melanoides tuberculatus*, in Guapimirim, Brazil (Giovanelli et al. 2005). *B. tenagophila* habitats were reported to show a tendency to cluster around urban centres with organic pollution (Santana Teles 2005). Kloos et al. (2004) also noted a strong association between pollution and mean density of *B. glabrata* in Minas Gerais, Brazil, which was due in part to vegetation density. *B. pfeifferi* seems to be more suited to the stable conditions of permanent water bodies such as isolated pools fed by perennial springs (De Kock and Van Eeden 1981). In the design of irrigation canals, it appears that occasional periods of strongly turbulent flow, rather than a steady rapid current, are better for eliminating snail populations (Brown 1994).

Water current was a key factor in determining the distribution of *B. alexandrina* in irrigation canals in Egypt (Yousif et al. 1993). The type of water body, together with temperature, was reported to be the major factor determining the distribution of *B. pfeifferi* in South Africa (De Kock et al. 2004), while temperature and water flow proved key determinants in the Zimbabwe high veld (Woolhouse 1992). In Richard Toll, Senegal, *B. pfeifferi* favoured smaller, man-made habitats, while *Bulinus* spp. favoured larger natural and man-made habitats (Sturrock et al. 2001), while in Kano State, Nigeria, *B. pfeifferi* was primarily lake dwelling, although also found in irrigation canals (Betterton et al. 1988). In the Awash Valley in Ethiopia, *B. pfeifferi* was most common in the medium-sized canals bypassing labour camps (Kloos and Lemma 1974), and its presence was correlated with the irrigation requirements of the crops being grown – thus, in banana and sugarcane plantations which require year-round irrigation, *B. pfeifferi* was common, while in seasonally irrigated cotton schemes, where the field canals dry out between irrigations, the snail was absent or rare (Kloos 1985). Primary and secondary canals were shown to play the most important role in transmission by *B. pfeifferi* in Mali because of the stability of these habitats (Dabo et al. 1994).

A study of *B. glabrata* in Minas Gerais, Brazil, found highest densities of the snail in overflow ponds, irrigation ponds, springs, canals and wells, and lowest densities in fishponds and water tanks, while streams, swamps, dams and cattle ponds had intermediate densities (Kloos et al. 2004). These authors also investigated a number of other parameters and reported various interrelationships between the presence of *B. glabrata* and vegetation, fish, pollution, substrate, water flow and turbidity.

3.2.4 Aquatic Plants

Plants provide snails with shelter from the sun and the water current as well as egg-laying sites and a source of food, although snails do not readily consume fresh tissues but feed mainly on decaying plant matter (Brown 1994). The aquatic plant trade has also played a key role in the spread of planorbid snails across the globe, as reviewed by Pointier et al. (2005). Thomas and co-workers (Thomas 1987; Thomas and Kowalczyk 1997; Eaton and Thomas 1999) have reported extensively on the interactions between freshwater pulmonate snails and macrophytes, addressing ecological, food-web, biochemical and co-evolutionary questions.

Biomphalaria spp. are reported to have differing preferences regarding vegetation, for example, in the Nile Valley in Egypt, *B. alexandrina* and a *B. alexandrina* × *B. glabrata* hybrid were found differently associated with aquatic plants in comparison with the *B. glabrata* parent (Yousif et al. 1998b), although both parents had the same major physiochemical requirements. Similar observations were made by Kader (2001) for *B. alexandrina* and *Bulinus truncatus* in Egypt, who reported different associations in relation to aquatic plants and snail species, including *L. natalensis* and *Physa acuta*. A correlation between the patchy distribution of *B. pfeifferi* and aquatic plants in a river habitat in Zimbabwe was reported by Woolhouse and Chandiwana (1989). In Nigeria, *B. pfeifferi* occurred in habitats dominated by the subaquatic macrophytes *Acroceras zizanioides*, *Alternanthera sessilis*, *Commelina* spp., *Ludwigia leptocarpa* and other members of the family Graminiae mainly *Paspalum* spp., while the predominant aquatic forms were *Lemna pausicostata* and *Nymphaea lotus* (Ndifon and Ukoli 1989).

In Venezuela, *B. glabrata* and *B. prona* were observed to be associated with distinct types of habitats in Lake Valencia (Amarista et al. 2001). Kloos et al. (2004) reported a significant association between vegetation density and snail occurrence in Minas Gerais, Brazil, with large numbers of *B. glabrata* snails and ova in water cress (*Nasturtium officinale*) and small numbers in the dense bulrush (*Typha* sp., locally known as "taboa") vegetation found in swamps and slow-flowing streams. However, they also found large populations of *B. glabrata* in calcium-rich limestone springs and wells with little or no macro vegetation. In St Lucia, Sturrock and Upatham (1973) recorded numerous marsh and terrestrial vascular plants, such as dasheen (*Colocasia esculuntu*), eddoes (*Xunthosoma sugittifoliu*), watergrass (*Commelinu difssu*) and several sedge species, which were frequently associated with snail-infested habitats. However, they concluded that these plants appeared to indicate favourable conditions for the snails rather than to attract them in any specific way.

The nature of vegetation may also impact on transmission of schistosomiasis. Higher prevalence of schistosomiasis was reported in areas with latossolo soil type and transitional vegetation, in Bahia, Brazil (Bavia et al. 1999). Although less numerous in ferruginous aquatic environments, *B. glabrata* occurred widely in areas of lateritic soil in Brazil (Kloos et al. 2004). Ndifon and Ukoli (1989) noted that *B. pfeifferi* seemed to prefer minimal (0–25%) shade in their habitats in Nigeria. Where plants give excessive shade, however, *B. pfeifferi* will be less likely to thrive, probably as a result of indirectly removing their diatomatious food; indeed, shade may be so unfavourable as to provide a potential means for biological control (Loreau and Baluku 1991).

3.3 Effects of Water Development Projects on *Biomphalaria*

There have been many extensive water development programmes across the globe that have had a significant impact on the distribution, prevalence and intensity of schistosomiasis, primarily through their impact on distribution and survival of the respective intermediate snail hosts. In Nigeria, for example, over 300 dams and reservoirs have been established since 1973 (Owojori et al. 2006). Irrigation schemes in the Awash Valley Ethiopia lead to an increase in the number of potential habitats for *B. pfeifferi*, especially in the medium-sized canals (Kloos and Lemma 1974), and the impact of such water resources development on schistosomiasis was subsequently evaluated by Kloos (1985). The building of the Aswan Dam in Egypt and resulting change from annual flooding (basin irrigation) to perennial irrigation, with elimination of the so-called winter closure, during which canals dried up, have meant that the irrigation canals and drains now harbour snail hosts throughout the year, providing permanent transmission sites (Malek 1975). Similarly, for the Gezira scheme in Sudan, small field irrigation canals known as "abu eshreens" now play a role in transmission of *S. mansoni* (Fenwick et al. 1981). A systematic review and meta-analysis of schistosomiasis and water resources development, and estimates of people at risk is given by Steinmann et al. (2006), who concluded that 779 million people were at risk of schistosomiasis in 2003, up 10.9% from the mid-1990s, and 106 million (13.6%) of whom lived in proximity to large dam reservoirs and irrigation schemes.

Ecological changes in the early 1980s as a result of sugar cane irrigation, followed by more widespread changes in the late 1980s, with the development and extension of rice culture, when dams were constructed across the Senegal River at Diama and Manantali for preventing seasonal rain-fed floods and sea water intrusion, with a resultant reduction in salinity and more stable water flow, have resulted in the rapid spread of *S. mansoni* around Richard Toll (Vercruysse et al. 1994; Sturrock et al. 2001). *B. pfeifferi* is the snail host responsible for transmission during the rainy season there and appears to be well adapted because of its ability to withstand changes in temperature and to aestivate (Picquet et al. 1996; Ernould et al. 1999).

Intestinal schistosomiasis was almost unheard of in Guiedougou, Burkina Faso, before 1987, but following the development of the hydroagricultural complex of Sourou, its prevalence ranged from 8 to 69% by 1998 (Poda et al. 2003). Similarly for the Kariba Dam in Zambia, *S. mansoni* prevalence increased from 16 to 56% over a 20-year period (while that for *S. haematobium* decreased from 69 to 17%) (Mungomba et al. 1993). The first finding of *S. mansoni* in Dhofar, Oman, was related to the occurrence of *B. arabica* in water bodies in the area (Moné et al. 2003). Human migration linked to water development projects and urbanisation was considered to be the principal factor associated with the spread of schistosomiasis in the mid-west area and Antananarivo plain of Madagascar, leading to rapid endemisation of the disease (Ollivier et al. 1999). Studies by Green et al. (1992) on the surfacing and water leaving behaviour of *B. glabrata* and *B. jousseaumei* could be relevant to the design and management of irrigation canals where there is a schistosomiasis risk.

Investigations of hydroelectric schemes in Brazil have highlighted potential foci for enhanced transmission of *S. mansoni*, such as by *B. amazonica* at the Manso Dam, Matto Grosso (*B. occidentalis* was not permissive) (Fernandez and Thiengo 2006), or by *B. straminea* at the Serra da Mesa Dam, Goias (Fernandez and Thiengo 2002), and around Sao Paulo (Santana Teles 1996).

3.4 Role of *Biomphalaria* in Transmission of *Schistosoma Mansoni*

3.4.1 Species Specificity

The two most important intermediate host species of *S. mansoni* are *B. glabrata* in the New World and *B. pfeifferi* in the Old World. All the African species of *Biomphalaria* tested have proved more or less compatible with *S. mansoni*, but *B. pfeifferi* is the most important intermediate host of *S. mansoni* in Africa because of its wide distribution. The degree of compatibility may vary widely with the origins of both the snail and parasite. For example, *B. alexandrina* proved permissive only for local *S. mansoni* from Egypt, while *B. camerunensis* was most compatible with the Central African schistosome. However, Greer et al. (1990) concluded that *B. camerunensis* was a poor host in comparison with *B. pfeifferi*, since the latter, although less common overall, occurred where *S. mansoni* infection rates were highest. *B. pfeifferi* from Senegal was shown to be much more compatible to the local strain of *S. mansoni* than to a strain from Cameroon (Tchuem Tchuenté et al. 1999; Southgate et al. 2000).

In contrast with Africa, in Brazil, although ten species and one subspecies of *Biomphalaria* are recognised, only three are found naturally infected with *S. mansoni*, namely *B. glabrata*, *B. tenagophila* and *B. straminea*, while *B. amazonica* and *B. peregrina* are known to be susceptible to infection in the laboratory and so represent potential intermediate hosts (reviewed by Caldeira et al. 2009). *B. occidentalis* proved refractory to infection in the laboratory (Fernandez and Thiengo 2006). A hybrid between *B. glabrata* and *B. alexandrina* has been found infected with *S. mansoni* in the Nile Valley in Egypt, and is assumed to be participating in disease transmission there (Yousif et al. 1998b; Lotfy et al. 2005).

Certain snail host lineages, such as *B. tenagophila* Taim lineage, are resistant to *S. mansoni* and will pass the resistance trait on when crossed with susceptible snails. The possibility of releasing such a snail host into the field for the purposes of interrupting schistosome transmission has been attempted by Coelho et al. (2008).

Other trematodes may also be harboured by *Biomphalaria*; for example, in a recent study in Sao Paulo, Brazil, *B. tenagophila* was reported to be shedding four morphologically distinct cercariae, including those of *S. mansoni* (De Moraes et al. 2009), and similarly for *B. glabrata* from Pernambuco, Brazil (De Souza et al. 2008).

3.4.2 Impact of Seasonality on Transmission

Strong seasonal influences on the production of schistosome cercariae have been reported, and it is possible therefore that the effect of climatic temperature on the intramolluscan stages could play a part in determining the parasite's distribution, which may not coincide exactly with that of its snail host (Brown 1994), as reported for *B. pfeifferi* and *S. mansoni* in Ethiopia, where the parasite is absent above altitudes of 2,200 m although the snail is widespread (Kloos et al. 1988).

Seasonality in the transmission of *S. mansoni* by *B. pfeifferi* at Richard Toll, Senegal, was reported by Sturrock et al. (2001) and Southgate et al. (2001) with most transmission occurring between May and August. *B. pfeifferi* population density and infection rate were higher in the dry season and lower during periods of heavy rainfall in Liberia (Dennis et al. 1983) – newer rice paddies contained few or no vector snails, whereas older paddies pre-dating a development project (BCADP) contained infected vectors, which reflected in higher disease prevalence. In a freshwater mangrove in Guadeloupe, *B. glabrata* population dynamics and *S. mansoni* transmission mainly depend on the alternating dry and rainy seasons, with infections of rats around 20% at the beginning of the rainy season but 100% by the end (Pointier and Theron 1979).

Coelho and Bezerra (2006) investigated the effects of temperature change on the infection rate of *B. glabrata* with *S. mansoni* and found that a decrease in temperature from 30 to 15°C had a significant effect on cercarial release. Similar temperature effects had been reported by Fried et al. (2002), where 12°C caused a significant decrease in cercarial release of both *S. mansoni* and *Echinostoma caproni*, and by Lwambo et al. (1987) who reported that no infection of *B. arabica* by *S. mansoni* occurred at 10°C, while highest infection rates were obtained between 28 and 34°C. De Souza et al. (1995) extended their observations on temperature effects to the ratio of male to female *S. mansoni* larvae in *B. glabrata*, both in the laboratory and in the field in Brazil, and concluded that infection of vertebrate hosts would be higher in the summer when higher levels of parasites and eggs were observed. Upatham (1973), studying the effect of water temperature on penetration and development of St Lucian *S. mansoni* miracidia in local *B. glabrata*, reported a peak of infection at 34°C (71.4%) and concluded that while excessively high temperatures during summer might reduce transmission in some habitats, they would be unlikely to eliminate it entirely. Anderson et al. (1982) reported experimental studies of miracidial survival and infectivity in relation to larval age, water temperature, host size and host age, in relation to transmission of *S. mansoni* from man to snail. Lee and Cheng (1971) found that *S. mansoni*-infected *B. glabrata* had a lowered higher thermal tolerance limit than uninfected snails.

Significantly more *S. mansoni*-infected *B. glabrata* were found in the dry season in the Pamparrao area of Brazil (there was a negative correlation between snails and rainfall), probably because the lower water volume at that time increased the chances of interactions between the intermediate host and parasite (Giovanelli et al. 2001). Enhanced transmission in the dry season was noted also in Ethiopia (Malone et al. 2001a, b). Similarly, the majority of infected *B. tenagophila* near Sao Paulo, Brazil, were collected during the dry season (Santana Teles et al. 2002), while the duration of the annual dry season was one of the most important determinants of prevalence of schistosomiasis in Bahia, Brazil (Bavia et al. 1999) – rainfall and temperature were not found to be significant.

Snails may aestivate to circumvent unfavourable environmental conditions, such as long periods of drought, and it has been shown in *B. glabrata* that aestivation does little to decrease the snail's ability to act as an intermediate host for *S. mansoni* or to interrupt the development of the parasite (Cooper et al. 1992).

Conductivity may also affect the infection rate with schistosomes. This was shown experimentally for a Saudi Arabian isolate of *S. mansoni* and *B. arabica*, where the infection rate decreased as salinity increased, up to 4,500 mg/L, above which no infection occurred (Lwambo et al. 1987). Sturrock and Upatham (1973) investigated the interactions of salinity, turbidity and pH on infection of *B. glabrata* and found that the infection rate was related curvilinearly to pH, but showed a negative relationship with the level of both turbidity and salinity. Increasing turbidity produced a smaller reduction in the infection rate at pH 5 than at 7 or 9, while increasing salinity produced a smaller reduction in the infection rate at pH 5 than at 7 or 9, and increasing turbidity caused a smaller reduction in the infection rate at high salinities than at low ones.

The thermal preferences of resistant and susceptible strains of *B. glabrata* exposed to *S. mansoni* were investigated by Lefcort and Bayne (1991). The mean temperature selected by resistant snail's post-exposure was significantly lower and is probably due to altered levels of endogenous cytokines in association with parasite activation of the snail's defence mechanism.

The effects of diurnally fluctuating temperatures on prepatency of *S. mansoni* in *B. glabrata* have been reported by Pflueger (1980, 1981). A decrease in temperature from 26 to 24°C during the prepatent period induced a drop of 33% of the cercarial production index (Theron 1981). Shiff et al. (1975) found that the prepatent period for *S. mansoni* in *B. pfeifferi* in Rhodesia was prolonged in winter but found no evidence of sporocyst dormancy. Valle et al. (1973) showed that when infected *B. glabrata* were kept at constant temperature but subjected to diurnal light variation, they shed cercariae in a characteristic rhythm, with about 98% between 6 a.m. and 6 p.m. A comparative study of circadial rhythms in cercarial emergence from *B. tenagophila* and *B. glabrata* in Brazil was reported by Favre et al. (1997), showing a diurnal pattern, with emergence concentrated in the afternoon. Asch (1972) concluded that cercarial emergence was light induced rather than by any change in temperature.

3.4.3 Focality of Transmission

Where schistosomiasis is endemic, there is a wide variation in the prevalence and intensity of infection locally, resulting in focal patterns of transmission, involving a complex of interacting factors (Sturrock and Upatham 1973; Southgate and Rollinson 1987), including environmental/climatic parameters as well as human behaviour/man-made water developments. Focal transmission of *S. mansoni*, despite the widespread occurrence of the *B. glabrata*, is typical in Guadeloupe (Golvan et al. 1977). Differing distributions of *B. sudanica* and *B. stanleyi* over small distances were reported by De Moira et al. (2007), and this, together with distinct spatial patterns of water contact, resulted in markedly variable exposure to *S. mansoni* between two tribes in a Ugandan fishing community. Focality and seasonality of *S. mansoni* transmission in the Gezira Irrigated Area of Sudan were reported by Babiker et al. (1985), who found that almost 90% of the infected *B. pfeifferi* were found in one minor canal near the village, while sites near smaller settlements some distance from the minor canals yielded few infected snails; factors influencing the prevalence of infection in snails were temperature, turbidity and human contact. Spatial analysis of *B. glabrata* and *S. mansoni* in Pernambuco, Brazil, also identified focal infections with two high intensity areas (Barbosa et al. 2004; Araújo et al. 2007), and these researchers had previously reported seasonal variations in snail density (Barbosa et al. 2000). Boissier et al. (2003) reported altered behaviour of *B. glabrata* as a result of infection with *S. mansoni*, with a tendency for infected snails to aggregate – if this occurs in the field, it may well lead to enhanced transmission.

The rise and fall of discharge from the Thamalakane River in the Okavango Delta of Botswana has been shown to correlate with the rise and fall of epidemic transmission of *S. mansoni* at Maun (Appleton et al. 2008).

3.5 Monitoring and Surveillance Systems

3.5.1 Use of Geographical Information System for Predicting Distribution of Snails and Disease

Detailed knowledge of the distribution patterns of schistosome intermediate hosts and their population dynamics and the factors affecting these patterns provide important information regarding the possibilities for and desirability of introducing control measures in various transmission scenarios (Madsen 1992). GIS technology provides a valuable tool for better understanding and predicting spatial disease distribution and has been used for modelling of schistosomiasis in the context of control programmes (Malone et al. 2004). GIS incorporates meteorological and other environmental factors as well as social variables and utilises recently launched orbital sensors such as the Moderate Resolution Imaging Spectroradiometer (MODIS) and the Shuttle Radar Topography Mission (SRTM). Prediction of schistosomiasis using GIS was first attempted in the Philippines and in the Caribbean by Cross et al. (1984). In 2001, a global network for the control of snail-borne disease using satellite surveillance and GIS was formed, which included a "virtual research group" through the Internet site www.GnosisGIS.org (GIS Network On Snail-borne Infections with special reference to schistosomiasis using GIS) (Malone et al. 2001a). In addition to using GIS models to predict the health impacts of water development projects, this initiative should facilitate evaluation of the consequences of both inter-annual climate variation and long-term global climate change, as well as the cost-effectiveness of hypothetical or proposed modifications of control strategies. The most recent review of remote sensing, GIS and spatial analysis for schistosomiasis in Africa is that of Simoonga et al. (2009), who highlight the need for more detailed information on socio-economic factors as well as data on intermediate snail host distribution.

In a recent study of schistosomiasis risk in Minas Gerais State, Brazil (Guimarães et al. 2008; De Toledo et al. 2008), the key variables selected as important for modelling purposes were topographic elevation, minimum summer temperature, the normalised difference vegetation index (NDVI) and the social index, Human Development Index, HDI_{91}. Geostatistical procedures such as indicator kriging have proved a robust tool in risk mapping in respect of *Biomphalaria glabrata*, *B. tenagophila* and/or *B. straminea* in Brazil (Guimarães et al. 2009). Earlier studies in Bahia, Brazil (Bavia et al. 1999, 2001), indicated that population density and the duration of the annual dry period were the most important determinants of prevalence of schistosomiasis, together with NDVI and diurnal temperature fluctuation, while maximum rainfall and max/min temperatures were not significant factors.

In Uganda, where GIS was used to study the *S. mansoni*-"*Biomphalaria pfeifferi*"-*B. sudanica* parasite-snail system, NDVI values of 151–174, day temperatures of 26–36°C and night temperatures of 15–20°C were used as criteria for the prediction model (Stensgaard et al. 2005). Species-specific ecological preferences were identified, and absence of the disease in the south-west highlands of Uganda could not be attributed to the absence of snail hosts but rather to low night-time temperature as one of the significant factors inhibiting *S. mansoni* transmission in that area. In Ethiopia, using data derived from the AVHRR (Advanced Very High Resolution Radiometer, on board the NOAA-11, National Oceanic and Atmospheric Administration satellite), the model-predicted endemic area for *S. mansoni* corresponded to values of NDVI 125–145 and T_{max} 20–33°C in the annual composite map (10–32°C for *B. pfeifferi* specifically), 125–145 and 18–29°C for the wet season map and 125–140 and 22–37°C for the dry season map (Kristensen et al. 2001; Malone et al. 2001b), and when used with FAO agroecologic zone climate data limits of <27°C for average annual mean temperature and annual moisture deficits (annual rain-annual potential evapotranspiration) of <–1,300 mm, the model accurately represented the regional distribution of *S. mansoni*–*B. pfeifferi* when extrapolated to East Africa. Thermal-hydrological data from AVHRR were also used successfully for identifying predilection sites of endemic *S. mansoni* and its snail host *B. alexandrina* in the Nile delta in Egypt (Abdel-Rahman et al. 2001; Yousif et al. 1998a). Infected *B. alexandrina* snails fluctuated seasonally, showing a minor peak in August and a higher peak in November (Yousif et al. 1999). Although *S. mansoni* infection in schools in northern Mozambique was found to be rare and had not previously been reported (Traquinho et al. 1998), the *Biomphalaria* vector was nevertheless discovered in the area – revised mapping of schistosomiasis is advised in view of probable changes after many years of conflict (similarly for schistosomiasis in Addis Ababa, Ethiopia, Erko et al. 1996).

3.5.2 Molecular Markers for Identification of Prepatent Infections in Snails

Detection of patent *S. mansoni* infection in snails has been routinely performed by observing cercarial shedding from snails exposed to artificial light, or by squeezing snails between two slides in order to observe cercariae and/or sporocysts in the digestive gland. However, because detection of cercarial shedding is only possible 30 days after infection, detection of prepatent infection and infection in dead snails, as well as differentiation of *S. mansoni* from other trematode infections, is more readily accomplished using molecular methods. The development of molecular approaches for the detection of *S. mansoni* and identification of transmission sites of schistosomiasis has been reported by groups in Brazil (Jannotti-Passos et al. 1997; Abath et al. 2006; Gomes et al. 2006; Jannotti-Passos et al. 2006; Melo et al. 2006). Hamburger et al. (1998) developed a PCR assay using a highly repeated *S. mansoni* sequence that allowed detection of infection 1 day after penetration by a single miracidium and showed the potential of the technique for large-scale determination of prepatent infection within snails. Similarly, a molecular detection system based on PCR amplification of mitochondrial DNA and the internal transcribed spacer (ITS2) region of *S. mansoni* was shown by Jannotti-Passos et al. (1997) to detect infection in *B. glabrata* during the prepatent period.

PCR amplification of the small subunit rRNA gene of *S. mansoni* proved to have a higher level of sensitivity than standard screening of intermediate snail hosts by cercarial shedding when DNA was purified from pools of snails collected from endemic areas in Brazil (Melo et al. 2006). Three variants of PCR were used, namely conventional PCR, nested PCR and single-tube nested PCR, with detection limits of parasite DNA of 10, 1 and 0.1 fg, respectively. The limit of detection of parasite DNA in pools of snails was 10 fg of purified genomic DNA, which is well below the equivalent of one parasite genome of 580 fg, thus providing a useful tool for indirectly quantifying parasite burden in snails and other samples such as cercarial infested water and human faeces. Earlier studies also reported on the use of the 18S rRNA of *S. mansoni* in *B. alexandrina* for determining whether snails in endemic areas have prepatent infections (Hanelt et al. 1997).

3.6 Concluding Remarks

The introduction of new molecular technologies for snail and parasite characterization together with the introduction of new mapping techniques will facilitate a much better understanding of the distribution of *Biomphalaria* spp. As resolution increases and fine scale mapping becomes more feasible, new malacological data will be of tremendous value to schistosomiasis control programme managers who need to target chemotherapy and pinpoint areas of high risk water contact and transmission hot spots. In their mathematical analysis of the transmission dynamics of schistosomiasis, Chiyaka and Garira (2009) showed that control strategies that target the transmission of the disease from snail to man will be more effective than those that block transmission from man to snail. We are fast approaching a time when the research pendulum needs to swing back to generate greater focus on the snail hosts, especially as the importance of transmission control becomes more widely recognised. It is clear that a full appreciation of the impact of water development projects and subtle changes in climate on disease transmission will only be achieved by a better understanding of the many factors governing the distribution of *Biomphalaria* spp.

References

Abath FGC, Gomes ALDV, Melo FL, Barbosa CS, Werkhauser RP (2006) Molecular approaches for the detection of *Schistosoma mansoni*: Possible applications in the detection of snail infection, monitoring of transmission sites, and diagnosis of human infection. Mem Inst Oswaldo Cruz 101:S145–S148

Abdel-Rahman MS, El-Bahy MM, Malone JB, Thompson RA, El-Bahy NM (2001) Geographic information systems as a tool for control program management for schistosomiasis in Egypt. Acta Trop 79:49–57

Amarista M, Niquil N, Balzan C, Pointier JP (2001) Interspecific competition between freshwater snails of medical importance: a Venezuelan example. C R Acad Sci III 324:143–148

Anderson RM, Mercer JG, Wilson RA, Carter NP (1982) Transmission of *Schistosoma mansoni* from man to snail: experimental studies of miracidial survival and infectivity in relation to larval age, water temperature, host size and host age. Parasitology 85:339–360

Appleton CC (1977a) The influence of above optimal constant temperatures on South African *Biomphalaria pfeifferi* (Krauss) (Mollusca: Planorbidae). Trans R Soc Trop Med Hyg 71:140–143

Appleton CC (1977b) The influence of temperature on the life-cycle and distribution of *Biomphalaria pfeifferi* (Krauss, 1948) in South-Eastern Africa. Int J Parasitol 7:335–345

Appleton CC, Bruton MN (1979) The epidemiology of schistosomiasis in the vicinity of Lake Sibaya, with a note on other areas of Tongaland (Natal, South Africa). Ann Trop Med Parasitol 73:547–561

Appleton CC, Eriksson IM (1984) The influence of fluctuating above-optimal temperature regimes on the fecundity of *Biomphalaria pfeifferi* (Mollusca: Planorbidae). Trans R Soc Trop Med Hyg 78:49–54

Appleton CC, Ellery WN, Byskov J, Moogkweetsinyana SS (2008) Epidemic transmission of intestinal schistosomiasis in the seasonal part of the Okavango Delta, Botswana. Ann Trop Med Parasitol 102:611–623

Araújo KCGM, Resendes APDC, Souza-Santos R, Silveira-Junior JC, Barbosa CS (2007) Análise espacial dos focos de *Biomphalaria glabrata* e de casos humanos de esquistossomose mansônica em Porto de Galinhas, Pernambuco, Brasil, no ano 2000. Cad Saude Publica 23:409–417

Asch HL (1972) Rhythmic emergence of *Schistosoma mansoni* cercariae from *Biomphalaria glabrata*: control by illumination. Exp Parasitol 31:350–355

Babiker A, Fenwick A, Daffalla AA, Amin MA (1985) Focality and seasonality of *Schistosoma mansoni* transmission in the Gezira Irrigated Area, Sudan. J Trop Med Hyg 88:57–63

Baptista DF, Jurberg P (1993) Factors conditioning the habitat and the density of *Biomphalaria tenagophila* (Orbigny, 1835) in an isolated schistosomiasis focus in Rio de Janeiro city. Mem Inst Oswaldo Cruz 88:457–464

Barbosa ND, Pimentel-Souza F, Sampaio IB (1987) The effect of seasonal temperature and experimental illumination on reproductive rate in the snail *Biomphalaria glabrata*. Braz J Med Biol Res 20:685–696

Barbosa CS, Pieri OS, Da Silva CB, Barbosa FS (2000) Ecoepidemiologia da esquistossomose urbana na ilha de Itamaracá, Estado de Pernambuco. Rev Saude Publica 34:337–341

Barbosa CS, Araújo KC, Antunes L, Favre T, Pieri OS (2004) Spatial distribution of schistosomiasis foci on Itamaracá Island, Pernambuco, Brazil. Mem Inst Oswaldo Cruz 99:79–83

Bavia ME, Hale LF, Malone JB, Braud DH, Shane HM (1999) Geographic information systems and the environmental risk of schistosomiasis in Bahia, Brazil. Am J Trop Med Hyg 60:566–572

Bavia ME, Malone JB, Hale L, Dantas A, Marroni L, Reis R (2001) Use of thermal and vegetation index data from earth observing satellites to evaluate the risk of schistosomiasis in Bahia, Brazil. Acta Trop 79:79–85

Betterton C, Ndifon GT, Bassey SE, Tam RM, Oyeyi T (1988) Schistosomiasis in Kano State, Nigeria. I. Human infections near dam sites and the distribution and habitat preferences of potential snail intermediate hosts. Ann Trop Med Parasitol 82:561–570

Boissier J, Rivera ER, Moné H (2003) Altered behavior of the snail *Biomphalaria glabrata* as a result of infection with *Schistosoma mansoni*. J Parasitol 89:429–433

Brown DS (1994) Freshwater snails of Africa and their medical importance. Taylor and Francis, London

Caldeira RL, Jannotti-Passos LK, Carvalho OS (2009) Molecular epidemiology of Brazilian *Biomphalaria*: a review of the identification of species and the detection of infected snails. Acta Trop 111:1–6

Campbell G, Jones CS, Lockyer AE, Hughes S, Brown D, Noble RL, Rollinson D (2000) Molecular evidence supports an African affinity of the neotropical freshwater gastropod, *Biomphalaria glabrata*, Say 1818, an intermediate host for *Schistosoma mansoni*. Proc Roy Soc B Biol Sci 267:2351–2358

Chieffi PP, Moretti IG (1979) Monthly variation of mollusc populations in a natural pond of Londrina county, state of Parana, Brazil. Rev Inst Adolfo Lutz 39:45–50

Chiyaka ET, Garira W (2009) Mathematical analysis of the transmission dynamics of schistosomiasis in the human-snail hosts. J Biol Syst 17:397–423

Christie JD, Edward J, Goolaman K, James BO, Simon J, Dugat PS, Treinen R (1981) Interactions between St. Lucian *Biomphalaria glabrata* and *Helisoma duryi*, a possible competitor snail, in a semi-natural habitat. Acta Trop 38:395–417

Ciocco NF, Scheibler EE (2008) Malacofauna of the littoral benthos of a saline lake in southern Mendoza, Argentina. Fundam Appl Limnol 172:87–98

Coelho JR, Bezerra FSM (2006) The effects of temperature change on the infection rate of *Biomphalaria glabrata* with *Schistosoma mansoni*. Mem Inst Oswaldo Cruz 101:223–224

Coelho PMZ, Rosa FM, Maciel E, Negrão-Correa DA, Carvalho OS, Caldeira RL, Jannotti-Passos LK, Moreira LA, Oliveira GC, Teles HM (2008) Transmission control of schistosomiasis mansoni by introduction of a resistant strain of *Biomphalaria tenagophila* in areas where transmission is maintained by this species. Acta Trop 108:245–248

Cooper LA, Richards CS, Lewis FA (1992) *Schistosoma mansoni* infection in diapausing *Biomphalaria glabrata* snails: Studies of temperature and genetic influences on diapausing behavior. Am J Trop Med Hyg 47:498–504

Coulibaly G, Madsen H (1990) Seasonal density fluctuations of intermediate hosts of schistosomes in two streams in Bamako, Mali. J Afr Zool 104:201–212

Cross ER, Sheffield C, Perrine R, Pazzaglia G (1984) Predicting areas endemic for schistosomiasis using weather variables and a Landsat data base. Mil Med 149:542–544

Da Silva PB, Barbosa CS, Pieri O, Travassos A, Florencio L (2006) Physico-chemical and biological aspects related to the occurrence of *Biomphalaria glabrata* in foci of schistosomiasis in coastal areas of the State of Pernambuco, Brazil. Quim Nova 29:901–906

Dabo A, Diop S, Doumbo O (1994) Distribution of intermediate mollusc hosts in human schistosomiasis in the Office of Niger (Mali). II. Role of different habitats. Bull Soc Pathol Exot 87:164–169

De Kock KN, van Eeden JA (1981) Life table studies on freshwater snails. The effect of constant temperature on the population dynamics of *Biomphalaria pfeifferi* (Krauss). Wetenskaplike Bydraes van die Potchefstroomse Universiteit, B. Natuurwetenskappe 107:1–17

De Kock KN, Wolmarans CT, Bornman M (2004) Distribution and habitats of *Biomphalaria pfeifferi*, snail intermediate host of *Schistosoma mansoni*, in South Africa. Water SA 30:29–36

De Moira AP, Fulford AJC, Kabatereine NB, Kazibwe F, Ouma JH, Dunne DW, Booth M (2007) Microgeographical and tribal variations in water contact and *Schistosoma mansoni* exposure within a Ugandan fishing community. Trop Med Intern Health 12:724–735

De Moraes J, Da Silva MPN, Ohlweiler FP, Pires F, Kawano T (2009) *Schistosoma mansoni* and other larval trematodes in *Biomphalaria tenagophila* (planorbidae) from Guarulhos, São Paulo State, Brazil. Rev Inst Med Trop Sao Paulo 51:77–82

De Souza CP, Jannotti-Passos LK, Ferreira SS, Sampaio IBM (1995) Influence of temperature on development of *Schistosoma mansoni* female cercariae in *Biomphalaria glabrata*. Rev Inst Med Trop São Paulo 37:319–324

De Souza MAA, Barbosa VS, Wanderlei TNG, Barbosa CS (2008) Criadouros de *Biomphalaria*, temporários e permanentes, em Jaboatão dos Guararapes, PE. Rev Soc Bras Med Trop 4:252–256

De Toledo F, Da Costa C, Dutra LV, Rodrigues F, Souza FJP, Santos R, Mourao AC, Dos Santos O (2008) Risk mapping of the schistosomiasis in the Minas Gerais state, Brazil, using modis and socioeconomic spatial data. In International geoscience and remote sensing symposium (IGARSS), vol 1, pp I268–I271 doi 10.1109/IGARSS.2008.4778845

Dejong RJ, Morgan JAT, Wilson WD, Al-Jaser MH, Appleton CC, Coulibaly G, D'Andrea PS, Doenhoff MJ, Haas W, Idris MA, Magalhães LA, Moné H, Mouahid G, Mubila L, Pointier JP, Webster JP, Zanotti-Magalhães EM, Paraense WL, Mkoji GM, Loker ES (2003) Phylogeography of *Biomphalaria glabrata* and *B. pfeifferi*, important intermediate hosts of *Schistosoma mansoni* in the New and Old World tropics. Mol Ecol 12:3041–3056

Dennis E, Vorkpor P, Holzer B (1983) Studies on the epidemiology of schistosomiasis in Liberia: the prevalence and intensity of schistosomal infections in Bong County and the bionomics of the snail intermediate hosts. Acta Trop 40:205–229

Eaton P, Thomas JD (1999) The accumulation and metabolism of amino acids by *Biomphalaria glabrata*, a freshwater pulmonate snail. Comp Biochem Physiol A Mol Integr Physiol 122:421–428

El Hassan AA (1974) Laboratory studies on the direct effect of temperature on *Bulinus truncatus* and *Biomphalaria alexandrina*, the snail intermediate hosts of schistosomes in Egypt. Folia Parasitol 21:181–187

El-Emam MA, Madsen H (1982) The effect of temperature, darkness, starvation and various food types on growth, survival and reproduction of *Helisoma duryi*, *Biomphalaria alexandrina* and *Bulinus truncatus* (Gastropoda: Planorbidae). Hydrobiologia 88:265–275

Erko B, Gemetchu T, Gemeda N, Dessie S (1996) Transmission of intestinal schistosomiasis in Addis Ababa, Ethiopia. East Afr Med J 73:732–734

Erko B, Balcha F, Kifle D (2006) The ecology of *Biomphalaria sudanica* in Lake Ziway, Ethiopia. Afr J Ecol 44:347–352

Ernould JC, Ba K, Sellin B (1999) The impact of the local water-development programme on the abundance of the intermediate hosts of schistosomiasis in three villages of the Senegal River delta. Ann Trop Med Parasitol 93:135–145

Favre TC, Bogea T, Rotenberg L, Silva HS, Pieri OS (1997) Circadian rhythms in the cercarial emergence of *Schistosoma mansoni* by *Biomphalaria tenagophila* at outdoors: a comparative study with *Biomphalaria glabrata*. Biol Rhythm Res 28:348–357

Fenwick A, Cheesmond AK, Amin MA (1981) The role of field irrigation canals in the transmission of *Schistosoma man*soni in the Gezira scheme, Sudan. Bull World Health Org 59:777–786

Fernandez MA, Thiengo SC (2002) Susceptibility of *Biomphalaria straminea* (Dunker, 1848) from Serra da Mesa Dam, Goiás, Brazil to infection with three strains of *Schistosoma mansoni* Sambon, 1907. Mem Inst Oswaldo Cruz 97:59–60

Fernandez MA, Thiengo SC (2006) Susceptibility of *Biomphalaria amazonica* and *Biomphalaria occidentalis* from Manso Dam, Mato Grosso, Brazil to infection with three strains of *Schistosoma mansoni*. Mem Inst Oswaldo Cruz 101:235–237

Freitas JR, Bedê LC, De Marco Júnior P, Rocha LA, Santos MB (1987) Population dynamics of aquatic snails in Pampulha reservoir. Mem Inst Oswaldo Cruz 82:299–305

Fried B, LaTerra R, Kim Y (2002) Emergence of cercariae of *Echinostoma caproni* and *Schistosoma mansoni* from *Biomphalaria glabrata* under different laboratory conditions. J Helminthol 76:369–371

Giovanelli A, Soares MS, D'Andréa PS, Gonçalves MM, Rey L (2001) Abundance and *Schistosoma mansoni* infection of the snail *Biomphalaria glabrata*, Brazil. Rev Saude Publica 35:523–530

Giovanelli A, Da Silva CLPA, Leal GB, Baptista DF (2005) Habitat preference of freshwater snails in relation to environmental factors and the presence of the competitor snail *Melanoides tuberculatus* (Müller, 1774). Mem Inst Oswaldo Cruz 100:169–176

Goll PH (1982) Seasonal changes in the distribution of *Biomphalaria sudanica sudanica* (Martens) in Lake Zwai, Ethiopia. Ann Trop Med Parasitol 76:159–164

Golvan YJ, Houin R, Combes C (1977) Natural transmission of *Schistosoma mansoni* in Guadeloupe (French Antilles). Preliminary note. Ann Parasitol Hum Comp 52:259–275

Gomes AL, Melo FL, Werkhauser RP, Abath FG (2006) Development of a real time polymerase chain reaction for quantitation of *Schistosoma mansoni* DNA. Mem Inst Oswaldo Cruz 101:133–136
Green P, Dussart GBJ, Gibson C (1992) Surfacing and water leaving behaviour of the freshwater pulmonate snails *Lymnaea peregra* (Müller), *Biomphalaria glabrata* (Say) and *Bulinus jousseaumei* (Dautzenberg). J Mollus Stud 58:169–179
Greer GJ, Mimpfoundi R, Malek EA, Joky A, Ngonseu E, Retard RD (1990) Human schistosomiasis in Cameroon. Distribution of the snail hosts. Am J Trop Med Hyg 42:573–580
Gretillat S, Gaston G (1975) Remarks on the ecological adaptation of the snail aquatic fauna in saline medium of the Dallol ponds (Republique du Niger). Ann Parasitol Hum Comp 50:595–601
Grisolia ML, de Freitas JR (1985) Physical and chemical characteristics of the habitat of *Biomphalaria tenagophila* (Mollusca, Planorbidae) | [Características físicas e químicas do habitat da *Biomphalaria tenagophila* (Mollusca, Planorbidae).]. Mem Inst Oswaldo Cruz 80:237–244
Guimarães RJ, Freitas CC, Dutra LV, Moura AC, Amaral RS, Drummond SC, Scholte RG, Carvalho OS (2008) Schistosomiasis risk estimation in Minas Gerais State, Brazil, using environmental data and GIS techniques. Acta Trop 108:234–241
Guimarães RJ, Freitas CC, Dutra LV, Felgueiras CA, Moura AC, Amaral RS, Drummond SC, Scholte RG, Oliveira G, Carvalho OS (2009) Spatial distribution of *Biomphalaria* mollusks at São Francisco River Basin, Minas Gerais, Brazil, using geostatistical procedures. Acta Trop 109:181–186
Hamburger J, He-Na XXY, Ramzy RM, Jourdane J, Ratard RC (1998) A polymerase chain reaction assay for detecting snails infected with Bilharzia parasites (*Schistosoma mansoni*) from very early prepatency. Am J Trop Med Hyg 59:872–876
Hanelt B, Adema CM, Mansour MH, Loker ES (1997) Detection of *Schistosoma mansoni* in *Biomphalaria* using nested PCR. J Parasitol 83:387–394
Ishii AI (1984) Effects of temperature on the larval development of *Angiostrongylus cantonensis* in the intermediate host, *Biomphalaria glabrata*. Z Parasitenkd 70:375–379
Jannotti-Passos LK, Vidigal TH, Dias-Neto E, Pena SD, Simpson AJ, Dutra WO, Souza CP, Carvalho-Parra JF (1997) PCR amplification of the mitochondrial DNA minisatellite region to detect *Schistosoma mansoni* infection in *Biomphalaria glabrata* snails. J Parasitol 83:395–399
Jannotti-Passos LK, Magalhães KG, Carvalho OS, Vidigal TH (2006) Multiplex PCR for both identification of Brazilian *Biomphalaria* species (Gastropoda: Planorbidae) and diagnosis of infection by *Schistosoma mansoni* (Trematoda: Schistosomatidae). J Parasitol 92:401–403
Joubert PH, Pretorius SJ, De Kock KN (1986) Survival of *Bulinus tropicus* (Krauss), *Lymnaea natalensis* Krauss and *Biomphalaria glabrata* (Say) at low temperatures. South Afr J Sci 82:322–323
Kader AA (2001) The effect of ecological parameters on the distribution of snail vectors of schistosomiasis. J Egypt Soc Parasitol 31:145–152
Kazibwe F, Makanga B, Rubaire-Akiiki C, Ouma J, Kariuki C, Kabatereine NB, Booth M, Vennervald BJ, Sturrock RF, Stothard JR (2006) Ecology of *Biomphalaria* (Gastropoda: Planorbidae) in Lake Albert, Western Uganda: snail distributions, infection with schistosomes and temporal associations with environmental dynamics. Hydrobiologia 568:433–444
King CH (2009) Global health: toward the elimination of schistosomiasis. New Engl J Med 360:106–109
Kloos H (1985) Water resources development and schistosomiasis ecology in the Awash Valley, Ethiopia. Soc Sci Med 20:609–625
Kloos H, Lemma A (1974) Bilharziasis in the Awash Valley. II. Molluscan fauna in irrigation farms and agricultural development. Ethiop Med J 12:157–173
Kloos H, Lemma A, Desole G (1978) *Schistosoma mansoni* distribution in Ethiopia: a study in medical geography. Ann Trop Med Parasitol 72:461–470
Kloos H, Lo CT, Birrie H, Ayele T, Tedla S, Tsegay F (1988) Schistosomiasis in Ethiopia. Soc Sci Med 26:803–827
Kloos H, Jannotti Passos LK, LoVerde P, Oliveira RC, Gazzinelli A (2004) Distribution and *Schistosoma mansoni* infection of *Biomphalaria glabrata* in different habitats in a rural area in the Jequitinhonha Valley, Minas Gerais, Brazil: environmental and epidemiological aspects. Mem Inst Oswaldo Cruz 99:673–681

Klutse A, Baleux B (1996) Survival of *Bulinus truncatus* and *Biomphalaria pfeifferi* in sewer water purified in stabilization ponds in a subsaharan zone. Med Trop 56:41–47

Kristensen TK, Malone JB, McCarroll JC (2001) Use of satellite remote sensing and geographic information systems to model the distribution and abundance of snail intermediate hosts in Africa: A preliminary model for *Biomphalaria pfeifferi* in Ethiopia. Acta Trop 79:73–78

Lee FO, Cheng TC (1971) *Schistosoma mansoni* infection in *Biomphalaria glabrata*: Alterations in heart rate and thermal tolerance in the host. J Invert Pathol 18:412–418

Lefcort H, Bayne CJ (1991) Thermal preferences of resistant and susceptible strains of *Biomphalaria glabrata* (Gastropoda) exposed to *Schistosoma mansoni* (Trematoda). Parasitology 103:357–362

Loreau M, Baluku B (1991) Shade as a means of ecological control of *Biomphalaria pfeifferi*. Ann Trop Med Parasitol 85:443

Lotfy WM, DeJong RJ, Black BS, Loker ES (2005) Specific identification of Egyptian *Biomphalaria* species and possible hybrids using the polymerase chain reaction based on nuclear and mitochondrial loci. Mol Cell Probes 19:21–25

Lwambo NJ, Upatham ES, Kruatrachue M, Viyanant V (1987) The host-parasite relationship between the Saudi Arabian *Schistosoma mansoni* and its intermediate and definitive hosts. 1. *S. mansoni* and its local snail host *Biomphalaria arabica*. Southeast Asian J Trop Med Public Health 18:156–165

Madsen H (1992) Ecological studies on the intermediate host snails and the relevance to schistosomiasis control. Mem Inst Oswaldo Cruz 87:249–253

Madsen H, Daffalla AA, Karoum KO, Frandsen F (1988) Distribution of freshwater snails in irrigation schemes in the Sudan. J Appl Ecol 25:853–866

Malek EA (1975) Effect of the Aswan High Dam on prevalence of schistosomiasis in Egypt. Trop Geogr Med 27:359–364

Malone JB, Bergquist NR, Huh OK, Bavia ME, Bernardi M, El Bahy MM, Fuentes MV, Kristensen TK, McCarroll JC, Yilma JM, Zhou XN (2001a) A global network for the control of snail-borne disease using satellite surveillance and geographic information systems. Acta Trop 79:7–12

Malone JB, Yilma JM, McCarroll JC, Erko B, Mukaratirwa S, Zhou X (2001b) Satellite climatology and the environmental risk of *Schistosoma mansoni* in Ethiopia and east Africa. Acta Trop 79:59–72

Malone JB, McNally KL, McCarroll JC, Corbett JD, Mkoji G (2004) Modeling the biocoenose of parasitic diseases using remote sensing and geographic information systems. Parassitologia 46:59–61

Mandahl-Barth, G. 1958. Intermediate hosts of *Schistosoma* – African *Biomphalaria* and *Bulinus*. World Health Monograph Series No. 37.

Martin DL, Fried B, Sherma J (2007) Effects of increased salinity on survival and lipid composition of *Helisoma trivolvis* (Colorado strain) and *Biomphalaria glabrata* in laboratory cultures. Veliger 49:101–104

Melo FL, Gomes AL, Barbosa CS, Werkhauser RP, Abath FG (2006) Development of molecular approaches for the identification of transmission sites of schistosomiasis. Trans R Soc Trop Med Hyg 100:1049–1055

Mohamed AM, el-Sharkawi FM, el-Fiki SA (1978) Prevalence of schistosomiasis among fishermen of Lake Maryut. Egypt J Bilharz 5:85–90

Moné H, Mouahid G, Shaban MA, Al Jabri AA, Boissier J, Ruppel A, Idris MA (2003) Ecological and molecular studies on emerging schistosomiasis mansoni in Dhofar Governorate, Sultanate of Oman. Trop Med Intern Health 8:269–276

Mungomba LM, Chandiwana SK, Madesen H (1993) Schistosomiasis around Siavonga, on the shores of Lake Kariba, Zambia. Ann Trop Med Parasitol 87:365–371

Ndifon GT, Ukoli FMA (1989) Ecology of freshwater snails in south-western Nigeria. I: distribution and habitat preferences. Hydrobiologia 171:231–253

Ollivier G, Brutus L, Cot M (1999) La schistosomose intestinale à *Schistosoma mansoni* à Madagascar: Extension et focalisation de l'endémie. Bull Soc Pathol Exot 92:99–103

Olofintoye LK, Odaibo AB (1996) Influence of ecological factors in the population and infection dynamics of *Bulinus globosus* and *Biomphalaria pfeifferi* in the river Odo-Ona, Ibadan, Nigeria. Helminthologia 33:81–86

Owojori OJ, Asaolu SO, Ofoezie IE (2006) Ecology of freshwater snails in Opa Reservoir and Research Farm Ponds at Obafemi Awolowo University Ile-Ife, Nigeria. J Appl Sci 6:3004–3015

Pflueger W (1978) Ecological studies in Madagascar of *Biomphalaria pfeifferi*, intermediate host of *Schistosoma mansoni*. 2. Biology and dynamics in the non-endemic area of Antananarivo. Arch Inst Pasteur Madagascar 46:241–269

Pflueger W (1980) Experimental epidemiology of schistosomiasis. I. The prepatent period and cercarial production of Schistosoma mansoni in Biomphalaria snails at various constant temperatures. Z Parasitenkd 63:159–169

Pflueger W (1981) Experimental epidemiology of schistosomiasis. II. Prepatency of *Schistosoma mansoni* in *Biomphalaria glabrata* at diurnally fluctuating temperatures. Z Parasitenkd 66:221–229

Picquet M, Ernould JC, Vercruysse J, Southgate VR, Mbaye A, Sambou B, Niang M, Rollinson D (1996) The epidemiology of human schistosomiasis in the Senegal river basin. Trans R Soc Trop Med Hyg 90:340–346

Pimentel-Souza F, Schall VT, Barbosa ND, Schettino M, Lautner R Jr (1988) Influence of experimental illumination and seasonal variation on crossbreeding mating in the snail *Biomphalaria glabrata*. Mem Inst Oswaldo Cruz 83:79–85

Pimentel-Souza F, Barbosa ND, Resende DF (1990) Effect of temperature on the reproduction of the snail *Biomphalaria glabrata*. Brazil J Med Biol Res 23:441–449

Poda JN, Sondo B, Parent G (2003) Influence des hydro-aménagements sur la distribution des bilharzioses et de leurs hôtes intermédiaires au Burkina Faso. Cah Sante 13:49–53

Pointier JP, Theron A (1979) La schistosomose intestinale dans les forêts marécageuses à Pterocarpus de Guadeloupe (Antilles françaises). Ecologie du mollusque vecteur, *Biomphalaria glabrata* et de son parasite *Schistosoma mansoni*. Ann Parasitol Hum Comp 54:43–56

Pointier JP, David P, Jarne P (2005) Biological invasions: the case of planorbid snails. J Helminthol 79:S249–S256

Rollinson D, Webster JP, Webster B, Nyakaana S, Jørgensen A, Stothard JR (2009) Genetic diversity of schistosomes and snails: implications for control. Parasitology 136(13):1801–1811

Rumi A, Hamann MI (1990) Potential schistosome-vector snails and associated trematodes in ricefields of Corrients province, Argentina. Preliminary results. Mem Inst Oswaldo Cruz 85:321–328

Rumi A, Hamann MI (1992) Spatial and seasonal trends of a natural population of *Biomphalaria occidentalis* in northeastern Argentina. Mem Inst Oswaldo Cruz 87:181–186

Rumi A, Bechara JA, Hamann MI, Ostrowski de Nuñez M (2002) Ecology of potential hosts of schistosomiasis in urban environments of Chaco, Argentina. Malacologia 44:273–288

Santana Teles HM (1996) Distribuição de *Biomphalaria straminea* ao Sul da Região Neotropical, Brasil. Rev Saude Publica 30:341–349

Santana Teles HM (2005) Distribuição geográfica das espécies dos caramujos transmissores de *Schistosoma mansoni* no Estado de São Paulo. Rev Soc Brasil Med Trop 38:426–432

Santana Teles HM, Ferreira CS, De Carvalho ME et al (2002) Schistosomiasis Mansoni in Bananal (State of São Paulo, Brazil). II. Intermediate hosts. Mem Inst Oswaldo Cruz 97(suppl 1):37–41

Shiff CJ (1960) Observations on the capability of freshwater vector snails to survive dry conditions. J Trop Med Hyg 63:89–63

Shiff CJ, Evans A, Yiannakis C, Eardley M (1975) Seasonal influence on the production of *Schistosoma haematobium* and *S. mansoni* cercariae in Rhodesia. Int J Parasitol 5:119–123

Simoonga C, Utzinger J, Brooker S, Vounatsou P, Appleton CC, Stensgaard AS, Olsen A, Kristensen TK (2009) Remote sensing, geographical information system and spatial analysis for schistosomiasis epidemiology and ecology in Africa. Parasitology 136:1683–1693

Southgate VR, Rollinson D (1987) The natural history of transmission and interactions. In: Rollinson D, Simpson AJG (eds) Biology of Schistosomes: from genes to latrines. Academic Press, London, pp 347–378

Southgate VR, Tchuem Tchuenté LA, Théron A, Jourdane J, Moncrieff CB, Gryseels B (2000) Compatibility of *Schistosoma mansoni* Cameroon and *Biomphalaria pfeifferi* Senegal. Parasitology 121:501–505

Southgate V, Tchuem Tchuenté LA, Sène M, De Clercq D, Théron A, Jourdane J, Webster BL, Rollinson D, Gryseels B, Vercruysse J (2001) Studies on the biology of Schistosomiasis with emphasis on the Senegal River Basin. Mem Inst Oswaldo Cruz 96:S75–S78

Steinmann P, Keiser J, Bos R, Tanner M, Utzinger J (2006) Schistosomiasis and water resources development: systematic review, meta-analysis, and estimates of people at risk. Lancet Infect Dis 6:411–425

Stensgaard AS, Jørgensen A, Kabatereine NB, Rahbek C, Kristensen TK (2005) Modeling the distribution of *Schistosoma mansoni* and host snails in Uganda using satellite sensor data and geographical information Systems. Parassitologia 47:115–125

Sturrock RF (1966) The influence of temperature on the biology of *Biomphalaria pfeifferi* (Krauss), an intermediate host of *Schistosoma mansoni*. Ann Trop Med Parasitol 60:100–105

Sturrock RF, Sturrock BM (1972) The influence of temperature on the biology of *Biomphalaria glabrata* (Say), intermediate host of *Schistosoma mansoni* on St. Lucia, West Indies. Ann Trop Med Parasitol 66:385–390

Sturrock RF, Upatham ES (1973) An investigation of the interactions of some factors influencing the infectivity of *Schistosoma mansoni* miracidia to *Biomphalaria glabrata*. Int J Parasitol 3:35–41

Sturrock RF, Diaw O-T, Talla I, Niang M, Piau JP, Capron A (2001) Seasonality in the transmission of schistosomiasis and in populations of its snail intermediate hosts in and around a sugar irrigation scheme at Richard Toll, Senegal. Parasitology 123:S77–S89

Tchuem Tchuenté LA, Southgate VR, Théron A, Jourdane J, Gryseels B (1999) Compatibility of *Schistosoma mansoni* and *Biomphalaria pfeifferi* in Northern Senegal. Parasitology 118:595–603

Theron A (1981) Dynamics of larval populations of *Schistosoma mansoni* in *Biomphalaria glabrata*. I. Rhythmic production of cercariae in monomiracidial infections. Ann Trop Med Parasitol 75:71–77

Thomas JD (1987) An evaluation of the interactions between freshwater pulmonate snail hosts of human schistosomes and macrophytes. Philos Trans R Soc Lond B Biol Sci 315:75–125

Thomas JD, Kowalczyk C (1997) Utilization of dissolved organic matter (DOM), from living macrophytes, by pulmonate snails: Implications to the "food web" and "module" concepts. Comp Biochem Physiol A Physiol 117:105–119

Traquinho GA, Quintó L, Nalá RM, Gama Vaz R, Corachán M (1998) Schistosomiasis in northern Mozambique. Trans R Soc Trop Med Hyg 92:279–281

Upatham ES (1973) The effect of water temperature on the penetration and development of St. Lucian *Schistosoma mansoni* miracidia in local *Biomphalaria glabrata*. Southeast Asian J Trop Med Public Health 4:367–370

Utzinger J, Tanner M (2000) Microhabitat preferences of *Biomphalaria pfeifferi* and *Lymnaea natalensis* in a natural and a man-made Habitat in Southeastern Tanzania. Mem Inst Oswaldo Cruz 95:287–294

Utzinger J, Mayombana C, Smith T, Tanner M (1997) Spatial microhabitat selection by *Biomphalaria pfeifferi* in a small perennial river in Tanzania. Hydrobiologia 356:53–60

Valle CM, Pellegrino J, Alvarenga N (1973) Rhythmic emergence of *Schistosoma mansoni* cercariae from *Biomphalaria glabrata*: influence of the temperature. Rev Inst Med Trop Sao Paulo 15:195–201

Vercruysse J, Southgate VR, Rollinson D, De Clercq D, Sacko M, De Bont J, Mungomba LM (1994) Studies on transmission and schistosome interactions in Senegal, Mali and Zambia. Trop Geograph Med 46:220–226

Williams NV (1970) Studies on aquatic pulmonate snails in Central Africa. I. Field distribution in relation to water chemistry. Malacologia 10:153–164

Woolhouse MEJ (1992) Population biology of the freshwater snail *Biomphalaria pfeifferi* in the Zimbabwe highveld. J Appl Ecol 29:687–694

Woolhouse MEJ, Chandiwana SK (1989) Spatial and temporal heterogeneity in the population dynamics of *Bulinus globosus* and *Biomphalaria pfeifferi* and in the epidemiology of their infection with schistosomes. Parasitology 98:21–34

Yousif F, Lämmler G (1975) The effect of some biological and physical factors on infection of *Biomphalaria glabrata* with *Angiostrongylus cantonensis*. Z Parasitenkd 47:191–201

Yousif F, Kamel G, el-Emam M, Mohamed S (1993) Ecology of *Biomphalaria alexandrina* the snail vector of *Schistosoma mansoni* in Egypt. J Egypt Soc Parasitol 23:29–42

Yousif F, el-Emam M, Abdel-Kader A, el-Din AS, el-Hommossany K, Shiff C (1998a) Schistosomiasis in newly reclaimed areas in Egypt. 1-distribution and population seasonal fluctuation of intermediate host snails. J Egypt Soc Parasitol 28:915–928

Yousif F, Ibrahim A, Abdel-Kader A, el-Bardicy S (1998b) Invasion of the Nile Valley in Egypt by a hybrid of *Biomphalaria glabrata* and *Biomphalaria alexandrina*, snail vectors of *Schistosoma mansoni*. J Egypt Soc Parasitol 28:569–582

Yousif F, el-Emam M, Abdel-Kader A, el-Din AS, el-Hommossany K, Shiff C (1999) Schistosomiasis in newly reclaimed areas in Egypt. 2 – Patterns of transmission. J Egypt Soc Parasitol 29:635–648

Chapter 4
Studies of *Biomphalaria* Snails Infected with Larval Trematodes Using Chromatography and Atomic Spectrometry

Joseph Sherma and Bernard Fried

Abstract This chapter describes chromatographic and atomic spectrometric methods used for the analysis of *Biomphalaria* snails infected with larval trematodes. Some of the analyses are concerned directly with larval schistosomes and echinostomes, whereas other analyses consider the snail tissues and organs infected with the larval schistosomes and echinostomes. Coverage includes the chromatographic methods used for the determination of different classes of organic compounds and atomic spectrometric methods for elemental analyses. The major chromatographic methods used have been thin-layer chromatography and high performance thin-layer chromatography. Atomic absorption spectrometry and inductivity coupled plasma-optical emission spectrometry have been used for determining certain elements, including metals. These studies have contributed to a better understanding of the chemical composition of larval schistosomes and echinostomes and of *Biomphalaria* infected by these digeneans.

4.1 Introduction

This chapter describes certain methods that have been used in the analysis of larval trematodes from *Biomphalaria* snails. Considered are also studies on the analyses of *Biomphalaria* tissues infected with larval trematodes, mainly schistosomes and echinostomes. Included are the chromatographic methods for the determination of different classes of organic compounds and atomic spectrometric methods for elemental analysis. Explanations of the abbreviations used in this chapter are given in Table 4.1. The major chromatographic methods applied

B. Fried (✉)
Department of Biology, Lafayette College, 111 Kunkel Hall, Easton, PA 18042, USA
e-mail: friedb@lafayette.edu

R. Toledo and B. Fried (eds.), Biomphalaria *Snails and Larval Trematodes*,
DOI 10.1007/978-1-4419-7028-2_4,

Table 4.1 The abbreviations used in this review

AAS	Atomic absorption spectrometry
ASW	Artificial spring water
C-18	Silica gel layer chemically bonded with octadecylsilyl groups
CE	Cholesteryl esters
DGG	Digestive gland–gonad complex
DI	Deionized water
U.S.A. EPA	United States of America Environmental Protection Agency
FAAS	Flame atomic absorption spectrometry
FFA	Free fatty acids
FS	Free sterols
GFAAS	Graphite furnace atomic absorption spectrometry
GLC	Gas–liquid chromatography
HPLC	Column high-performance liquid chromatography
HPTLC	High-performance thin-layer chromatography
ICP-AES	Inductively coupled plasma-atomic emission spectrometry
L-T	Lettuce-tetramin
MALDI-MS	Matrix assisted laser desorption ionization mass spectrometry
nano-HPLC/ESI-IT-MS	HPLC/electrospray ionization-ion trap mass spectrometry
NMID	Nonmethylene-interrupted diene
PC	Phosphatidycholine
PE	Phosphatidylethanolamine
PI	Postinfection
PIn	Phosphatidylinositol
PS	Phosphatidylserine
RP	Reversed phase
SCW	Snail-conditioned water
SE	Steryl esters
SPE	Solid phase extraction
TG	Triacylglycerols
TLC	Thin-layer chromatography
UV	Ultraviolet
Y-L	Yolk-lipid

have been TLC and HPTLC, with a few papers reporting analyses by GLC and column HPLC; AAS and ICP-AES have been reported for the determination of elements, including metals. Earlier studies of the TLC analysis of larval trematodes in *Biomphalaria* from the 1970s through the mid-1990s were reviewed by Fried and Haseeb (1996). Some TLC information on *Biomphalaria* infected with larval echinostomes has been given in Sherma and Fried (2009). This chapter includes mainly TLC-related publications from 1995 through 2009. Publications earlier than 1995 are included if they are related to our 2005–2009 coverage. No previous review of analyses of column chromatographic and spectrometric methods is available. Therefore, all references on these topics are covered here.

4.2 Thin Layer Chromatography Principles and Methods

4.2.1 Principles

TLC is a type of liquid chromatography in which the stationary phase is a layer of sorbent on a glass plate, aluminum foil, or plastic sheet. In the basic TLC procedure, standard compound and sample solutions are applied to the origin line at the bottom of the plate, and the plate is developed by placing it in a closed chamber containing a pool of mobile phase, which is a mixture of two or more solvents. As the mobile phase rises through the layer, the compounds move at different rates and are separated, forming the chromatogram. The separated zones are then visualized by their natural color in daylight or under UV light, or by an application of a detection reagent. Quantification is performed by instrumental densitometry.

Only those materials, instruments, and techniques of TLC that have been applied to studies of larval trematode infected *Biomphalaria* are described in Sects. 2.2–2.10, and the references given in these sections are examples rather than comprehensive. Additional details on TLC are available in Sherma and Fried (2003).

4.2.2 Snail Maintenance and Sample Preparation

Biomphalaria glabrata snails were maintained in aerated glass containers, 20 per container, in 800 mL of ASW. The formulation of ASW was given in Ulmer (1970). Snails were fed Romaine lettuce ad libitum, and cultures were changed twice a week. The details of snail maintenance are given in Schneck and Fried (2005). For estivation studies, snails were maintained on moist paper towels in a closed container at a relative humidity of 98% and a temperature of 24–25°C. The details of the estivation studies are given in White et al. (2007). *B. glabrata* snails maintained under conditions of estivation survived for up to 7 days. In starvation experiments, snails were maintained without exogenous food for up to 1 week. Most snails survived these conditions for the week (White et al. 2007).

Snails were infected with *Schistosoma mansoni* miracidia as described in Fried et al. (2001). Usually snails 5–8 mm in shell diameter were infected in masse with about five miracidia per snail. Infected snails were maintained as described in Fried et al. (2001). *B. glabrata* snails were infected with *Echinostoma caproni* miracidia or eggs as described in Idris and Fried (1996), and infected snails were maintained as described in that paper. Additional information on the maintenance of *B. glabrata* snails infected with echinostome larvae was given in Fried and Peoples (2009).

In our laboratory, when *B. glabrata* samples were prepared for analyses by the techniques described in this review, most of the snail tissue that was used consisted of the DGG. However, in some studies whole bodies were used, and we also occasionally used shells for analyses. The details of the procedures used for obtaining *B. glabrata* samples are described in Fried and Sherma (1990). In brief, shells were lightly cracked with a hammer and removed to obtain either the entire snail or the DGG. For the DGG, the posterior region of the snail was severed from the visceral mass, and the DGG was placed in a Petri dish half filled with Locke's solution. The formulation of Locke's solution was given in Fried (1994). When whole bodies were used, the shell was removed, and the entire body was placed in Locke's solution. Whole snail bodies or DGGs were rinsed with several changes of Locke's solution to remove debris. When hemolymph was needed, it was obtained by gently cracking the snail's shell and allowing hemolymph to drain into the bottom half of a Petri dish. The hemolymph was removed with a Pasteur pipet, placed in an Eppendorf tube, and briefly centrifuged to separate the plasma (snail hemolymph minus the hemocytes is referred to as plasma) from the hemocytes and debris. The details of this procedure are given in Fried and Sherma (1990). In most cases, plasma was used for analysis. All tissues and fluids were used immediately after collection or within a few hours following storage at 4°C. However, some tissues were stored at −20°C and used from 2 to 14 days later.

4.2.3 Sample Extraction and Purification

The classic Folch procedure was used to extract lipids and phospholipids (polar lipids) from the DGG of *B. glabrata* infected with *S. mansoni* (White et al. 2007; Fried et al. 2001); from whole bodies, DGG, and hemolymph of snails patently infected with *Echinostoma caproni* (Bandstra et al. 2006); from rediae, cercariae, and encysted metacercariae of *E. caproni* removed from experimentally infected *B. glabrata* (Marsit et al. 2000); and from the feces of *B. glabrata* infected with *E. caproni* (Schneck et al. 2005). The Folch procedure involves extraction of the sample with chloroform-methanol (2:1) in a glass homogenizer, usually in the ratio of 20 parts solvent to 1 part sample, and then filtration through glass wool if necessary. The extraction is performed up to three times. With a ratio of four parts sample volume to one part salt solution, Folch wash (0.88% KCl in DI water) is added, and the upper, aqueous layer is discarded. The combined extracts are evaporated to dryness in a warm water bath under nitrogen gas and then reconstituted in chloroform-methanol (2:1). The reconstitution volume is chosen so that the scan areas of zones in sample chromatograms are bracketed within the scan areas of the standard zones of the calibration graph for densitometric quantification.

Neutral lipids were extracted from SCW collected from snails infected with *E. caproni* using chloroform-methanol (2:1) in a centrifuge tube, and the lower layer was filtered through glass wool, evaporated, and reconstituted for TLC analysis (Schneck et al. 2005). Neutral lipids and phospholipids were isolated from cercariae of *S. mansoni* from *B. glabrata* snails by vortexing a sample of cercariae and ASW with chloroform-methanol (2:1), refrigerating overnight to obtain two layers, and evaporating and reconstituting the lower layer for TLC analysis (Schariter et al. 2002).

Glycolipids were extracted from *B. glabrata* infected with *S. mansoni* with 20 volumes of chloroform-methanol (1:1), the extract was centrifuged (1,000 g for 10 min), and the residue was re-extracted in 10 volumes of chloroform-methanol (1:2). The extract was again centrifuged and the supernatants combined and taken to dryness on a rotary evaporator. Glycolipids were separated from other lipid classes of the total extract by Florisil column chromatography of their acetylated derivatives (Maloney et al. 1990).

Prior to TLC determination, extraction was performed using acetone in a glass homogenizer for the pigments lutein and beta-carotene from whole bodies or DGGs of *B. glabrata* infected with *E. caproni* and *Helisoma trivolvis* infected with *E. trivolvis* (Evans et al. 2004); by vortexing with 70% aqueous ethanol for free pool amino acids in cercariae, rediae, encysted and excysted metacercariae from *B. glabrata* snails experimentally infected with *E. caproni* (Ponder et al. 2003); and by homogenization in a glass tissue grinder for sugars in the DGG of estivated *B. glabrata* and those infected with *S. mansoni* (Jarusiewicz et al. 2006). *S. mansoni* cercariae from *B. glabrata* were vortexed for 10 min with absolute ethanol-DI water (7:3) to extract sugars and amino acids, followed by centrifugation, evaporation of the supernatant, and reconstitution for TLC analysis (Wagner et al. 2002).

4.2.4 Layers and Mobile Phases

Neutral and polar lipids have been analyzed on 10×20 cm HPTLC silica gel plates with diatomaceous earth preadsorbent (or concentrating zone) and 19 scored lanes (HPTLC-HLF, Analtech, Newark, DE; White et al. 2007 and LHP-KDF, Whatman Inc., Florham Park, NJ; Schariter et al. 2002 and Bandstra et al. 2006). The preadsorbent zone adjacent to the main analytical layer facilitates manual sample application of relatively high volumes of biological sample extracts, and the lanes serve as a guide for positioning of the initial zones and the densitometer source slit for scanning chromatograms. Glycolipids were analyzed on silica gel 60 HPTLC plates without preadsorbent from EMD Chemicals, Inc. (an affiliate of Merck KGaA, Darmstadt, Germany) (Maloney et al. 1990).

The usual mobile phase for separating neutral lipid classes on these plates was petroleum ether-diethyl ether-glacial acetic acid (80:20:1); steryl esters and methyl esters were better separated from each other and from unidentified nonpolar hydrocarbons using the less polar mobile phase hexane-petroleum ether-diethyl ether-glacial ether (50:25:5:1). The mobile phase chloroform-methanol-water (65:25:4) was used to separate polar lipids and glycolipids.

Lutein and beta-carotene were determined on unlaned EMD Chemicals, Inc. 10×20 cm C-18 chemically bonded silica gel RP-HPTLC plates with a preadsorbent zone. The mobile phase was petroleum ether-acetonitrile-methanol (1:1:2) (Evans et al. 2004).

Sugars were analyzed on Whatman Inc. LK5DF silica gel 20×20 cm TLC plates containing a preadsorbent zone and 19 lanes. The mobile phase was ethyl acetate-glacial acetic acid-methanol-water (60:15:15:10) (Jarusiewicz et al. 2006; Wagner et al. 2002).

The analysis of amino acids required four layers having different separation mechanisms (Ponder et al. 2003, 2004; Wagner et al. 2002): EMD Chemicals, Inc. silica gel HPTLC with *n*-butanol-acetic acid-water (3:1:1) mobile phase (normal phase adsorption), EMD Chemicals, Inc. cellulose HPTLC with the same mobile phase (normal phase partition), Whatman Inc. 20×20 cm preadsorbent C-18 bonded silica gel with *n*-propanol-0.5 M NaCl (4:6) mobile phase (RP), and Polygram Ionex-25 SA-Na sheets (Macherey-Nagel, Bethlehem, PA) with pH 3.3 citrate buffer mobile phase (strong acid cation exchange).

4.2.5 Standard and Sample Solution Application

The neutral lipid, phospholipid, carbohydrate, amino acid, and pigment standards were purchased from commercial sources, such as Matreya, Inc. (Pleasant Gap, PA), Nu-Check Prep (Elysian, MN), or Sigma (St. Louis, MO), and dissolved in an appropriate solvent to prepare the standard solutions.

Standard solutions and reconstituted sample extracts were generally applied to the layer in 1.00–16.0 μL aliquots. It is important that the sample zone weight applied is within the weight range of the standard calibration graph for quantitative analysis. Drummond (Broomall, PA) Digital Microdispensers (10 and 25 μL) with disposable glass pipets were used to apply the initial zones of samples and standards. Solutions are applied as diffuse vertical streaks to plates containing a preadsorbent zone, and tight band-shaped initial zones are automatically formed at the preadsorbent-analytical sorbent interface. Solutions must be applied as compact spots to layers without a preadsorbent. In the analysis of glycolipids (Maloney et al. 1990), a Camag (Wilmington, NC) Linomat IV semiautomatic applicator with 100 μL syringe was used to apply bands by the spray-on technique.

4.2.6 Plate Development with the Mobile Phase

Isocratic, linear, ascending development was carried out in a large volume, covered glass TLC or HPTLC chamber (normal chamber or N-chamber) that was lined with a saturation pad (Analtech, Newark, DE) or piece of chromatography paper and pre-equilibrated with the vapors of the mobile phase for 10–15 min before insertion of the plate with the initial zones. Most used was the Camag twin trough chamber, a special N-chamber with an inverted V-shaped ridge on the bottom dividing it into two sections that allow development with a very low volume of mobile phase. The ambient temperature in the laboratory was typically 22°C and the humidity was 50%.

4.2.7 Methods for Detection of Separated Zones

After removing the mobile phase from the plate by drying inside a fumehood with cool air from a hair dryer, neutral lipids were detected as blue zones on a yellow background by spraying with PMA detection reagent (5%, w/v, in ethanol) and heating on a plate heater (Camag) for 10 min at 115°C. Polar lipids were detected as brown–black zones on a white background by spraying with a 10% solution of cupric sulfate in 8% phosphoric acid and heating at 140°C for 15 min (White et al. 2007). Glycolipids were visualized using an orcinol reagent specific for carbohydrate residues (Maloney et al. 1990).

Lutein and beta-carotene appeared as natural yellow zones in daylight (Evans et al. 2004). Sugars were detected as dark purple zones against a light yellow background by spraying with alpha-naphthol-sulfuric acid reagent and heating at 110°C for 5 min (Jarusiewicz et al. 2006), and amino acids as purple to blue zones on a pale background by spraying with ninhydrin reagent (0.3 g ninhydrin and 3 mL of glacial acetic acid in 100 mL of *n*-butanol) and heating for 10 min at 110°C (Ponder et al. 2003).

4.2.8 Documentation of Chromatograms

Chromatograms on plates were documented by scanning with a videodensitometer (VideoScan, Camag). Pigment zones fade rapidly and must be photographed (and quantified, if required) immediately after plate development.

4.2.9 Identification of Zones

Compounds in sample chromatogram zones have been identified by comparison of R_F values and detection colors with standard zones developed on adjacent lanes of the same plate. The R_F value is defined as the distance from the origin to the center of the zone divided by the distance from the origin to the mobile phase front.

4.2.10 Quantification of Analytes

Accurate and precise quantification was performed using a Camag TLC Scanner II slit scanning densitometer with which standard calibration graphs were calculated by linear regression and sample weights interpolated automatically under computer control. Typical settings of the TLC Scanner II used for measuring the bands formed on the preadsorbent plates by the aliquot scanning method were slit width 4, slit length 4, and scanning speed 4 mm/s. Scanning wavelengths found to provide optimum quantitative results for the separated and detected standard and sample zones were 610 nm for neutral lipids, 370 nm for polar lipids, 610 nm for amino acids (except 495 nm for histidine), 515 nm for sugars, 448 nm for lutein, and 455 nm for beta-carotene. All wavelengths were provided by the tungsten densitometer source except for 390 nm (deuterium source).

The interpolated weight of analyte in a zone, extract reconstitution volume, volume applied, and sample weight (solid samples) or volume (liquid samples) were used in appropriate equations to calculate the concentration of the analyte in various samples. Up to 15 different samples could be analyzed with four standards on a single plate.

4.2.11 Statistical Analysis of Data

The accuracy and precision of densitometric quantitative data are high because calibration graphs relating scan areas vs. standard weights are established under the same TLC conditions as for sample zones separated on the same plate. In most studies, Student's *t*-test (*P* equal or less than 0.05) was used to determine the significance of data based on the mean ± standard error values of analytes from different sample populations.

White et al. (2007) did statistical analysis using a single factor analysis of variance (ANOVA) to determine whether there was a significant difference in the lipid percentage of the DGG among infected-estivated, infected-starved, and control snails. If a significant difference ($P<0.05$) was found, the data were subjected to the Bonferroni method to determine among which populations the difference occurred. The SPSS version 13.0 software was used for all data analyses.

4.3 Thin Layer Chromatography Applications

4.3.1 Introduction to Thin Layer Chromatography Applications

This section examines TLC studies on lipids, amino acids, carbohydrates, and lipophilic pigments isolated in larval stages of schistosomes and echinostomes. It also examines these analytes in *Biomphalaria* infected with the intramolluscan stages of these larval trematodes.

4.3.2 Lipids in Biomphalaria *Infected with Larval Trematodes*

This section examines the TLC of lipids in *Biomphalaria* snails infected with larval trematodes, mainly schistosomes and echinostomes. The TLC analyses are of analytes from larval stages and also include studies on effects of larval trematodes on the lipid content of the snail hosts. Most studies are concerned with neutral lipids and phospholipids (polar lipids). Unless otherwise noted, the TLC and HPTLC studies have used silica gel layers, the Mangold solvent system, detection by spraying with PMA or cupric sulfate, and quantification by densitometry. Unless otherwise noted, the studies are arranged in chronological order.

Fried et al. (1989) determined the effects of larval *E. caproni* infection on the neutral lipid content of the DGG of *B. glabrata* using semiquantitative HPTLC. The TG fraction was greater in control than in infected DGG, as also were the SE and FFA fractions. The FS fraction did not differ between control and infected DGG samples. This study showed that larval echinostomes can reduce depot fats in the DGG of the *B. glabrata* snail.

Pérez et al. (1995) analyzed the phospholipids in the DGG of *B. glabrata* infected with *E. caproni*. The mean weight percentages of PE and PC in the DGG of the two populations of snails were not significantly different, but the mean weight percentage of PS was 1.5 times greater in the DGG of the infected snails than that of the uninfected snails. This study showed that larval echinostomes can alter the concentration of some phospholipids in the host snail.

Chaffee et al. (1996) used TLC to study neutral lipids in SCW from *B. glabrata* snails. A major lipid fraction in SCW at 2 and 4 h after snail incubation contained FFA, and the amounts of these acids were quantified by densitometry. The amount of FFA released at 4 h was significantly greater than at 2 h. Snails also released other lipids into the water. The potential of these neutral lipids to serve as chemoattractants for larval echinostomes and schistosomes remains to be determined.

Haas et al. (1997) used TLC to obtain surface lipids from the skin of humans and pigs. Such lipids provide chemical signals that stimulate the secretion of acetabular gland contents of the cercariae of *S. mansoni*. They showed that skin lipids such as FFA, phospholipids, and glucosylceramides play a role in stimulating acetabular secretions of *S. mansoni* cercariae.

Marsit et al. (2000) used HPTLC to analyze the neutral lipids in the rediae, cercariae, and encysted metacercariae of *E. caproni* removed from experimentally infected *B. glabrata* snails. Visual observations showed that the most abundant lipid fraction in all stages was FS. The concentrations of FS were determined in all stages by densitometry.

Thompson et al. (1991) reported that *B. glabrata* infected with *S. mansoni* and maintained on a diet of hen's egg yolk produced fully developed cercariae in about one half of the time taken by snails fed romaine lettuce. Fried et al. (2001) reexamined the Thompson et al. (1991) claim of rapid cercarial development as a factor of snail diet, and also examined the nutritional effects of the high fat diet on larval schistosome development. Infected snails maintained at 26°C and fed either diet produced fully developed cercariae by 4 week PI. The earlier contention that the yolk diet enhanced the time to cercarial patency was not confirmed. HPTLC analysis of neutral lipids show that the DGG of infected snails fed the yolk diet contained significantly higher concentrations of FS and CE, but not TG, than those of the infected snails fed the lettuce diet.

Schariter et al. (2002) used HPTLC to analyze neutral lipids and phospholipids in the cercariae of a Puerto Rican strain of *S. mansoni*. Visual observations of the chromatograms showed that the most abundant neutral lipids in the cercariae were FFA and FS, and the most abundant phospholipids were PC and PE. Densitometry was used to quantify the above-mentioned lipid fractions. This was the first study to provide quantitative data on neutral lipids and phospholipids on a per cercaria basis for *S. mansoni*. The possible functions of lipids in schistosome cercariae were discussed.

Schneck et al. (2004) used HPTLC to analyze neutral lipids in chemically excysted metacercariae of *E. caproni*. The encysted metacercariae (cysts) were obtained from the kidney/pericardium of experimentally infected *B. glabrata* snails. The cysts were chemically excysted, and the excysted metacercariae were maintained in Locke's solution. The excysted metacercariae released neutral lipids into the medium, and densitometry was used to quantify these neutral lipids. Neutral lipid release in echinostomes is related to worm mediated chemoattraction (Haseeb and Fried 1988).

Schneck et al. (2005) used HPTLC to determine neutral lipids in SCW and feces from *B. glabrata* snails infected with *E. caproni*. Analysis of SCW showed the presence of FS, FFA, and TG, but significant differences in the concentrations of these lipids in SCW from infected vs. uninfected snails were not detected. The FFA fraction was the major neutral lipid detected in the feces, and the amount of this lipid was significantly lower in the uninfected samples vs. the infected samples as determined by quantitative densitometry.

Bandstra et al. (2006) examined the effects of larval echinostome infection on the neutral lipid and phospholipid content *B. glabrata* patently infected with rediae of *E. caproni*. Uninfected snails were used as controls. The major neutral lipids in whole bodies and DGGs of both groups were FS, FFA, and TG. The major phospholipids were PC and PE. Densitometric analysis showed that the concentration of TG in infected DGGs was significantly less than that of the uninfected snails.

4.3.3 Effects of Diet and Larval Trematodes on Lipids in Biomphalaria *Snails*

Beers et al. (1995) determined the effects of diet and *E. caproni* parasitism on the neutral lipids in the DGG of *B. glabrata* snails. An analysis of the snails fed on the Y-L diet showed that TG were significantly reduced in the DGG of infected snails whereas levels of FS were significantly elevated in the infected snails. In the snails fed the L-T diet, the TG level was reduced in the DGG of infected snails, but the difference was not significant. The FS level was also elevated as in snails on the Y-L diet, but again this finding was not significant. There were significant differences between TG and FS levels in both infected and uninfected snails maintained on the Y-L diet, but there were no significant differences in any lipid class in the infected vs. uninfected snails. The results of this study showed that both diet and larval trematode parasitism can influence the lipid constituents of *Biomphalaria* snails.

Thompson et al. (1991) used two-dimensional TLC to analyze neutral lipids and phospholipids in the DGG of uninfected *B. glabrata* snails and those infected with *S. mansoni*. Both snail populations were maintained on different diets, as follows. Some snails were fed either egg yolk or lettuce whereas others were not fed (starved). The neutral lipids were separated by two-dimensional TLC on 20×20 cm activated 250-μm layers of silica gel G. The mobile phase for the first direction was hexane-diethyl ether, 80:20, and that for the second direction was hexane-diethyl ether-methanol, 70:20:10. Each mobile phase contained 1% acetic acid. The zones were visualized by spraying the plate with 50% sulfuric acid and heating to char the lipids. Identities were established by use of co-chromatography and comparison of R_F values. The phospholipids were identified by the use of two-dimensional TLC on activated layers of silica gel G. Plates were developed with chloroform-methanol-water-40% aqueous ammonia, 86:28:1:3, in the first direction and chloroform-methanol-water-acetic acid, 76:9.5:2.5:12, in the second direction. Spots were visualized by spraying with molybdenum blue reagent, and the choline-containing phospholipids were identified with Dragendorff's reagent (Thompson and Lee 1987). The major neutral lipid fraction was TG, and this fraction was greatest in the DGG of infected or uninfected snails maintained on the egg yolk diet. FFA were present in trace amounts in all populations; likewise, the major phospholipid fraction was PC, but other polar lipids were also present. Reduced levels of PC and some other phospholipids were found in the DGG of snails infected with *S. mansoni*. These reduced lipid levels coincided with the idea that membrane phospholipids were used by the developing sporocysts and cercariae. In one trial, increased levels of lipids and a shortened period of patency (4 or 5 weeks for snails maintained on egg yolk compared with 8 or 9 weeks for snails maintained on lettuce) were observed for snails maintained on the egg yolk diet. Thompson et al. (1991) were not able to confirm this finding in a subsequent trial. Fried et al. (2001), using HPTLC-densitometry, repeated the experiment of Thompson et al. (1991) and found that maintaining *B. glabrata* infected with *S. mansoni* on a Y-L diet did not increase the time of patency compared with those maintained solely on lettuce. An analysis of snails at 6 weeks PI revealed a significant increase in the FS and SE fractions in the DGG of infected snails maintained on a Y-L diet compared with those of infected snails maintained on a lettuce diet.

White et al. (2007) examined the effects of estivation or starvation on the neutral lipid and phospholipid content of *B. glabrata* patently infected with *S. mansoni* by HPTLC densitometry. Infected-estivated snails were maintained in a moist chamber at 24±1°C and a relative humidity of 98±1%. Infected-starved snails were maintained in ASW at 23±1°C without exogenous food. Infected snails (the controls) were maintained in ASW at 23±1°C and fed lettuce ad libitum. The three groups were maintained in the laboratory for 7 days, and then the lipids from the DGG were extracted and analyzed by class. Infected-estivated snails exhibited a greater mortality rate and weight loss after 7 days than did the infected-starved snails. The SE concentration in the infected-starved snails was significantly increased ($P=0.010$) compared with the controls but not compared with infected-estivated snails; the concentration of PC in infected-estivated snails was significantly decreased ($P=0.007$) compared with the controls but not when compared with the infected-starved snails. Estivation or starvation had a significant effect on the concentration of certain lipid classes in the DGG of *B. glabrata* infected with *S. mansoni*.

4.3.4 *Amino Acids in* Biomphalaria *Snails Infected with Larval Trematodes*

Wagner et al. (2002) used TLC and HPTLC to analyze the amino acids present in the cercariae of *S. mansoni*. Visual observations of the amino acid chromatograms confirmed the presence of histidine, tryptophan, isoleucene, alanine, and proline. Several other unknown ninhydrin-positive zones were also detected. Quantification of the histidine and tryptophan zones in cercarial samples by densitometry gave mean values of 3.7 and 0.59 ng/cercariae, respectively. This is the first study to report concentrations of amino acids in *S. mansoni* cercariae.

Pachuski et al. (2002) used TLC and HPTLC to determine the amino acid content in the DGG of *B. glabrata* snails infected with larval *S. mansoni* for 8 weeks. DGG of infected and control snails were pooled, extracted in ethanol, and chromatographed using various sorbent-mobile phase combinations applicable to amino acid analysis. Zones were detected with ninhydrin and quantified by densitometry. Qualitative analysis showed the presence of histidine, lysine, alanine, methionine, threonine, asparagine, proline, and leucine/isoleucine in both the infected and uninfected DGG. Quantitative analysis was done on histidine, lysine, alanine, and methionine, but only lysine showed a significant reduction in concentration (Student's t-test, $P<0.05$) in the DGG of infected snails compared to the controls. Lysine, an essential amino acid for the developing schistosome larvae, is obtained in significant amounts from the snail DGG.

Ponder et al. (2003) used TLC and HPTLC to determine the free pool amino acids in the rediae, cercariae, encysted metacercariae, and excysted metacercariae of *E. caproni*. Larvae of each type were pooled and extracted in ethanol, and their free pool amino acids separated using four types of layers with different separation mechanisms. The zones were detected with ninhydrin spray reagent and quantified by densitometry. Qualitative analysis revealed the presence of valine, leucine, lysine, histidine, and alanine in rediae; histidine in cercariae; histidine and alanine in encysted metacercariae; and leucine in excysted metacercariae. Quantitative analysis showed that rediae contained 0.76 ng of lysine per organism, and excysted metacercariae contained 0.16 ng of leucine per organism.

Ponder et al. (2004) used TLC and HPTLC to analyze the free pool amino acids in the DGG of *B. glabrata* infected with *E. caproni* vs. uninfected control snails. Qualitative analysis showed the presence of histidine, lysine, serine, alanine, valine, and isoleucine in all samples. Quantitative analysis of lysine and valine gave mean weight percentages of 0.007 and 0.002, respectively, in the DGG of uninfected snails and 0.005 and 0.003, respectively in the DGG of infected snails.

However, the differences in these values between infected and uninfected snails were not statistically significant.

4.3.5 Carbohydrates in Biomphalaria *Snails Infected with Larval Trematodes*

Wagner et al. (2001) used TLC to analyze the effects of larval trematode parasitism by *E. caproni* on carbohydrates in the whole body of *B. glabrata* snails infected for 10 weeks. The major sugars found in snail whole bodies were glucose, maltose, and raffinose. At 10 weeks PI, infection by larval *E. caproni* caused a significant reduction in the concentrations of both maltose and raffinose in the whole bodies of *B. glabrata*. No significant reduction was observed in the concentration of glucose in the infected vs. control snails at 10 weeks PI. This paper reviewed earlier studies on the effects of larval trematode parasitism on the carbohydrate content of host snails.

Wagner et al. (2002) used TLC to analyze the carbohydrates present in the cercariae of *S. mansoni* released from *B. glabrata*. Visual observation of the chromatograms showed the presence of glucose and raffinose and three unidentified alpha-naphthol-positive zones. Quantification of the glucose zones by densitometry in three cercarial samples gave a mean value of 0.49 ng per cercaria. Raffinose was present at lower levels in different cercarial batches. This was the first quantitative study to report sugar values in the *S. mansoni* cercaria.

Jarusiewicz et al. (2006) used TLC to analyze the glucose and maltose concentrations of the DGG of uninfected-estivated *B. glabrata* snails and estivated *B. glabrata* patently infected with *S. mansoni*. All snails were estivated in a moist chamber at a relative humidity of 98 ± 1% and a temperature of 23 ± 1°C for 14 days. Carbohydrates were extracted from the DGG with 70% aqueous ethanol, and extracts were analyzed on silica gel preadsorbent plates using ethyl acetate-glacial acetic acid-methanol-water (60:15:15:10) mobile phase, alpha-naphthol-sulfuric acid detection reagent, and quantification by densitometry. The concentrations of glucose and maltose were significantly reduced in both uninfected-estivated snails and infected-estivated snails.

4.3.6 Lipophilic Pigments in Biomphalaria *Snails Infected with Larval Trematodes*

Evans et al. (2004) used HPTLC to quantify the concentration of beta-carotene and lutein in several planorbid snails infected with larval trematodes. An important combination was laboratory raised *B. glabrata* snails experimentally infected with *E. caproni*. Uninfected snails served as controls. The pigments were extracted from the snail whole bodies and DGG, separated by reversed phase HPTLC, and quantified by densitometry. *B. glabrata* infected with *E. caproni* showed no significant differences in the concentrations of lutein and beta-carotene compared to the uninfected cohorts. Differences in this study compared with other previous studies on this topic reflected intrinsic differences in the larval trematode snail systems used.

4.4 Gas Chromatography

4.4.1 Principles

GLC is a method in which a liquid sample is injected through a rubber septum into a port and vaporized onto the front end of a column. Isothermal or programmed temperature GLC on a packed column with an inert solid support onto which a nonvolatile liquid stationary phase is adsorbed or on a wall coated open tubular (WCOT) capillary column has been used for studies of infected *B. glabrata*. As the inert carrier gas (mobile phase) flows through the column, the sample components move at different rates and reach the detector separated in time. In general, the compound with the lowest boiling point elutes first. The injection port, column, and detector are heated to keep the sample from condensing during the analysis. The flame ionization detector has been used in all analyses reported below.

The chromatogram is a display of the detector signal as a series of peaks at different retention times that are the basis of qualitative analysis for the separated compounds. Quantitative analysis is performed by generating a calibration graph of peak area vs. weight of a series of standards for each analyte, by dividing the peak area of a compound by the total area of all peaks if the response factors for all analytes are equal, or correcting the detector response using predetermined response factors for each analyte.

See the book by Grob and Barry (2004) for information on the theory, instrumentation, and techniques of GLC.

4.4.2 Methods and Applications

Furlong and Caulfield (1988) analyzed the sterol composition of cercariae, schistosomules, and adult *S. mansoni* from *B. glabrata* by GLC. It was found that cercariae and schistosomules contained cholesterol, desmosterol, campesterol, stigmasterol, and beta-sitosterol while adults contained only cholesterol. In all stages, cholesterol comprised greater than 50% of the total sterols, and in cercariae and schistosomules desmosterol comprised 38 and 21% of the total sterols, respectively. The other three sterols made up approximately 10% of the total. The same five sterols found in cercariae and schistosomules were present in the hepatopancreas of uninfected snails but with a much higher desmosterol concentration in the parasite (38%) than in the snail (2%). As in cercariae and schistosomules, the three minor sterols comprised approximately 10%. Thus, the sterol composition of cercariae and schistosomules was similar but not identical to the snail host. Lipids were extracted from samples by the Folch et al. method, and neutral lipids and phospholipids were separated by silicic acid liquid chromatography. Neutral lipids were saponified and acetylated derivatives were prepared and analyzed using a PerkinElmer (Waltham, MA) Model 3920 gas chromatograph with an FID. Both free sterols and acetylated derivatives were injected onto either 3% SE-30 or 3% OV-17 packed columns operated isothermally at 244 or 262°C, respectively. Retention times and peak areas of sterol peaks were calculated with a PerkinElmer Model LCI-100 integrator operating in the default gas chromatography mode.

Shetty et al. (1992) studied the sterols in the DGG complex of *B. glabrata* infected with the patent larval stages of *E. caproni* vs. uninfected controls. The major sterol present was cholesterol at levels of 59% in the infected snails and 51% in controls. Both populations contained the phytosterols desmosterol, campesterol, stigmasterol, and beta-sitosterol, but the percentage composition was reduced in the infected snails. A Hewlett-Packard (Avondale, PA) Model 5890 gas chromatograph with a 30 m×0.32 mm id SPB-1 methylsilicone WCOT column programmed from 200 to 280°C and an FID was used. Sterols were identified based on retention times compared to standards, and percentage composition was calculated by dividing the peak area of a particular sterol by the total area of all peaks in the chromatogram.

Fried et al. (1993) did GLC studies to determine the fatty acid composition of *B. glabrata* snails experimentally infected with the intramolluscan stages of *E. caproni*. The infection was found to reduce the amounts of saturated fatty acids in whole snail bodies but increased the amounts of these acids in the DGG complex of the snails. In both whole bodies and DGGs, infection increased markedly the amounts of palmitoleic acid (16:1n-9) and 20:2 non-methylene-interrupted diene (NMID), but reduced the amounts of 16:1n-7, 20:1n-11+9, and 22:1n-11+13; docosahexaenoic acid (DHA) concentration was markedly decreased in infected whole snail bodies. Lipids were extracted by the Folch method and then treated with 7% BF_3-methanol to convert acyl lipids to methyl esters. GLC of the fatty acid methyl esters was carried out on a PerkinElmer model 8240 gas chromatograph equipped with a 30 m×0.25 mm id Supelcowax-10 polyethylene glycol fused silica capillary column programmed from 185 to 230°C and an FID. Peak areas were converted to % fatty acids by correcting the FID response using a previously developed computer program.

4.5 Column High-Performance Liquid Chromatography

4.5.1 Principles

HPLC involves high pressure flow of a liquid mobile phase through a metal tube (column) containing the stationary phase. The sample is applied to the column through a loop injector, and separated mixture components are eluted into a detector.

HPLC columns have a short length and small internal diameter (id) compared to classical LC. The differences of HPLC compared to GLC are that the former is usually performed at ambient temperature, and the HPLC mobile phase affects the selectivity but the GLC carrier gas does not. Detectors for HPLC include fluorescence, electrochemical, or UV. Mechanisms of separation, depending upon the type of column used, include adsorption, normal and RP phase partition, bonded phase, ion exchange, and size exclusion.

See the book by Dong (2006) for information on the theory, instrumentation, and techniques of HPLC.

4.5.2 Methods and Applications

Furlong and Caulfield (1988) used two HPLC procedures to determine the phospholipid composition of *S. mansoni* cercariae, schistosomules, and adults from *B. glabrata*. It was found that PC was the major phospholipid of all three stages (50%). The remaining phospholipids consisted of PE, PS, and PIn. In addition, in adults there were small quantities of sphingomyelin and lysophosphatidylcholine. The percentage of each phospholipid was similar among stages except for a slight increase in PS in adults compared to the other two stages. In the first protocol, HPLC was carried out on a Waters (Milford, MA) automated gradient control system. Lipid from approximately 25,000 cercariae or schistosomules or 50 adult pairs was eluted from a Waters Z-module cartridge with acetonitrile-methanol-phosphoric acid (130:1.8:1.25) at a flow rate of 5 mL/min. Phospholipids were detected by UV absorbance at 214 nm, and the major phospholipid classes were identified by comparison of retention times with standards. Quantification was performed by cutting out and weighing peaks recorded on a strip chart recorder; the total weight of all peaks exclusive of the solvent front was taken as 100%, and the percentage of a given peak was calculated accordingly. The second HPLC protocol separated and determined neutral lipids and phospholipids in a single run using a Spherisorb 10 µm silica column; a ternary gradient system with hexane-THF (99:1), chloroform-isopropanol (1:4), and isopropanol-water (1:1); and a mass selective detector using individual calibration graphs for quantification of each compound.

Manger et al. (1996) measured the biogenic monoamines serotonin, dopamine, and L-dopa using HPLC in the extracts of the central nervous system and plasma of infected *B. glabrata* in snails at 7, 14, 21, and 28 days post exposure to the miracidia of *S. mansoni*. The findings suggested that serotonin acts as a stimulant for egg production in *B. glabrata*, and that parasitic castration may be due, at least in part, to larval induced suppression of serotonin in the snails' central nervous system and plasma during the course of infection with *S. mansoni*. The HPLC system included a pump, injector with 20 µL sample loop, electrochemical (amperometric) detector with a glassy carbon working electrode and Ag/AgCl reference electrode, and BAS 3 µm ODS column (10 cm × 3.2 mm id). The mobile phase consisted of

0.1 M citrate buffer (pH 3) containing 0.7 mM Na_2EDTA, 0.25 mN octyl sulfate, and 10% acetonitrile with a flow rate of 0.6 mL/min.

During the intramolluscan life cycle stages of *S. mansoni* within the *B. glabrata* intermediate host, the larval parasites synthesize a wide array of glycoconjugates exhibiting, in part, unique carbohydrate structures. In addition, the larval parasites express definitive host-like sugar epitopes, such as Lewis X determinants, supporting the concept of carbohydrate-mediated molecular mimicry as an invasion and survival strategy. Lehr et al. 2006 investigated whether common carbohydrate determinants occur also at the level of the intermediate host by performing structural characterization of hemolymph glycoprotein-*N*-glycans of *B. glabrata*. *N*-glycans were released from tryptic peptides and labeled with 2-aminopyridine. Sugar chains serologically cross reacting with *S. mansoni* glycoconjugates were isolated by immunoaffinity chromatography using a polyclonal antiserum directed against schistosomal egg antigens, and fractionated by *Aleuria aurantia* lectin affinity chromatography and HPLC. Obtained glycans were analyzed by MALDI-MS and nano-HPLC/ESI-IT-MS as well as by monosaccharide constituent and linkage analysis. The results revealed a highly heterogeneous oligosaccharide pattern. Cross-reacting species represented about 5% of the total glycans and exhibited a terminal Fuc(alpha1-3) GalNAc unit, a (1–2)-linked xylosyl residue, or both types of structural motifs. The study demonstrated the presence of common carbohydrate epitopes in larval *S. mansoni* and in the intermediate host *B. glabrata*. HPLC was carried out at 40°C using a TSK-Amide-80 column (4×250 mm) and fluorescence detection; the mobile phase was a gradient of 0–100% of 3% aqueous triethylamine/acetic acid, pH 7.3-acetonitrile (1:1) in 50 min.

Bezerra et al. (1997) examined the profile of carboxylic acids in *S. mansoni* resistant and susceptible strains of *B. glabrata*. The acids were extracted from the hemolymph of two susceptible strains (PR, Puerto Rico and Ba, Jacobina-Bahia from Brazil) and from two resistant strains (13-16-R_1 and 10R_2) using SPE followed by HPLC. The carboxylic acids identified in all hemolymph samples by comparison with known standards were pyruvic, lactic, succinic, malic, fumaric, acetic, propionic, beta-hydroxybutyric, and acetoacetic. Under standard conditions, the concentration of each acid varied among the strains tested and appeared to be specific for each strain. Only the concentration of fumarate was consistently different ($P<0.05$) between resistant and susceptible strains. Organic acids were recovered from centrifuged hemolymph by direct application onto SAX anion exchange quaternary amine Bond-Elut SPE column (Varian Inc., Walnut Creek, CA) and elution with 0.5 M sulfuric acid, and the eluate was subjected to HPLC on a Biorad (Hercules, CA) Aminex ion exclusion HPX-87H column (300×7.8 mm) with 0.5 mM sulfuric acid mobile phase and UV detection at 210 nm.

Massa et al. (2007) used HPLC to determine the effects of a patent *S. mansoni* infection on certain carboxylic acids in the DGG and hemolymph of *B. glabrata*. Hemolymph was analyzed for carboxylic acids using the same anion exchange SPE and ion exclusion HPLC-UV techniques as described above (Bezerra et al. 1997) with an Agilent 1100 series system (Wilmington, DE) except the mobile phase was 5.0 mM instead of 0.50 mM sulfuric acid to achieve better and more consistent analyte separations. An analysis of DGG samples was done using extraction with 50% Locke's solution, cleanup of the extract by anion exchange SPE, and ion exclusion HPLC-UV. Acetic, fumaric, malic, and pyruvic acids were detected, confirmed, and quantified at concentrations ranging from 12 to 280 ppm in the DGG and 124–8,000 μg/dL in the hemolymph. Infection with *S. mansoni* caused a significant reduction in concentrations of acetic, fumaric, malic, and pyruvic acids in the DGG but not in the hemolymph of *B. glabrata* compared to uninfected cohort snails. Reductions in the infected DGG suggest these acids are utilized by the sporocysts and cercariae in the snail tissue, or that infection stimulates reduced production or increased utilization by the snail tissue.

4.6 Inductively Coupled Plasma-Atomic Emission Spectrometry and Atomic Absorption Spectrometry

4.6.1 Principles

ICP-AES and AAS are atomic spectrometric methods used to determine elemental analytes in samples. ICP-AES is an emission method in which the source is very hot plasma of ionized argon, into which samples and standards are introduced. Atoms are ionized and emissions are measured using a sequential, monochromator design or a simultaneous multichannel, polychromator or array-based design to isolate different wavelengths, and one or more PMT detectors. The wavelength is characteristic of a particular element, and the intensity of the emission indicates the concentration of the element in the sample.

In AAS a liquid sample is aspirated into a flame (FAAS) or pipetted into a graphite furnace (GFAAS) at high temperature, and the resulting unexcited gaseous analyte atoms absorb light emitted from a hollow cathode lamp and passed through the flame or furnace. The lamp emits exactly the best wavelength required for the analysis. A photomultiplier tube detector converts the amount of light reaching it into an electrical signal that is read out as absorption. The analyte concentration in the sample is determined from its amount of absorption using a calibration graph prepared using standards of known concentration.

See the paper by Sherma (2008) for further information on the procedures and instrumentation of atomic spectrometry.

4.6.2 Methods and Applications

Mohammad and Mostafa (2007) used FAAS to determine the alteration of Ca concentration in the soft parts and shells of *B. alexandrina*, a planorbid species related to *B. glabrata*, and in *Bulinus truncatus* due to infection with *S. mansoni* and *S. haematobium*, respectively. The results showed significant lowering in the Ca content in the shells of infected *B. alexandrina* and *B. truncatus* relative to the shells of uninfected snails. In contrast, the Ca content in the soft parts of snails that were releasing cercariae was significantly higher than in the soft parts of uninfected snails. Generally, Ca content was significantly higher in the shells than in the soft parts of the snails, whether infected or uninfected. Samples were digested with concentrated nitric acid to extract Ca, and extracts were analyzed using the source wavelength and sample aspiration rate recommended by the manufacturer of the instrument, which was not identified. Four aqueous standards having analyte concentrations within the linear range of the instrument and containing the same amount of nitric acid as samples were used for calibration. Each standard, sample, and reagent blank was analyzed using three 10s integrations, and the blank value was subtracted to give the final concentration.

Ong et al. (2004) employed ICP-AES to study element ions in whole bodies of uninfected *B. glabrata* snails and those experimentally infected with larval stages of *S. mansoni*. Infected and cohort uninfected snails were analyzed 8 weeks PI. Of the 28 elements measured, 16 (Al, B, Ba, Ca, Cd, Cu, Fe, K, Mg, Mn, Na, Ni, Pb, Se, Sn, and Zn) were identified in infected and uninfected whole bodies at concentrations above the detection limit of ICP-AES analysis. Of these, Ca, Cd, Mn, and Na were present in significantly higher amounts in whole infected vs. whole uninfected snails. Ca was extracted from snail bodies using 2% nitric acid, and analysis was done with a Thermo Jarrell Ash (Franklin, MA) simultaneous reading spectrometer with an autosampler. The instrument was calibrated using U.S.A. EPA Method 6010B, which employs a three point calibration graph generated from the measurement of a calibration blank and two multielement standards. Interelement correction factors were applied to minimize interference between elements in samples. The samples, standards, and a reagent blank were analyzed using three 30s integrations.

4.7 Concluding Remarks

Most of the studies reported here have used TLC and HPTLC for the analyses of larval schistosomes and echinostomes and *Biomphalaria* snails infected with these digeneans. These studies have provided considerable information on the qualitative and quantitative analyses of various organic compounds, including neutral lipids, phospholipids, glycolipids, amino acids, sugars, and lipophilic pigments such as lutein and beta-carotene. Changes in *Biomphalaria* tissues and hemolymph infected with schistosomes and echinostomes have documented changes in the pathobiochemical effects of the infection on this host. The atomic spectrometry studies have provided new information on elemental analyses of schistosomes and echinostomes and on the pathochemical effects of schistosome and echinostome infection on *Biomphalaria* snails. Less information is available on GLC and HPTLC analysis of schistosomes and echinostomes associated with *Biomphalaria* snails. Further work using these analytical tools may provide much needed information on the molecular species of compound classes in *Biomphalaria* snails, information which is not easily obtained by TLC and HPTLC alone.

Acknowledgment We thank Ms. Karen E. LeSage for assistance in preparing this manuscript.

References

Bandstra SR, Fried B, Sherma J (2006) High performance thin layer chromatographic analysis of neutral lipids and phospholipids in *Biomphalaria glabrata* patently infected with *Echinostoma caproni*. Parasitol Res 99:414–418

Beers K, Fried B, Fujino T, Sherma J (1995) Effects of diet on the lipid composition of the digestive gland-gonad complex of *Biomphalaria glabrata* (Gastropoda) infected with larval *Echinostoma caproni* (Trematoda). Comp Biochem Physiol 110B:729–737

Bezerra JCB, Becker W, Zelck UE (1997) A comparative study of the organic acid content of the hemolymph of *Schistosoma mansoni* resistant and susceptible strains of *Biomphalaria glabrata*. Mem Inst Oswaldo Cruz 92:421–425

Chaffee LA, Fried B, Sherma J (1996) Neutral lipids in snail conditioned water from *Biomphalaria glabrata* (Gastropoda: Planorbidae). J Chem Ecol 22:231–235

Dong MW (2006) Modern HPLC for practicing scientists. Wiley-Interscience, Malden

Evans RT, Fried B, Sherma J (2004) Effect of diet and larval trematode parasitism on lutein and beta-carotene concentrations in planorbid snails as determined by quantitative high performance thin layer chromatography. Comp Biochem Physiol 137B:179–186

Fried B (1994) Metacercarial excystment of trematodes. Adv Parasitol 33:91–144

Fried B, Haseeb MA (1996) Thin layer chromatography in parasitology. In: Fried B, Sherma J (eds) Practical thin layer chromatography. CRC, Boca Raton, pp 51–70

Fried B, Peoples RC (2009) Maintenance, cultivation and excystation of echinostomes: 2000-2007. In: Fried B, Toledo R (eds) The biology of echinostomes: from the molecule to the community. Springer, New York, pp 111–128

Fried B, Sherma J (1990) Thin layer chromatography of lipids found in snails. J Planar Chromatogr 3:290–299

Fried B, Schater S, Kim S (1989) Effects of *Echinostome caproni* infection on the lipid composition of *Biomphalaria glabrata*. Int J Parasitol 19:353–354

Fried B, Sherma J, Rao KS, Ackman RG (1993) Fatty acid composition of *Biomphalaria glabrata* (Gastropoda: Planorbidae) experimentally infected with the intramolluscan stages of *Echinostoma caproni* (Trematoda). Comp Biochem Physiol 104B:595–598

Fried B, Muller EE, Broadway A, Sherma J (2001) Effects of diet on the development of *Schistosoma mansoni* in *Biomphalaria glabrata* and on the neutral lipid content of the digestive gland gonad complex of the snail. J Parasitol 87:223–225

Furlong ST, Caulfield JP (1988) *Schistosoma mansoni*: sterol and phospholipid composition of cercariae, schistosomula, and adults. Exp Parasitol 65:222–231

Grob RL, Barry EF (2004) Modern practice of gas chromatography. Wiley, Malden

Haas W, Diekhoff D, Koch K, Schmalfuss G, Loy C (1997) *Schistosoma mansoni* cercariae: stimulation of acetabular gland secretion is adapted to the chemical composition of mammalian skin. J Parasitol 83:1079–1085

Haseeb MA, Fried B (1988) Chemical communication in helminths. Adv Parasitol 27:169–207

Idris N, Fried B (1996) Development, hatching, and infectivity of *Echinostoma caproni* (Trematoda) eggs, and histologic and histochemical observations on the miracidia. Parasitol Res 82:136–142

Jarusiewicz JA, Sherma J, Fried B (2006) Thin layer chromatographic analysis of glucose and maltose in estivated *Biomphalaria glabrata* snails and those infected with *Schistosoma mansoni*. Comp Biochem Physiol 145B:346–349
Lehr T, Geyer H, Maass K, Doenhoff MJ, Geyer R (2006) Structural characterization of *N*-glycans from the freshwater snail *Biomphalaria glabrata* cross reacting with *Schistosoma mansoni* glycoconjugates. Glycobiology 17:82–103
Maloney MD, Semprevivo LH, Coles GC (1990) A comparison of the glycolipid compositions of the cercarial and adult *Schistosoma mansoni* and their associated hosts. Int J Parasitol 20:1091–1093
Manger P, Li J, Christensen BM, Yoshino TP (1996) Biogenic monoamines in the freshwater snail, *Biomphalaria glabrata*: influence of infection by the human blood fluke *Schistosoma mansoni*. Comp Biochem Physiol 114A:227–234
Marsit CJ, Fried B, Sherma J (2000) Neutral lipids in cercariae, encysted metacercariae, and rediae of *Echinostoma caproni*. J Helminthol 74:365–367
Massa DR, Chejlava MJ, Fried B, Sherma J (2007) High performance column liquid chromatographic analysis of selected carboxylic acids in *Biomphalaria glabrata* patently infected with *Schistosoma mansoni*. J Parasitol 101:925–928
Mohammad O, Mostafa S (2007) Effects of *Schistosoma mansoni* and *Schistosoma haematobium* infections on calcium content in their intermediate hosts. Parasitol Res 101:963–966
Ong JHL, Chejlava M, Fried B, Koehnlein KM, Bosavage GL, Sherma J (2004) Effects of *Schistosoma mansoni* infection on inorganic elements in the snail *Biomphalaria glabrata*. J Helminthol 78:343–346
Pachuski J, Fried B, Sherma J (2002) HPTLC analysis of amino acids in *Biomphalaria glabrata* infected with *Schistosoma mansoni*. J Liq Chromatogr Relat Technol 25:2345–2349
Pérez MK, Fried B, Sherma J (1995) Comparison of mobile phase and HPTLC qualitative and quantitative analyses, on preadsorbent silica gel plates, of phospholipids in *Biomphalaria glabrata* (Gastropoda) infected with *Echinostoma caproni* (Trematoda). J Planar Chromatogr 7:340–343
Ponder EL, Fried B, Sherma J (2003) Thin layer chromatographic analysis of free pool amino acids in cercariae, rediae, encysted metacercariae, and excysted metacercariae of *Echniostoma caproni*. J Liq Chromatogr Relat Technol 26:2697–2702
Ponder EL, Fried B, Sherma J (2004) Free pool amino acids in *Biomphalaria glabrata* infected with *Echniostoma caproni* as determined by thin layer chromatography. J Parasitol 90:665–666
Schariter JA, Pachuski J, Fried B, Sherma J (2002) Determination of neutral lipids and phospholipids in the cercariae of *Schistosoma mansoni* by high performance thin layer chromatography. J Liq Chromatogr Relat Technol 25:1615–1622
Schneck JL, Fried B (2005) Growth of *Biomphalaria glabrata* (NMRI strain) and *Helisoma trivolvis* (Colorado Strain) under laboratory conditions. Am Malacol Bull 20:71–73
Schneck JL, Fried B, Sherma J (2004) Effects of tonicity on the release of neutral lipids in *Echinostoma caproni* adults and observations on lipids in excysted metacercariae. Parasitol Res 92:285–288
Schneck JL, Bandstra JR, Fried B, Sherma J (2005) Thin layer chromatographic analysis of neutral lipids in snail conditioned water and feces of *Biomphalaria glabrata* infected with *Echinostoma caproni*. J Liq Chromatogr Relat Technol 27:2039–2045
Sherma J (2008) Instruments for atomic spectrometry. J AOAC Int 91:17A–24A
Sherma J, Fried B (eds) (2003) Handbook of thin layer chromatography, 3rd edn. Marcel Dekker, New York
Sherma J, Fried B (2009) Studies of echinostomes using chromatography and atomic spectrometry. In: Fried B, Toledo R (eds) The biology of echinostomes: from the molecule to the community. Springer, New York, pp 229–248
Shetty PH, Fried B, Sherma J (1992) Effects of *Echinostoma caproni* infection on the sterol composition of the digestive gland-gonad complex of *Biomphalaria glabrata* as determined by gas liquid chromatography. J Helminthol 66:68–71

Thompson SN, Lee RW-K (1987) Characterization of the ^{31}P NMR spectrum of the schistosome vector *Biomphalaria glabrata* and of the changes following infection by *Schistosoma mansoni*. J Parasitol 73:64–68

Thompson SN, Mejia-Scales V, Borchardt DB (1991) Physiologic studies of snail-schistosome interactions and potential for improvement of in vitro culture of schistosomes. In Vitro Cell Dev Biol 27A:497–504

Ulmer MJ (1970) Notes on rearing of snails in the laboratory. In: MacInnes AJ, Voge M (eds) Experiments and techniques in parasitology. Freeman, California, pp 143–144

Wagner SD, Kim Y, Fried B, Sherma J (2001) Thin layer chromatographic analysis of glucose, maltose, and raffinose in *Biomphalaria glabrata* (Gastropoda) infected with *Echinostoma caproni* (Trematoda). J Planar Chromatogr 14:459–461

Wagner SD, Pachuski J, Fried B, Sherma J (2002) Thin layer chromatographic analysis of carbohydrates and amino acids in *Schistosoma mansoni* (Trematoda) cercariae. Acta Chromatogr 12:159–169

White MM, Fried B, Sherma J (2007) Effects of aestivation and starvation on the neutral lipid and phospholipid content of *Biomphalaria glabrata* infected with *Schistosoma mansoni*. J Parasitol 93:1–3

Chapter 5
Effects of Larval Schistosomes on *Biomphalaria* Snails

Judith Humphries

Abstract The aim of this chapter is to provide a clearer understanding of the complex relationship between *Biomphalaria* snails and larval schistosome parasites. This chapter describes the numerous changes in host physiology, biochemistry, and behavior brought about by infection. Specifically, the effects of larval schistosomes on host mortality, growth, metabolism, reproduction, organic and inorganic elements, and behavior are focused on. The chapter concentrates on three *Biomphalaria* species; *B. glabrata*, *B. alexandrina*, and *B. pfeifferi* and one species of schistosome, *Schistosoma mansoni*.

5.1 Introduction

The long history of research on the relationship between *Biomphalaria* snails and larval *Schistosoma* parasites spans from the mid 1900s, and is active and ongoing today. Up to 18 different *Biomphalaria* species are thought to act as the intermediate host of the trematode parasite *S. mansoni* over a wide range of tropical and subtropical countries (DeJong et al. 2001). However, just a few *Biomphalaria* species have been researched in depth, in particular *B. glabrata*, *B. pfeifferi* and *B. alexandrina*, and this disparity is reflected in this chapter. The numerous species and strains of *Biomphalaria* vary in their compatibility as a schistosome host such that some display resistance to infection while others are susceptible. When resistant snails are exposed to *S. mansoni* miracidia, hemocytes (snail blood cells) migrate towards the recently transformed sporocysts and enclose them in a multilayered cellular encapsulation. Soon after, the sporocysts are killed by a cytotoxic reaction which most likely involves free radicals such as hydrogen peroxide and/or nitric oxide (Hahn et al. 2001). In contrast, susceptible snails are not able to

J. Humphries (✉)
Biology Department, Lawrence University,
711 East Boldt Way, Appleton, WI 54911, USA
e-mail: judith.humphries@lawrence.edu

R. Toledo and B. Fried (eds.), Biomphalaria *Snails and Larval Trematodes*,
DOI 10.1007/978-1-4419-7028-2_5, © Springer Science+Business Media, LLC 2011

successfully defend against *S. mansoni* larvae and an infection will develop following miracidial exposure. After penetrating the head/foot region of susceptible snails, schistosome miracidia transform into sporocysts. Then, second generation sporocysts known as daughters develop inside the first generation sporocysts through polyembryony. Larvae remain in the head/foot region for approximately 2–3 weeks, after which they can be found in the digestive gland-gonad (DGG) complex where they continue to multiply. When the infection reaches patency at approximately 4 weeks postinfection, cercariae are released from daughter sporocysts and are released from the snail. The specific timing of the above events differs with various host–parasite combinations and maintenance conditions.

The comparison of uninfected and infected snails has allowed any changes occurring within infected snails to be more evident. However, caution must be exercised when interpreting the changes observed in the infected host. The *Biomphalaria*-schistosome relationship is highly complex and it is difficult to determine whether the effects of infection are direct and whether they benefit the host, the parasite or perhaps neither.

This chapter describes the nonimmunological effects of *S. mansoni* larvae on *Biomphalaria* snails, such as host physiology, biochemistry, and behavior. Specifically, the effects of infection on mortality, growth, metabolism, organic and inorganic elements, reproduction, and behavior are discussed in the following sections. Host growth, metabolism, and reproduction have received more attention in the text because of the relative abundance of available literature focusing on these areas.

5.2 Survival and Mortality

5.2.1 Mortality

Several studies have suggested that schistosome-infected *Biomphalaria* snails show decreased survival compared to uninfected snails (Sorensen and Minchella 2001). Few of these studies were carried out in the field but the work of Woolhouse (1989) suggested that laboratory findings correspond closely to that in the field. *B. pfeifferi* were collected from a river in Zimbabwe and when monitored in the laboratory; schistosome-infected snails exhibited higher mortality rates than the uninfected snails as well as a significant (>60%) reduction in expected life span (Woolhouse 1989). *B. glabrata* likewise displayed an elevated mortality rate during a schistosome infection (Schwanbeck et al. 1986; Blair and Webster 2007), such that 80% of a group of infected *B. glabrata* died during the first 12 weeks of infection compared to only 15% of the control group (Meier and Meier-Brook 1981).

The time at which a significant increase in mortality is evident, differs between studies but often it coincides with the onset of cercarial release, otherwise known as patency. Ibrahim (2006) found no significant differences in survival rates between uninfected and infected *B. alexandrina* snails during the first 3 weeks of infection but survival was significantly reduced during weeks 4–9 when infections had reached patency. In contrast, Meuleman (1972) reported that the onset of increased mortality occurred at 10 weeks postinfection. The later onset of increased mortality was not due to a prolonged prepatent period as cercarial shedding was observed at 22–24 days postinfection. Variation in the timing of events in these two studies may be due to the different miracidial dosages used and also that snails were originally from different geographical regions, South Africa and the Sinai.

5.2.2 Miracidial Dose

A number of studies suggest that there is a relationship between miracidial dose and the increase in mortality rate of infected snails. Makanga (1981) studied the effect of increasing doses of *S. mansoni* miracidia on the survival of *B. pfeifferi* and results displayed dose dependence as there was 5% mortality in controls and 10, 40, and 75% in snails exposed to 1 2, 4 or 4 miracidia, respectively. In contrast, Blair and Webster (2007) exposed snails to increasing numbers of miracidia but did not observe a dose-dependent relationship with mortality. Unexpectedly, resistant snails that had been exposed to miracidia did not develop an infection but showed an increased mortality rate which was proportional to the miracidial dose (Blair and Webster 2007). This is particularly interesting as it demonstrates that not only does the mortality rate of susceptible snails increase with infection, but also that of resistant *Biomphalaria* snails following exposure to schistosomes. These data suggest that the antischistosome response of resistant snails is costly in terms of survival.

5.2.3 Snail Age

In some cases, an inverse relationship is apparent between mortality rate and the age of snails when infected. Meier and Meier-Brook (1981) showed that snails infected while 1 week old displayed a 77% mortality rate which lowered progressively as the age of snail when infected increased. The more severe effect of infection on younger snails may be due in part to an elevated parasitemia as those exposed while 1–3 weeks old showed a significantly higher rate of infection than those exposed at 4 weeks of age and older (Meier and Meier-Brook 1981). Uninfected juvenile snails showed a relatively high rate of mortality during their first 5 weeks but this was still significantly lower than that seen in infected snails of the same age (Meier and Meier-Brook 1981). Similar results were reported by Sturrock and Sturrock (1970) when infecting the first generation progeny of field collected *B. glabrata*. However, snails infected when 44 weeks old also showed a higher mortality rate similar to that seen in 2 week old infected snails. Control uninfected snails displayed an increase in mortality rate about this age. Therefore, it seems that the highest mortality rates occurred in snails infected at an age when mortality would have been elevated in uninfected *Biomphalaria* regardless.

5.2.4 *Susceptibility*

Webster and Woolhouse (1999) examined whether the effects of schistosomes on *Biomphalaria* correlated with the degree of susceptibility. A genetically polymorphic and moderately susceptible *B. glabrata* strain was subjected to selection for either resistance or susceptibility. As expected, snails selected for susceptibility showed higher mortality when infected than both the uninfected resistant selected and uninfected and unselected snails. However, there was no significant difference in the mortality rates between the more susceptible selected snails and the original unselected snails when both lines were infected (Webster and Woolhouse 1999) indicating that the degree of susceptibility does not affect mortality rate when infected.

5.3 Growth and Metabolism

5.3.1 *Growth*

The majority of studies examining the effect of a schistosome infection on the growth of *Biomphalaria* snails have focused on *B. glabrata,* and only a few have looked at *B. alexandrina* and *B. pfeifferi*. An examination of the literature shows that in most studies, schistosome infected *B. glabrata* show a significant decrease in growth at some point in the life cycle (Sturrock and Sturrock 1970; Thompson and Mejia-Scales 1989; Théron et al. 1992). In one such study, Théron et al. (1992) observed no significant changes during the first 3 weeks of infection but later on there was a significant decrease in the growth of infected snails to the extent that they were 1.4 times smaller than controls by 9 weeks postinfection. Occasionally, the reduction in growth rate was preceded by a short growth surge (Sturrock and Sturrock 1970; Gerard and Théron 1997). For example, Sturrock and Sturrock (1970) measured an initial increase in growth rate in *S. mansoni*-infected *B. glabrata* but thereafter growth rapidly declined and ultimately, the controls grew larger than the infected snails. These findings are in accord with Sorensen and Minchella (2001) who concluded that planorbid genera such as *Biomphalaria*, *Bulinus*, and *Helisoma* more commonly display stunting of growth during trematode infection rather than gigantism.

5.3.2 *Snail Age*

The results of several studies suggest the change in growth rate of infected *B. glabrata* varies depending on the age of the host when infected. In one study, infected juveniles (3–4 mm in diameter) demonstrated a significant growth spurt during the prepatent period, while in contrast infected adults displayed a reproductive boost (Gerard and Théron 1997). Furthermore, during the patent period, the growth rate of juveniles was significantly reduced while that of adults was not affected (Gerard and Théron 1997). Meier & Meier-Brook (1981) reported similar findings; snails infected at 1, 2 or 4 weeks of age showed significant decreases in growth, which contrasted the lack of change seen in those infected when 6 or 8 weeks old. Similarly, Sturrock and Sturrock (1970) observed that snails infected at a younger age were overtaken by controls while those infected when older were not.

5.3.3 B. alexandrina *and* B. pfeifferi

The few studies using *B. alexandrina* vary such that both increases and decreases in growth have been reported. Ibrahim (2006) observed an initial increase in the growth rate of *B. alexandrina* but then growth declined until the controls outgrew them. In contrast, Mohamed and Ishak (1979) measured an increase in the growth rate of infected snails which remained elevated for the duration of the experiment. Likewise, Sturrock (1966) reported an increase in the growth rate of *S. mansoni*-infected *B. pfeifferi* which lasted 4 weeks then paralleled that of controls but due to the initial surge, infected snails remained larger for the remainder of the experiments. These findings are supported by Meuleman's work (1972) which also recorded an increase in the shell diameter of infected *B. pfeifferi* and moreover measured an accompanying increase in the total weight of infected snails. The increased growth of infected *B. pfeifferi* and *B. alexandrina* contrasts with what has been reported for *B. glabrata*. Sturrock and Sturrock (1970) suggest the smaller size of *B. pfeifferi* compared to *B. glabrata* may be a contributing factor. Yousif et al. (1996) likewise found *B. alexandrina* to be smaller than *B. glabrata*. A positive correlation between cercarial output and shell diameter has been reported for both *B. alexandrina* and *B. pfeifferi*; therefore, larger snails are advantageous for the parasite (Sturrock 1966; Yousif et al. 1996). Perhaps the direction of change in growth, whether it increases or decreases, is determined by the size of snails.

Because of the complex nature of the *Biomphalaria*-schistosome relationship, it is unlikely that one specific mechanism is solely responsible for the altered growth of infected snails. Moreover, the particular mechanisms affecting growth of infected snails probably differ within and between *Biomphalaria* species. A number of studies report alterations in eating, food assimilation, and carbohydrate levels, all of which have the potential to change growth.

5.3.4 Glucose and Glycolytic Enzymes

Freshwater snails represent a particularly rich energy resource for larval trematodes as they contain a high concentration of carbohydrates (Veldhuijzen 1975). That said, it is not surprising that numerous species of freshwater snails are parasitized by trematode larvae and consequently show diminished carbohydrate stores when infected (Cheng and Snyder 1963). This has been seen in *S. mansoni*-infected *B. glabrata* as a number of studies have measured significantly reduced glucose levels in host hemolymph (Cheng and Lee 1971; Stanislawski and Becker 1979; Rupprecht et al. 1989). Interestingly, hemolymph glucose levels in *S. mansoni*-infected *B. glabrata* 1 month into the infection are very similar to those seen in snails subjected to starvation (Stanislawski and Becker 1979).

Corresponding to the depressed glucose levels, El-Ansary et al. (2000) reported increased activities of glycolytic enzymes such as hexokinase, pyruvate kinase and glucose phosphate isomerase in *S. mansoni*-infected *B. alexandrina* when measured 2 weeks postinfection. It is not clear whether the host, parasite or both produce these enzymes. As host glucose has been localized inside *Glypthelmins* sporocysts during infection of *Helisoma trivolvis* (Cheng and Snyder 1963), it is possible that glycolysis is occurring inside schistosomes and sporocyst derived enzymes contribute to the increased levels of glycolytic enzymes within schistosome-infected *Biomphalaria*. The importance of glycolysis to larval schistosome development was demonstrated by treating infected *B. alexandrina* with a molluscicide known to interfere with glycolytic enzymes (El-Ansary et al. 2000). Molluscicide treatment significantly delayed the development of intramolluscan larval schistosomes and in addition, the number of cercariae produced was reduced by more than 50% (El-Ansary et al. 2000). In a subsequent study, El-Ansary et al. (2003) reported that molluscicide treatment also significantly reduced the pathogenicity of cercariae in their mammalian host.

Biomphalaria glucose content has not been unequivocally traced to sporocysts as yet but a possible scenario is that the larvae internalize host glucose for use as an immediate energy source while also converting some into glycogen stores within the enclosed cercariae (Cheng and Snyder 1963). This proposition is based on a study by Cheng and Snyder (1963) where they detected radioactively labeled host glucose both attached to, and inside the tegument of *Glypthelmins* sporocysts. In further support, Mohamed and Ishak (1979) reported that cercarial glycogen levels increased progressively during schistosome infection of *B. alexandrina*. Interestingly, sporocysts do not seem to store glycogen for their own needs as the only glycogen stores identified inside sporocysts have been those within the enclosed cercariae.

5.3.5 Glycogen

Not surprisingly, *Biomphalaria* glycogen levels are also affected by a schistosome infection as described by Mohamed and Ishak (1981) who measured significantly diminished glycogen levels in the female accessory sexual organs (ASO), mantle, and DGG of *B. alexandrina* 45 days following infection with *S. mansoni*. The highest concentration of glycogen was also measured in the DGG followed by the mantle (Mohamed and Ishak 1981). During infection, *Biomphalaria's* glycogen is likely to be converted into a more readily usable form of energy such as glucose. This assumption is supported by Schwartz and Carter (1982) who found that glycogenolysis was significantly increased in infected snails at 5 weeks postinfection. Moreover, Cheng and Snyder (1963) identified glucose on and in the tegument of *Glypthelmins* sporocysts and not glycogen, possibly because the former is easier to transport. It is not known whether schistosome sporocysts or the host are responsible for breaking down host glycogen into glucose.

5.3.6 Maltose

Maltose is another carbohydrate present in the digestive gonad gland of *Biomphalaria* snails at concentrations comparable to that of glucose (Cline et al. 1999; Jarusiewicz et al. 2006). However, this carbohydrate source does not appear to be accessed by schistosome larvae either, as Jarusiewicz et al. (2006) measured similar concentrations of maltose in infected and control snails. Moreover, Wagner et al. (2002) could not detect maltose in cercariae shed from *B. glabrata* snails. Perhaps maltose stores represent a host exclusive nutritional resource.

5.3.7 Anaerobic Metabolism

The production of lactic acid by *S. mansoni* sporocysts during infections of both *B. alexandrina* and *B. glabrata* has led several researchers to suggest that the majority of metabolic pathways inside infected snails have transitioned from aerobic to anaerobic (Coles 1972; Ishak et al. 1975). Corresponding to the increases in lactic acid, increased lactate dehydrogenase activity has also been demonstrated in infected *B. alexandrina* (Nabih et al. 1990). This enzyme is responsible for the interconversion of pyruvate and lactate. Moreover, Nabih et al. (1990) compared the ratio of H and M lactate dehydrogenase isoenzymes as this ratio reveals whether aerobic or anaerobic metabolic pathways are in use. A high H/M ratio is indicative of aerobic metabolism while a low ratio suggests anaerobic pathways. Results showed that the H/M ratio was decreased at 21 days postinfection but returned to a normal ratio 5 months later. The low H/M ratio further supports the findings of other studies that suggest that mainly anaerobic respiration is occurring in infected snails (Nabih et al. 1990). The latter study also indicates that infected snails regain the ability for aerobic respiration once cercarial shedding has ceased (Nabih et al. 1990). If the majority of metabolic pathways within infected snails were indeed aerobic, a concomitant increase in oxygen consumption by the snail would be expected. In correlation with the use of anaerobic pathways, a number of studies have reported no differences in the oxygen consumption rate of infected snails compared to uninfected (Von Brand and Files 1947; Van Aardt et al. 2003). In one particular study however, regression analysis demonstrated elevated oxygen consumption by infected snails, but this only reached statistical significance at 6 and 8 weeks postinfection (Lee and Cheng 1971).

5.3.8 Feeding Rate

One might predict that infected snails increase their feeding rate in order to compensate for the depletion of host carbohydrates by schistosomes. A small number of studies have investigated this matter but their findings are quite varied. For instance, Williams and Gilbertson (1983) reported that *S. mansoni*-infected *B. glabrata* spent more time feeding than controls while Gerard and Théron (1996) found no difference. Interestingly, results in the latter study were normalized for the total milligrams of snail dry weight which probably leads to more accurate data (Gerard and Théron 1996). After carrying out what appears to be the only study addressing this issue in *B. pfeifferi*, Meuleman (1972) reported a reduction in feeding rate during infection. As an explanation, McClelland and Bourns (1969) proposed that infected and therefore castrated snails consume less food because cercarial production requires less energy than producing eggs. Gerard and Théron (1996) also calculated food assimilation and found that conversion efficiency was decreased in infected snails (Gerard and Théron 1996), similar to results reported previously by Thompson and Mejia-Scales (1989).

5.4 Lipids and Organic Acids

Unfortunately, studies investigating the effects of schistosome infection on the lipid content of *Biomphalaria* are relatively lacking. One of the few available studies examined lipid levels in the DGG of *S. mansoni*-infected *B. glabrata* and found that triacylglycerol levels were three times that of the controls and total lipid levels were increased although not so significant (Thompson 1987). In addition, a particular unidentified lipid showed a significant reduction with infection. When compared with starved snails, it is interesting to see that triacylglycerols were not affected by starvation implying that infected *Biomphalaria* do not metabolize triacylglycerols as an energy source. The elevated triacylglycerol levels are difficult to explain especially in the absence of additional data. *Biomphalaria* lipid levels might be expected to decrease with infection as has been observed in *Fasciola hepatica*-infected *Lymnaea* snails (Humiczewska and Rajski 2005) and also because schistosomes cannot synthesize fatty acids (Meyer et al. 1970). In addition, Mandlowitz et al. (1960) showed that *S. mansoni* cercariae contain lipases suggesting they can utilize host lipids. Perhaps lipid stores in other regions of the snail body are significantly reduced and the increased triacylglycerol levels in the DGG of *S. mansoni*-infected snails are actually from the sporocysts and cercariae. In support, triacylglycerols are relatively abundant in adult schistosomes and constitute more than 40% of the total lipid content (Brouwers et al. 1997). The role of these triacylglycerol stores in adult schistosomes is not clear, though it does not appear that they can be used for ATP synthesis; so far the β oxidative pathway has not been found to be active in schistosomes, even though the genes for this pathway were identified through the schistosome genome project (Brouwers et al. 1997; Berriman et al. 2009).

It would be interesting to examine the levels of fatty acids in both infected snails and schistosome larval stages as it is known that fatty acids can be incorporated into phospholipids in adults (Brouwers et al. 1997). It is clear that more studies are necessary to fully understand the role of lipids in larval schistosomes during infection of *Biomphalaria* and to allow comparisons to be made with adult worms.

Likewise, organic acids within infected *Biomphalaria* snails have not been focused on to any great extent. A study carried out by Massa et al. (2007) found that acetic, fumaric, malic, and pyruvic acids were all significantly reduced in the DGG of *S. mansoni*-infected *B. glabrata* but no changes were seen in the hemolymph. It was suggested that these acids may be released from the DGG through physical damage caused by the sporocysts and subsequently be utilized by them, or alternatively, the host may be using the acids as an energy source to compensate for the increased metabolism associated with infection (Massa et al. 2007).

5.5 Proteins

Infection with schistosome larvae has been shown to alter the protein levels in both the hemolymph and body tissues of *Biomphalaria* snails (Gilbertson et al. 1967; Rupprecht et al. 1989; Crews and Yoshino 1991). A number of studies all report a significant reduction in *Biomphalaria* protein levels during the first 4 weeks of infection (Gilbertson et al. 1967; Stanislawski et al. 1979; Rupprecht et al. 1989). However, in one study, the initial reduction in hemolymph protein levels was followed by a brief surge at 5 weeks postinfection (Rupprecht et al. 1989). This surge was soon followed by a decline and then during the following weeks, protein levels were seen to increase progressively (Rupprecht et al. 1989). The sudden increase in protein levels at week 5 is difficult to explain with certainty, but it does coincide with cercarial shedding suggesting that once the infection is patent, the larval schistosomes place less of a demand on host proteins allowing the host to replenish stores. Rupprecht et al. (1989) suggested that the gradual increase in protein levels during the following weeks represent the establishment of equilibrium between host and parasite.

As protein levels are altered by infection, several groups have also investigated the amino acid profile of infected *Biomphalaria* and in general the data indicates that amino acids are significantly decreased in infected snails (Gilbertson et al. 1967; Stanislawski et al. 1979; Schnell et al. 1985). For example, Gilbertson et al. (1967) reported a decrease in free amino acids in *B. glabrata* hemolymph 16 days following exposure to *S. mansoni* and levels were further reduced to approximately half of that of the controls by 32 days postinfection. Infection also affects the amino acid levels of body tissues such as the DGG where Schnell et al. (1985) measured a 43% reduction in total free amino acids and amines as well as a 20% reduction in all other tissues. However, not all amino acids are affected to the same extent by an infection, and the amino acid profile observed during an infection, varies between studies. For instance, Schnell et al. (1985) found the levels of asparagine, glutamine, phenylalanine, and isoleucine in the DGG were more dramatically reduced than the remaining amino acids and similarly serine and glutamine were decreased the most in the remaining tissues. Likewise, Stanislawski et al. (1979) reported significant reductions in the hemolymph levels of the majority of amino acids except for a few, of which citrulline, isoleucine, and ornithine are noteworthy. It is assumed that such specific changes do not occur by chance, but the mechanisms underlying such changes have not been established. That said, Stanislawski et al. (1979) suggested the reduction in hemolymph arginine levels measured in their study may be due to the increased levels of arginase in *S. mansoni*-infected *B. glabrata* that have previously been reported (Schmale and Becker 1977). Ornithine is a product of arginine hydrolysis; therefore the relatively stable ornithine concentrations correlate with the proposed increase in arginase activity. Furthermore, concentrations of urea are elevated in infected snails and as a consequence, ornithine and citrulline levels may remain consistent due to their involvement in the urea cycle (Rupprecht et al. 1989).

Glutamine was also extremely reduced in infected snail tissues and in the past this was accounted for by glutamine's role as a store for ammonia (Stanislawski et al. 1979). More recent research suggests that sporocysts can utilize glutamine as a substrate for glycerol synthesis (glyceroneogenesis) and possibly also glucose synthesis (gluconeogenesis) (Coppin et al. 2003). In addition, evidence suggests the presence of sporocysts may actually cause snails to upregulate their expression of glutamine synthase, while in turn sporocysts may increase their levels of glutaminyl-tRNA synthetase in response to a snail-derived signal (Coustau et al. 2003).

Histidine, tryptophan, isoleucine, alanine, and proline have all been detected in *S. mansoni* cercariae (Wagner et al. 2002); therefore, it is expected that the levels of these amino acids in particular would be reduced in infected snails. However, only isoleucine has shown dramatic reductions with a schistosome infection (Schnell et al. 1985).

The effects of *S. mansoni* on *B. glabrata* protein levels are similar to that seen in snails starved for 12 days (Stanislawski and Becker 1979) suggesting the reduction in host protein and amino acids is most likely due to increased protein catabolism in infected snails. This correlates with the elevated levels of urea that have been reported by Thompson and Lee (1987) who measured a nearly tenfold increase in the levels of urea in the hemolymph of infected *Biomphalaria.* A similar increase in urea was seen in starved snails (Thompson and Lee 1987). Interestingly, a study by Smith et al. (1994) showed that elevated urea levels inhibited growth and reproduction, and may contribute to the castration and reduced growth of infected snails.

5.6 Inorganic Elements

Inorganic elements are integral to a multitude of processes within living organisms, such as respiration, metabolism, oxygen transport, cell signaling, and osmotic balance. For this reason, sporocysts and cercariae require these elements but they are restricted to obtaining them from the snail host. The DGG is a storage site for inorganic elements (Bebiano and Langston 1995) and while sporocysts are located there, they may cause physical damage, consequently releasing stores. Therefore, it is not surprising that the levels of inorganic elements within *Biomphalaria* are affected by a schistosome infection. In particular, changes in calcium levels in both the shell and soft body tissue have been most commonly reported. Three types of calcium cells labeled A, B and C have been identified in *B. glabrata* and morphological changes in cell type A were found during schistosome infection (Davies and Erasmus 1984). In addition, changes in the shell have been reported, such as erosion of the nacreous layer, pitting of the underlying surface of the shell, and increased fragility, all of which suggest the shells of infected snails have reduced calcium levels. This implies that *B. glabrata* does not show hypercalcification of the shell as has been observed in other trematode-snail interactions (Pinheiro and Amato 1995). However, White et al. (2005) did not detect a change in the calcium levels of shells of *S. mansoni*-infected *B. glabrata*, but this may be due to differences in snail maintenance or the experimental conditions used in the two studies. The effect of infection on *B. alexandrina* shells was also examined and results showed that calcium levels were also reduced in the shell (Mostafa 2007).

In contrast to the reduction in shell calcium, increased levels of calcium were measured in the soft tissue of *Biomphalaria* snails during schistosome infection (Ong et al. 2004; Mostafa 2007). Similar findings have been reported for other trematode-infected snails such as *E. revolutum*-infected *Lymnaea stagnalis* (Gabrashanska et al. 1991). The opposing effects of infection on shell and body calcium levels may imply that calcium from the shell and cellular stores in body tissues, especially in the digestive gland, are released during infection, raising the levels of free calcium (Shaw and Erasmus 1987; Bebiano and Langston 1995). Davies and Erasmus (1984) suggest that the calcium changes recorded in infected *Biomphalaria* may occur to counteract the condition of acidosis which results from the high metabolism of schistosome larvae. Alternatively, several authors suggest that cercariae are taking up host calcium as they are known to harbor stores of calcium in their preacetabular glands. Dresden and Edlin (1975) found the calcium concentration in preacetabular glands to be between 8 and 10 M, and suggest calcium is important as an inhibitor of protease activity until such activity is required. Shaw and Erasmus (1987) found that the effects of schistosome infection on host calcium could be detected as early as 2 days postinfection, long before cercariae are present, implying that host calcium is released for the purpose of balancing the acid/base levels in hemolymph associated with acidosis. Studies on the effects of other stressors such as starvation, on *Biomphalaria* are very useful for comparative purposes. Davies and Erasmus (1984) compared the effects of starvation, infection, and mechanical damage on calcium within *B. glabrata* and discovered that infection induced similar symptoms to that seen with mechanical damage and acidosis but differed from starvation. This suggests that acidosis and damage caused by schistosome larvae are at least partially responsible for the calcium changes brought about in *Biomphalaria*. However, this does not rule out the possibility that cercariae sequester host calcium into their preacetabular stores once the infection reaches patency.

As calcium is critical to multiple processes within animals, it is possible that the other physiological changes occurring in infected *Biomphalaria* are downstream effects of the changes in calcium levels. Mazuran et al. (1999) showed that high environmental calcium concentrations significantly reduced egg laying in *Planorbarius* snails, implying that the reduction in infected *Biomphalaria* egg laying may be partially due to the altered calcium levels.

5.7 Behavioral Effects

Numerous studies have examined a variety of behaviors displayed by *Biomphalaria* snails but few have focused on the behavior of schistosome-infected snails. One of the first studies in this area implied that snail chemosensitivity was affected by larval schistosomes (Etges 1963). Control and infected *B. glabrata* were monitored for their attraction to food bait and results showed that infected snails selected food less often than the controls (Etges 1963). A more recent study compared the behavior of infected and uninfected *B. glabrata* snails and discovered that all snails, irrespective of infection status, were more attracted to infected snails than to controls (Boissier et al. 2003). If this behavior is indeed replicated in the field, it might lead to aggregation of infected snails. In fact, Sire et al. (1999) carried out a field study in Guadeloupe which found infected snails to be aggregated and interestingly, infected snails tended to harbor single sex infections. In addition, the prevalence of infected snails was surprisingly low (0.21–4.76%) suggesting that aggregation increases the probability that a human host will get infected by both sexes of schistosomes and also by a genetically diverse population (Sire et al. 1999).

When comparing the behavior of infected and uninfected snails, Boissier et al. (2003) also examined several parameters associated with exploration and movement such as the length of time snails rested between travels, the size of area explored, and the rate and distance of travel. The results were interesting as snails infected with only male schistosome larvae showed significant differences with the controls in all of the above parameters. In contrast, snails harboring an all female infection showed significant differences only in the amount of time they rested between travels. Therefore, a female infection seems to alter *Biomphalaria* behavior to a lesser extent than a male infection, although the difference was not statistically significant. The significance of the difference between the behavior of male-infected and female-infected snails is unknown as yet.

5.8 Reproduction

5.8.1 Host Castration

Of all the aspects of *Biomphalaria* biology affected by schistosome infection, reproductive success has probably been of most interest within the field. Consequently, numerous studies have focused on this area yet the data have not resulted in many clear conclusions as the effects on reproduction vary with the age of snails, miracidial dose, and the various strains of snails and schistosomes studied. Infection affects several different components of *Biomphalaria* reproduction including egg and sperm production, size, growth, and metabolism of sexual organs such as the ovotestis and albumen glands, and even sexual behavior.

Numerous studies have demonstrated that schistosome infection of *B. glabrata* and *B. pfeifferi* leads to host castration, i.e., a reduction or termination of egg and/or sperm production (Sturrock 1966; Meuleman 1972; Looker and Etges 1979; Makanga 1981; Meier and Meier-Brook 1981; Minchella and Loverde 1981; Crews and Yoshino 1989; Cooper et al. 1996). The timing of this event varies between studies but the earliest reported decrease in egg production occurred at 2 and 3 weeks postinfection (Crews and Yoshino 1989; Meier and Meier-Brook 1981). This is commonly followed 1–2 weeks later by the complete cessation of egg production (Looker and Etges 1979; Crews and Yoshino 1989). Clearly, this reduces the host's lifetime reproductive success, the extent of which depends on the host's age at the time of infection. Snails which become infected when they are already sexually mature will have some reproductive success through mating events that took place prior to infection. Moreover, mature snails may still reproduce during the first few weeks of infection as they can still produce some eggs during this period. In contrast, infected juvenile snails are deprived entirely of any potential reproductive success. Juveniles have not yet developed sexually and if infected at this age they are incapable of becoming sexually mature and consequently cannot reproduce. Therefore, it can be said that *S. mansoni* infection is more virulent in juvenile snails than adults.

In some host–parasite relationships, hosts compensate for parasitic castration by increasing their reproductive output during the prepatent period, following parasite exposure. Whether this phenomenon, known as reproductive compensation, commonly occurs in schistosome-infected *Biomphalaria* has not been conclusively established. One of the first studies to report reproductive compensation in infected *B. glabrata* was carried out by Minchella and Loverde (1981). In their study, *B. glabrata* showed a significant increase in egg laying during the first 3 weeks of infection which was subsequently followed by a reduction (Minchella and Loverde 1981). This compensatory strategy has likewise been observed in a number of other *B. glabrata* studies (Gerard and Théron 1997) including one carried out by Thornhill et al. (1986) which interestingly suggests that reproductive compensation is age-dependent as it was observed in infected adults but not juveniles. The issue of whether reproductive compensation occurs in schistosome-infected *Biomphalaria* was recently revisited by Blair and Webster (2007). Their data did not show an increase in the reproductive effort of infected *B. glabrata* during prepatency. Curiously though, exposed but uninfected snails demonstrated an increase in reproductive output following exposure (Blair and Webster 2007). Therefore, when examining the cumulative data on reproductive compensation in *Biomphalaria*, no firm conclusions can be drawn as yet. This is not surprising as there is much variation between the aforementioned studies, including miracidial dose, age of snail, parasite, and host strains. However, it seems clear that schistosomes cause castration of *Biomphalaria* snails, irrespective of whether there is a preceding reproductive surge. Unfortunately, field studies are lacking and therefore not available for comparative purposes.

The exact mechanism(s) responsible for castration of schistosome-infected *Biomphalaria* are unknown as yet. Parasitic castration of hosts in general may occur through a number of mechanisms, such as direct ingestion of the sexual organs as in the case of redial larval stages, the redirection of resources that were allocated to host reproduction or through modulation of the host neuroendocrine system (De Jong-Brink 1995; Warr et al. 2006). Interestingly, host castration may be beneficial for both schistosomes and the host as castration releases nutritional resources which are presumably utilized by both the parasite and the host. Moreover, if castration did not take place, perhaps the host would die before cercariae are shed due to an extreme lack of resources, which would benefit neither in the relationship. This raises the question whether *Biomphalaria* are partly responsible for their own castration in order to increase their lifespan.

5.8.2 Nutritional Resources

The redirection of host nutritional resources in infected *Biomphalaria* snails is expected to contribute to the castration of snails during infection. Diet and nutrition have been shown to be very important for *Biomphalaria* egg production as evidenced by the severe negative impact of starvation and nutrient depletion on egg output (Stanislawski and Becker 1979; Thompson and Mejia-Scales 1989). Furthermore, the specific dietary composition is also important as high carbohydrate and high lipid diets were shown to reduce *B. glabrata* egg laying, while it was significantly increased when snails were fed a high protein diet (Stanislawski and Becker 1979). Schistosome infection and starvation both have been reported to reduce egg output although it is worth noting that in a comparative study, starvation caused a more significant reduction than infection (Stanislawski and Becker 1979). Furthermore, the effects of infection on *Biomphalaria* reproduction are not mirrored exactly by that seen in response to starvation. Therefore, redirection of nutritional resources such as glucose, glycogen, and proteins is most likely one of a number of factors contributing to castration of infected *Biomphalaria* snails.

5.8.3 Proteins

A variety of proteins are essential for *Biomphalaria* egg development and production, however, total protein levels throughout the body, including the reproductive organs, are significantly reduced during infection. Crews and Yoshino (1991) reported a significant reduction in the levels of specific but unidentified proteins in the ovotestis and albumen gland during infection and it is likely that these particular proteins play important roles in egg production. Likewise, when studying *S. mansoni*-infected *B. glabrata*, Looker and Etges (1979) measured a significant reduction in galactogen, a glycoprotein that is secreted by the albumen gland during egg development.

Furthermore, phenol oxidase activity, important for oxidizing egg shell proteins, was significantly inhibited in the albumen gland (Bai et al. 1997). The reduced levels of certain host proteins such as galactogen and phenol oxidase are most likely, having direct and specific negative effects on egg production.

5.8.4 Physical Damage

It is possible that sporocysts cause physical damage to the sexual organs as daughter sporocysts migrate from the headfoot region to the DGG between 2 and 3 weeks postinfection (Crews and Yoshino 1989). The number of schistosomes continues to increase and soon thereafter the DGG is packed full with sporocysts and cercariae (Théron et al. 1992). In addition to any physical damage that might be caused by the sporocysts, schistosomes can affect the sexual organs and gamete production through a number of other means.

5.8.5 Accessory Sexual Organs

Infection clearly stunts or inhibits the growth of the sexual organs such as the ovotestis and albumen gland (Meier and Meier-Brook 1981; Théron and Gerard 1994). Théron and Gerard (1994) examined the growth and development of male and female ASO in both immature and mature *B. glabrata* infected with schistosomes. Schistosome infection either delayed or inhibited the growth and development of ASO in both age groups of snails, although immature snails were more significantly affected than mature. Notably, the most dramatic effect was seen in the albumen gland, the growth of which was stopped completely in infected snails irrespective of age (Théron and Gerard 1994).

5.8.6 Bioamines

Involvement of the bioamines serotonin and dopamine in *B. glabrata* egg production has been demonstrated in several studies (Manger et al. 1996; Bai et al. 1997; Boyle and Yoshino 2002). Specifically, dopamine is required for the egg shell and/or the egg mass membrane and both dopamine and serotonin may regulate the albumen gland secretion of perivitelline fluid around the egg (Santhanagopalan and Yoshino 2000; Boyle and Yoshino 2002). Manger et al. (1996) demonstrated that the treatment of infected snails with serotonin increased egg laying and in doing so reversed schistosome-induced castration. This study strongly implies that serotonin plays an important regulatory role in *Biomphalaria* egg laying. Consequently, several researchers have investigated whether the aforementioned bioamines, which are associated with egg production, might be affected during infection. Manger et al. (1996) recorded lower levels of serotonin and dopamine in infected *B. glabrata* compared to uninfected controls in both the central nervous system and hemolymph and notably reductions in serotonin could be detected as early as 7 days postinfection. Sporocysts may be directly responsible for this reduction as recent evidence suggests sporocysts have the ability to take up exogenous serotonin via high-affinity surface transporters (Boyle and Yoshino 2002). This report together with the cloning of a schistosome serotonin transporter, provide a mechanism by which the infection-related decrease in snail serotonin levels could be occurring (Patocka and Ribiero 2007).

5.8.7 Neuroendocrine Interference

Until recently, schistosomin, a peptide originally isolated from *Lymnaea* snails, was proposed as a candidate responsible at least in part, for the castration of schistosome-infected *Biomphalaria*. The schistosome parasite *Trichobilharzia* decreases egg laying in *Lymnaea stagnalis* by targeting host secretion of a peptide called schistosomin (De Jong-Brink 1995). As a schistosomin homologue was recently identified from *B. glabrata,* it was proposed that a similar mechanism was occurring in *S. mansoni*-infected *Biomphalaria*. However, this was recently disproved by Zhang et al. (2009). The possibility that *S. mansoni* regulates host reproduction by another means of neuroendocrine interference or through a secreted factor(s) should not be discounted, but as yet, such strategies have not been identified.

5.8.8 Male Role in Mating

It has been suggested that even when egg production has ceased, infected snails may still gain some reproductive success by mating as males, during the following few weeks. Cooper et al. (1996) observed that infected *B. glabrata* were capable of fertilizing the eggs of uninfected snails at 9 weeks postinfection. However, it cannot be stated with certainty that the sperm used was produced postinfection and was not from stores. In contrast to the previous study, Webster et al. (2003) found that infected *B. glabrata* were not allowed to fertilize the eggs of uninfected snails when mating. The ability to mate successfully as males once egg production has ceased would obviously be extremely beneficial to the infected *Biomphalaria*.

5.8.9 Egg Viability

There is evidence to suggest that infection has a significant impact on the viability of any eggs that may be laid during an infection. Blair and Webster (2007) found that during patency, infected snails showed a decrease in the number of viable offspring that hatched. It is not surprising that egg viability or quality would decrease as the proteins necessary for egg production, such as peroxidase and galactogen are also significantly reduced.

5.8.10 Miracidial Dose

Several studies imply that the negative impact of schistosomes on *Biomphalaria* reproduction increases in severity relative to miracidial dose. Makanga (1981) compared the effects of 1, 2, 3 or 4 miracidia on *B. pfeifferi* reproduction and results showed that the number of young produced by infected snails was inversely proportional to the miracidial dose. However, dose-dependent effects are not always observed and within an individual study, some of the parasite-induced changes may be dose-dependent while others are not (Blair and Webster 2007). If infection intensity correlates closely with miracidial dose, it would be expected that any changes in host reproduction would also be dose dependent. Such a relationship between infection intensity and miracidial dose was demonstrated by Sturrock and Sturrock (1970) who reported that when snails were exposed to 2, 4 or 8 miracidia, the cercarial output increased in proportion.

5.8.11 Behavior

Schistosome infection not only affects the physiological aspects of *Biomphalaria* reproduction but also the behavioral. Webster et al. (2003) showed that uninfected *B. glabrata*, whether resistant or susceptible, acted equally as males and females when copulating with other uninfected snails. However, when resistant (uninfected) snails mated with infected snails, the resistant snails refused to act as females forcing the infected snails to mate as females (Webster et al. 2003). Possibly snails are weakened by schistosome infection and hence are less competitive in the precopulation conflicts that determine which gender each snail takes on while mating. A possible outcome of this pairing would be the insemination of infected snails by uninfected snails. It is assumed this sexually selective behavior is in the interest of the uninfected snails as it is more expensive to produce and nurture embryos than sperm and it would not be cost effective for uninfected snails to invest in embryos that share genes coding for a susceptible phenotype. Therefore, schistosome infection is extremely detrimental to the reproductive success of the host. It seems possible that uninfected (resistant) snails may fertilize the eggs of infected (susceptible) snails which could be beneficial to the progeny as they would inherit genes associated with host resistance. However, these eggs may either not be laid or may not be viable. This possibility may be seen as unprofitable to the resistant parent as their offspring will inherit part of a susceptible genome but perhaps passing on genes without the cost of egg production makes it worthwhile.

5.9 Gene Regulation

In recent years a number of studies have investigated gene transcription levels in *Biomphalaria*, most often to compare resistant and susceptible snails and also to uncover the genes involved in immune responses. Despite the foci of these previous studies, some of the resulting data refers to the more general effects of schistosomes on *Biomphalaria* snails. Lockyer et al. (2008) compared gene expression of hemocytes from resistant and susceptible snails within 24 h after exposure to schistosomes. The microarray data showed the upregulation of numerous transcripts in resistant snails which displayed homology to ornithine decarboxylase I, ADP/ATP carrier, lactate/malate dehydrogenase, glutamyl-prolyl-tRNA synthase, histidyl-tRNA synthetase, and tyrosyl-tRNA synthetase. Most of these transcripts are associated with protein synthesis and metabolism and perhaps are involved in the increased egg laying observed in exposed and uninfected (resistant) *B. glabrata* (Blair and Webster 2007). A few transcripts were also very significantly upregulated in susceptible snails but unfortunately they remain unidentified. The completion of the *Biomphalaria* genome project may aid identification of these transcripts and provide a clearer understanding of the effect of schistosomes on *Biomphalaria* snails during the first hours of infection.

5.10 Concluding Remarks

The phenotype of *S. mansoni*-infected *Biomphalaria* snails is determined by both the host and parasite genomes and as such differs dramatically from the uninfected phenotype. Numerous changes in host growth, metabolism, reproduction, and behavior have been observed and almost all changes are viewed as detrimental. Schistosome larvae are metabolically dependent on their intermediate hosts and this is evident in infected snails by the substantial reductions in host glucose and glycogen levels. Infection is especially costly for *Biomphalaria* in terms of lifetime reproductive success as depending on the age of the snail host, infection either reduces or completely negates the host's lifetime reproductive success. The redirection of host nutritional resources is almost certainly a major determinant of *Biomphalaria* castration but other factors are most likely involved as well, such as the potential uptake of host 5-HT by schistosome larvae and the increased levels of urea measured in infected hosts (Thompson and Lee 1987; Boyle et al. 2003). It seems that noncarbohydrate sources of energy are tapped into as well, such as proteins and perhaps carbonic acids. The use of host proteins presumably for metabolic purposes might have very direct effects on host physiological processes such as reproduction depending on which proteins are decreased (Looker and Etges 1979; Bai et al. 1997). Recent studies indicate that resistant snails also are affected by exposure to miracidia even though they do not develop an infection (Blair and Webster 2007). For instance, exposure to miracidia triggered significant changes in the egg laying and mortality rate of resistant snails (Blair and Webster 2007).

In general, many of the effects observed in infected *Biomphalaria* snails appear to be less severe in older mature snails, such as the effects on mortality and reproduction for example (Meier and Meier-Brook 1981; Thornhil et al. 1986). This corresponds with the observation that some *Biomphalaria* species, including *B. glabrata* and *B. alexandrina,* become more refractory to infection with age (Fernandez 1997; Fernandez and Pieri 2001; Jamjoom and Banaja 2007).

The mechanisms by which schistosome related changes are induced in *Biomphalaria* snails are not understood and because of the complexity of the relationship, they are not easy to decipher. Conclusions must be arrived at with caution as it is challenging to determine clearly whether such changes benefit the host or parasite or whether they are a side effect of infection. For instance, if a schistosome secreted molecule which caused host castration were to be identified, it would be evidence that the trait was selected for by evolution and was a beneficial adaptation of the parasite. As yet, no such schistosome molecule appears to have been identified in *S. mansoni*-infected *Biomphalaria* snails.

Comparative studies on the effects of starvation on the physiological processes within *Biomphalaria* snails are valuable as they allow the role of nutritional resources to be more clearly determined. Schistosome-infected and starved snails generally display common symptoms; however, it is clear that the effects seen in infected snails are not due solely to the redirection of host resources (Stanislawski and Becker 1979).

Despite the abundance of literature describing the effects of schistosome infection on *Biomphalaria* species, this chapter emphasizes the need for further detailed study of this relationship, including the utilization of additional *Biomphalaria* species. Furthermore, review of the available literature has led to few definite conclusions because of variation between studies, such as the host and parasite strains used, experimental setup and miracidial dose. Additionally, field studies are desirable to establish whether laboratory derived findings are truly representative of field situations. Nevertheless, it is recognized that such field studies would be difficult to perform.

References

Bai G, Johnston LA, Watson CO, Yoshino TP (1997) Phenoloxidase activity in the reproductive system of *Biomphalaria glabrata*: role in egg production and effect of schistosome infection. J Parasitol 83:852–858

Bebiano MJ, Langston WJ (1995) Induction of metallothionen synthesis in the gill and kidney of *Littorina littoerea* exposed to cadmium. J Mar Biol Assoc UK 75:173–186

Berriman M, Haas BJ, LoVerde PT, Wilson RA, Dillon GP, Cerqueira GC, Mashiyama ST, Al-Lazikani B, Andrade LF, Ashton PD, Aslett MA, Bartholomeu DC, Blandin G, Caffrey CR, Coghlan A, Coulson R, Day TA, Delcher A, DeMarco R, Djikeng A, Eyre T, Gamble JA, Ghedin E, Gu Y, Hertz-Fowler C, Hirai H, Hirai Y, Houston R, Ivens A, Johnston DA, Lacerda D, Macedo CD, McVeigh P, Ning Z, Oliveira G, Overington JP, Parkhill J, Pertea M, Pierce RJ, Protasio AV, Quail MA, Rajandream MA, Rogers J, Sajid M, Salzberg SL, Stanke M, Tivey AR, White O, Williams DL, Wortman J, Wu W, Zamanian M, Zerlotini A, Fraser-Liggett CM, Barrell BG, El-Sayed NM (2009) The genome of the blood fluke *Schistosoma mansoni*. Nature 460:352–360

Blair L, Webster JP (2007) Dose-dependent schistosome-induced mortality and morbidity risk elevates host reproductive effort. J Evol Biol 20:54–61
Boissier J, Rivera ER, Mone H (2003) Altered behavior of the snail *Biomphalaria glabrata* as a result of infection with *Schistosoma mansoni*. J Parasitol 89:429–433
Boyle JP, Yoshino TP (2002) Monoamines in the albumen gland, plasma, and central nervous system of the snail *Biomphalaria glabrata* during egg-laying. Comp Biochem Physiol A 132:411–422
Boyle JP, Hillyer JF, Yoshino TP (2003) Pharmacological and autoradiographical characterization of serotonin transporter-like activity in sporocysts of the human blood fluke, *Schistosoma mansoni*. J Comp Physiol A Neuroethol Sens Neural Behav Physiol 189:631–641
Brouwers JFHM, Smeenk IMB, Van Golde LMG, Tielens AGM (1997) The incorporation, modification and turnover of fatty acids in adult *Schistosoma mansoni*. Mol Biochem Parasitol 88:175–185
Cheng TC, Snyder RW (1963) Studies on host-parasite relationships between larval trematodes and their hosts. IV. A histochemical determination of glucose and its role in the metabolism of molluscan host and parasite. Trans Am Microsc Soc 82:343–346
Cheng TC, Lee FO (1971) Glucose levels in the mollusc *Biomphalaria glabrata* infected with *Schistosoma mansoni*. J Invertebr Pathol 18:395–399
Cline DJ, Fried B, Sherma J (1999) TLC and GC-MS identification of glucose and maltose in *Biomphalaria glabrata* (Gastropoda), and use of quantitative TLC to determine the effect of starvation on the amounts of these carbohydrates. Acta Chromatogr 9:79–86
Coles GC (1972) Carbohydrate metabolism of larval *Schistosoma mansoni*. Int J Parasitol 2:341–352
Coppin JF, Lefebvre C, Caby S, Cocquerelle C, Vicogne J, Coustau C, Dissous C (2003) Gene expression changes in *Schistosoma mansoni* sporocysts induced by *Biomphalaria glabrata* embryonic cells. Parasitol Res 89:113–119
Cooper LA, Larson SE, Lewis FA (1996) Male reproductive success of *Schistosoma mansoni*-infected *Biomphalaria glabrata*. J Parasitol 82:428–431
Coustau C, Mitta G, Dissous C, Guillou F, Galiner R, Alliene JF, Modat S (2003) *Schistosoma mansoni* and *Echinostoma caproni* excretory-secretory products differentially affect gene expression in *Biomphalaria glabrata* embryonic cells. Parasitology 127:533–542
Crews AE, Yoshino TP (1989) *Schistosoma mansoni*: effect of infection in reproduction and gonadal growth in *Biomphalaria glabrata*. Exp Parasitol 68:326–334
Crews AE, Yoshino TP (1991) *Schistosoma mansoni*: influence of infection on levels of translatable mRNA and on polypeptide synthesis in the ovotestis and albumen gland of *Biomphalaria glabrata*. Exp Parasitol 72:368–380
Davies TW, Erasmus DA (1984) An ultrastructural study of the effect of parasitism by larval *Schistosoma mansoni* on the calcium reserves of the host, *Biomphalaria glabrata*. Cell Tissue Res 236:643–649
DeJong RJ, Morgan JAT, Paraense WL, Pointier JP, Amarista M, Ayeh-Kumi PFK, Babiker A, Barbosa CS, Bremond P, Canese AP, de Souza CP, Dominguez C, File S, Gutierrez A, Incani RN, Kawano T, Kazibwe F, Kpikpi J, Lwambo NJS, Mimpfoundi R, Poda JN, Sene M, Velasquez LE, Yong M, Adema CM, Hofkin BV, Mkoji GM, Loker ES (2001) Evolutionary relationships and biogeography of *Biomphalaria* (Gastropoda: Planorbidae) with implications regarding its role as host of the human blood fluke *Schistosoma mansoni*. Mol Biol Evol 18:2225–2239
De Jong-Brink M (1995) How schistosomes profit from the stress responses they elicit in their hosts. Adv Parasitol 35:177–256
Dresden MH, Edlin EM (1975) *Schistosoma mansoni*: calcium content of cercariae and its effects on protease activity in vitro. J Parasitol 61:398–402
El-Ansary A, Sammour EM, Mohamed AM (2000) Susceptibility of *Biomphalaria alexandrina* to infection with *Schistosoma mansoni*: correlation with the activity of certain glycolytic enzymes. J Egypt Soc Parasitol 30:547–560
El-Ansary A, Mohamed AM, Mamoud SS, El-Bardicy S (2003) On the pathogenicity of attenuated *Schistosoma mansoni* cercariae released from metabolically disturbed *Biomphalaria alexandrina*. J Egypt Soc Parasitol 33:777–794

Etges FJ (1963) Effects of *Schistosoma mansoni* infection on chemosensitivity and orientation of *Australorbis glabratus*. Am J Trop Med Hyg 12:696–700
Fernandez MA (1997) *Schistosoma mansoni* infections in the first three months of life of sympatric intermediate hosts from Brazil. Mem Inst Oswaldo Cruz 92:27–29
Fernandez MA, Pieri OS (2001) Infection by *Schistosoma mansoni* Sambon 1907 in the first four months of life of *Biomphalaria straminea* (Dunker, 1848) in Brazil. Mem Inst Oswaldo Cruz 96:185–192
Gabrashanska M, Damyanova A, Kanev I (1991) Mineral composition of *Echinostoma revolutum* (Froelich, 1802) and its host *Lymnaea stagnalis* (L.). Khelmintology 31:3–7
Gerard C, Théron A (1996) Altered nutrition and assimilation of the snail host (*Biomphalaria glabrata*) as a consequence of the parasitic spatial constraint (*Schistosoma mansoni*). Acta Trop 61:51–55
Gerard C, Théron A (1997) Age/size- and time-specific effects of *Schistosoma mansoni* on energy allocation patterns of its snail host *Biomphalaria glabrata*. Oecologia 112:447–452
Gilbertson DE, Etges FJ, Ogle JD (1967) Free amino acids of *Australorbis glabratus* hemolymph: comparison of four geographic strains and effect of infection by *Schistosoma mansoni*. J Parasitol 53:545–568
Hahn UK, Bender RC, Bayne CJ (2001) Killing of *Schistosoma mansoni* sporocysts by hemocytes from resistant *Biomphalaria glabrata*: role of reactive oxygen species. J Parasitol 87:292–299
Ibrahim MM (2006) Energy allocation patterns in *Biomphalaria alexandrina* snails in response to cadmium exposure and *Schistosoma mansoni* infection. Exp Parasitol 112:31–36
Humiczewska M, Rajski K (2005) Lipids in the host-parasite system: digestive gland of *Lymnaea truncatula* infected with the developmental stages of *Fasciola hepatica*. Acta Parasitol 50:235–239
Ishak MM, Mohamed AM, Sharaf AA (1975) Carbohydrate metabolism in uninfected and trematode-infected snails *Biomphalaria alexandrina* and *Bulinus truncatus*. Comp Biochem Physiol 51B:499–505
Jamjoom MB, Banaja AEA (2007) Comparative studies on the susceptible and non-susceptible *Biomphalaria alexandrina* the intermediate snail host of *Schistosoma mansoni* in western Saudi Arabia. World J Med Sci 2:108–114
Jarusiewicz JA, Sherma J, Fried B (2006) Thin layer chromatographic analysis of estivated *Biomphalaria glabrata* snails and those infected with *Schistosoma mansoni*. Comp Biochem Physiol B 145:346–349
Lee FO, Cheng TC (1971) *Schistosoma mansoni*: respirometric and partial pressure studies in infected *Biomphalaria glabrata*. Exp Parasitol 30:393–399
Lockyer AE, Spinks J, Kane RA, Hoffman KF, Fitzpatrick JM, Rollinson D, Noble LR, Jones CS (2008) *Biomphalaria glabrata* transcriptome: cDNA microarray profiling identifies resistant- and susceptible-specific gene expression in hemocytes from snail strains exposed to *Schistosoma mansoni*. BMC Genomics 9:634
Looker DL, Etges FJ (1979) Effect of *Schistosoma mansoni* infection on fecundity and perivitelline fluid composition in *Biomphalaria*. J Parasitol 65:880–885
McClelland G, Bourns TKR (1969) Effects of *Trichobilharzia ocellata* on growth, reproduction, and survival of *Lymnaea stagnalis*. Exp Parasitol 24:137–146
Makanga B (1981) The effect of varying the number of *Schistosoma mansoni* miracidia on the reproduction and survival of *Biomphalaria pfeifferi*. J Invertebr Pathol 37:7–10
Mandlowitz A, Dusanic D, Lewert RM (1960) Peptidase and lipase activity of extracts of *Schistosoma mansoni* cercariae. J Parasitol 46:89–90
Manger P, Li J, Christensen BM, Yoshino TP (1996) Biogenic monoamines in the freshwater snail, *Biomphalaria glabrata*: influence of infection by the human blood fluke, *Schistosoma mansoni*. Comp Biochem Physiol 114:227–234
Massa DR, Chejlava MJ, Fried B (2007) High performance column liquid chromatographic analysis of selected carboxylic acids in *Biomphalaria glabrata* patently infected with *Schistosoma mansoni*. Parasitol Res 101:925–928

Mazuran N, Hrsak M, Tomic PD (1999) Effects of $CaCl_2$ and $CaBr_2$ on the fecundity of *Planorbarius corneus* L. Chemosphere 38:2345–2355

Meier M, Meier-Brook C (1981) *Schistosoma mansoni*: effect on growth, fertility, and development of distal male organs. Z Parasitenkd 66:121–131

Meuleman EA (1972) Host-parasite interrelationships between the freshwater pulmonate *Biomphalaria pfeifferi* and the trematode *Schistosoma mansoni*. Neth J Zool 22:355–427

Meyer F, Meyer H, Bueding E (1970) Lipid metabolism in the parasitic and free-living flatworms *Schistosoma mansoni* and *Dugesia dorotocephala*. Biochim Biophys Acta 210:257–266

Minchella DJ, Loverde PT (1981) A cost of increased early reproductive effort in the snail *Biomphalaria glabrata*. Am Nat 118:876–881

Mohamed AM, Ishak MM (1979) Growth raste and changes in tissue carbohydrates during schistosome infection of the snail *Biomphalaria alexandrina*. Hydrobiologia 76:17–21

Mostafa OMS (2007) Effects of *Schistosoma mansoni* and *Schistosoma haematobium* infections on calcium content in their intermediate hosts. Parasitol Res 101:963–966

Nabih I, El Dardiri Z, El-Ansary A, Rizk M (1990) Measurement of some selected enzymatic activities in infected *Biomphalaria alexandrina* snails. Cell Mol Biol 36:637–642

Ong JHL, Chejlava M, Fried B, Koehnlein KM, Bosavage GL, Sherma J (2004) Effects of *Schistosoma mansoni* infection on inorganic elements in the snail *Biomphalaria glabrata*. J Helminthol 78:343–346

Patocka N, Ribiero P (2007) Characterization of a serotonin transporter in the parasitic flatworm, *Schistosoma mansoni*: cloning, expression and functional analysis. Mol Biochem Parasitol 154:125–133

Pinheiro J, Amato S (1995) Calcium determination in the shell of *Lymnaea columella* (Mollusca, Gastropoda) infected with *Fasciola hepatica* (Platyhelminthes, Digenea). Braz Arch Biol Technol 38:761–767

Rupprecht H, Becker W, Schwanbek A (1989) Alterations in *Biomphalaria glabrata* components during long term infection with *Schistosoma mansoni* infection. Parasitol Res 75:233–237

Santhanagopalan V, Yoshino TP (2000) Monoamines and their metabolites in the freshwater snail *Biomphalaria glabrata*. Comp Biochem Physiol A Mol Integr Physiol 125:469–478

Schmale H, Becker W (1977) Studies on the urea cycle of *Biomphalaria glabrata* during normal feeding activity, in starvation and with infection of *Schistosoma mansoni*. Comp Biochem Physiol 58A:321–330

Schnell S, Becker W, Winkler A (1985) Amino acid metabolism in the freshwater pulmonate *Biomphalaria glabrata* infected with the trematode *Schistosoma mansoni*. Comp Biochem Physiol 81B:1001–1008

Schwanbeck A, Becker W, Rupprecht H (1986) Quantification of parasite development in the host-parasite system *Biomphalaria glabrata* and *Schistosoma mansoni*. Z Parasitenkd 72:365–373

Schwartz CFW, Carter CE (1982) Effect of *Schistosoma mansoni* on glycogen synthase and phosphorylase from *Biomphalaria glabrata* (Mollusca). J Parasitol 68:236–242

Shaw MK, Erasmus DA (1987) *Biomphalaria glabrata*: changes in calcium reserves following parasitism by larval *Schistosoma mansoni*. Parasitology 95:267–276

Sire C, Durand P, Pointier JP, Théron A (1999) Genetic diversity and recruitment pattern of *Schistosoma mansoni* in a *Biomphalaria glabrata* snail population: a field study using random-amplified polymorphic DNA markers. J Parasitol 85:436–441

Smith ME, Steiner SA, Isseroff H (1994) Urea: inhibitor of growth and reproduction in *Bulinus truncatus*. Comp Biochem Physiol 108A:569–577

Sorensen RE, Minchella DJ (2001) Snail-trematode life history interactions: past trends and future directions. Parasitology 123:S3–S18

Stanislawski E, Becker W (1979) Influences of semi-synthetic diets, starvation and infection with *Schistosoma mansoni* (Trematoda) on the metabolism of *Biomphalaria glabrata* (Gastropoda). Comp Biochem Physiol 63A:527–533

Stanislawski E, Becker W, Muller G (1979) Alterations of the free amino acid content in the hemolymph of *Biomphalaria glabrata* (Pulmonata) in starvation and after infection with *Schistosoma mansoni* (Trematoda). Comp Biochem Physiol 63B:477–482

Sturrock BM (1966) The influence of infection with *Schistosoma mansoni* on the growth rate and reproduction of *Biomphalaria pfeifferi*. Ann Trop Med Hyg 60:187–197

Sturrock BM, Sturrock RF (1970) Laboratory studies of the host-parasite relationship of *Schistosoma mansoni* and *Biomphalaria glabrata* from St. Lucia, West Indies. Ann Trop Med Parasitol 64:357–363

Théron A, Mone H, Gerard C (1992) Spatial and energy compromise between host and parasite: the *Biomphalaria glabrata-Schistosoma mansoni* system. Int J Parasitol 22:91–94

Théron A, Gerard C (1994) Development of accessory sexual organs in *Biomphalaria glabrata* (Planorbidae) in relation to timing of infection by *Schistosoma mansoni*: consequences for energy utilization patterns by the parasite. J Molluscan Stud 60:25–31

Thompson SN (1987) Effect of *Schistosoma mansoni* on the gross lipid composition of its vector *Biomphalaria glabrata*. Comp Biochem Physiol 87B:357–360

Thompson SN, Lee RWK (1987) Characterization of the ^{31}P NMR spectrum of the schistosome vector *Biomphalaria glabrata* and of the changes following infection by *Schistosoma mansoni*. J Parasitol 73:64–76

Thompson SN, Mejia-Scales V (1989) Effects of *Schistosoma mansoni* on the nutrition of its intermediate host, *Biomphalaria glabrata*. J Parasitol 75:329–332

Thornhill JA, Jones JT, Kusel JR (1986) Increased oviposition and growth in immature *Biomphalaria glabrata* after exposure to *Schistosoma mansoni*. Parasitology 93:443–450

Van Aardt WJ, de Kock KN, Naudé K (2003) The respiratory properties of *Biomphalaria glabrata* exposed to *Schistosoma mansoni* infection, starvation, CO and choices of different oxygen concentrations. Exp Parasitol 103:93–101

Veldhuijzen JP (1975) Glucose-tolerance in the pond snail *Lymnaea stagnalis* as affected by temperature and starvation. Neth J Zool 25:206–218

Von Brand T, Files VS (1947) Chemical and histological observations on the influence of *Schistosoma mansoni* on *Australorbis glabratus*. J Parasitol 33:476–482

Wagner SD, Pachuski J, Fried B, Sherma J (2002) Thin layer chromatographic analysis of carbohydrates and amino acids in *Schistosoma mansoni* (Trematoda) cercariae. Acta Chromatogr 12:159–169

Warr E, Meredith JM, Nimmo DD, Basu S, Hurd H, Eggleston P (2006) A tapeworm molecule manipulates vitellogenin expression in the beetle *Tenebrio molitor*. Insect Mol Biol 15:497–505

Webster JP, Hoffmann JI, Berdoy M (2003) Parasite infection, host resistance and mate choice: battle of the genders in a simultaneous hermaphrodite. Proc R Soc Lond B 270:1481–1485

Webster JP, Woolhouse MEJ (1999) Cost of resistance: relationship between reduced fertility and increased resistance in a snail-schistosome host-parasite system. Proc R Soc Lond B 266:391–396

White MW, Chejlava M, Freid B, Sherma J (2005) Effects of various larval digeneans on the calcium carbonate content of the shells of *Helisoma trivolvis*, *Biomphalaria glabrata* and *Physa* sp. Parasitol Res 95:252–255

Williams CL, Gilbertson DE (1983) Altered feeding response as a cause for the altered heartbeat rate and locomotor activity of *Schistosoma mansoni*-infected *Biomphalaria glabrata*. J Parasitol 69:671–676

Woolhouse MEJ (1989) The effect of schistosome infection on the mortality rates of *Bulinus globosus* and *Biomphalaria pfeifferi*. Ann Trop Med Parasitol 83:137–141

Yousif F, Haroun N, Ibrahim A, El-Bardicy S (1996) *Biomphalaria glabrata*: a new threat for schistosomiasis transmission in Egypt. J Egypt Soc Parasitol 26:191–205

Zhang SM, Nian H, Wang B, Loker ES, Adema CN (2009) Schistosomin from the snail *Biomphalaria glabrata*: Expression studies suggest no involvement in trematode-mediated castration. Mol Biochem Parasitol 165:79–86

Chapter 6
Effects of Nonschistosome Larval Trematodes on *Biomphalaria* Snails

Jose-Guillermo Esteban, Carla Muñoz-Antoli, Maria Trelis, and Rafael Toledo

Abstract This chapter examines the most salient literature on the development of nonschistosome larval digeneans in *Biomphalaria* spp. (members of the superfamilies Clinostomoidea, Diplostomoidea, Echinostomatoidea, Paramphistomoidea, and Pronocephaloidea). For this purpose, the general biology of the nonschistosome trematode species that use *Biomphalaria* spp. as intermediate hosts are analyzed. Thereafter, the main features of these parasitic infections are studied with particular emphasis on topics such as the effect of the infections on the snail, the antagonism and/or competition between different digenean species in the snail, the immunology of the infection, and the genomic and proteomic aspects of these host–parasite interactions.

6.1 Introduction

Snails belonging to the genus *Biomphalaria* are commonly studied in relation to their capacity to transmit the agent causing human schistosomiasis. However, *Biomphalaria* spp. also act as intermediate hosts of other relevant digenean species. The study of the interactions between *Biomphalaria* spp. and the nonschistosome trematodes is of importance since it may be useful for the analysis of several topics of great interest: (1) the population dynamics of snails and trematodes; (2) the transmission dynamics of trematodes; (3) the developmental biology of larval digeneans, (4) the immunology, genomic, and proteomic aspects of snail–larval trematode interactions; (5) effect of trematodes on wildlife; and also (6) for the development of strategies for the biological control of schistosomiasis. In this context, this chapter reviews the most salient literature on several issues in relation to different *Biomphalaria*–nonschistosome digenean combinations.

R. Toledo (✉)
Departamento de Biología Celular y Parasitología, Facultad de Farmacia,
Universidad de Valencia, Av. Vicent Andrés Estellés s/n, 46100 Burjassot-Valencia, Spain
e-mail: Rafael.Toledo@uv.es

R. Toledo and B. Fried (eds.), Biomphalaria *Snails and Larval Trematodes*,
DOI 10.1007/978-1-4419-7028-2_6, © Springer Science+Business Media, LLC 2011

6.2 *Biomphalaria* spp. as Intermediate Host of Nonschistosome Larval Trematodes

Freshwater pulmonate snails of the genus *Biomphalaria* act as intermediate hosts of schistosomes, which cause schistosomiasis, affecting more than 80 million people around the world (Crompton 1999). However, this group of freshwater snails is involved in the transmission of other trematode species belonging to the superfamilies Clinostomoidea, Diplostomoidea, Echinostomatoidea, Paramphistomoidea, and Pronocephaloidea. Moreover, different types of cercariae without specific descriptions infecting *Biomphalaria* snails have been recorded in field studies from several countries. However, certain reports on *Biomphalaria* specimens involved in the life cycles of some trematodes are probably controversial (Goodman and Panesar 1976).

Table 6.1 shows a list of the superfamilies, families, genera, and most relevant species of nonschistosome trematodes for which different species of *Biomphalaria* act as first and/or second intermediate hosts, both in nature as well as experimentally in the laboratory.

The species listed in Table 6.1 have an aquatic indirect life cycle involving, typically, two or three different hosts: a vertebrate definitive host; *Biomphalaria* spp. snails as aquatic first intermediate hosts; and often a second intermediate host carrying an encysted metacercarial stage (Fig. 6.1).

In all these life cycles, the eggs are emitted through the definitive host's feces and require some time for embryonation in the aquatic medium. When the eggs are fully embryonated, they hatch in water and release the miracidia that swim searching for their *Biomphalaria* first intermediate host snails. Within the *Biomphalaria* snail, a miracidium develops into different larval stages depending on the trematode species involved until reaching the cercarial stage (Fig. 6.2).

The free-swimming cercariae escape from the *Biomphalaria* snail. In Clinostomatidae, Diplostomidae, Strigeidae, Echinostomatidae, and Psilostomatidae, the cercariae reach a compatible second intermediate host, often invertebrate or even a vertebrate. Cercariae actively penetrate the host's body and encyst as metacercariae; but in some Echinostomatidae species, the cercariae may re-enter the same snail for encystment. In the life cycles of other trematode species belonging to the families Fasciolidae, Cladorchiidae, Zygocotylidae, and Notocotylidae, the cercarial encystment may also take place on aquatic vegetation, on the water surface, or even on snail shells.

The definitive host is a vertebrate, becoming infected when metacercariae are ingested. Within the definitive host, the encysted metacercaria excysts and gradually matures into an adult in its definitive microhabitat.

6.2.1 *As First Intermediate Host*

Among the reports based on field studies of *Biomphalaria* as first intermediate hosts, including life cycles, the studies concerning trematode species of the genus *Ribeiroia* are particularly noteworthy for they have shed light on the important

Table 6.1 Trematodes species that use *Biomphalaria* species as first and second intermediate hosts (I.H.) either in nature (N) or experimentally (E), with a list of the most relevant literature

Superfamily / Family / Genus	Parasite species	*Biomphalaria* spp.	First I.H. N/E	Second I.H. N/E	Most relevant references
Clinostomoidea					
Clinostomidae					
Clinostomum	*C. complanatum*	*B. glabrata*	–/+	–/–	Dias et al. (2003)
		B. peregrina	+/+	–/–	Dias et al. (2003)
	C. golvani	*B. glabrata*	+/–	–/–	Nassi and Bayssade-Dufour (1980)
Diplostomoidea					
Diplostomidae					
Diplostomum	*D. compactum*	*D.* cf. *Havanensis*	+/–	–/–	Violante-González et al. (2009)
Strigeidae					
Apatemon	*A. graciliformis*	*B. glabrata*	+/–	–/–	Dubois and Nassi (1977)
Echinostomatoidea					
Echinostomatidae					
Echinostoma	*E. caproni*	*B. alexandrina*	+/+	+/+	Yousif and Haroun (1986), Fried and Huffman (1996)
		B. camerunensis	–/+	–/–	Fried and Huffman (1996)
		B. glabrata	–/+	–/+	Fried and Huffman (1996), Jeyarasasingam et al. (1972), Richard and Brygoo (1978), Christensen et al. (1980), Kuris and Warren (1980), Keeler and Huffman (2009)
		B. pfeifferi	+/+	–/+	Fried and Huffman (1996), Keeler and Huffman (2009)
		B. straminea	–/–	–/+	Fried and Huffman (1996)
	E. echinatum	*B. glabrata*	+/+	–/+	Lie and Basch (1966), Kostadinova and Gibson (2000), Carney (1991)
		B. straminea	–/–	–/+	Lie and Basch (1966), Lie et al. (1967)
		B. tenagophila	–/–	–/+	Lie and Basch (1966)
	E. deserticum	*B. glabrata*	–/–	–/+	Kechemir et al. (2002)
		B. pfeifferi	–/–	–/+	Kechemir et al. (2002)

(continued)

Table 6.1 (continued)

Superfamily / Family / Genus	Parasite species	*Biomphalaria* spp.	First I.H. N/E	Second I.H. N/E	Most relevant references
	E. friedi	*B. glabrata*	–/+	–/+	Muñoz-Antoli et al. (2006)), Muñoz-Antolí et al (2007), Muñoz-Antolín et al. (unpublished)
	E. luisreyi	*B. glabrata*	–/–	–/+	Maldonado et al. (2003)
	E. malayanum	*Biomphalaria* spp.	–/–	–/+	Lo (1973)
	E. paraensei	*B. glabrata*	+/+	–/+	Lie and Basch (1967), Stein and Basch (1977), Maldonado et al. (2001), Maldonado and Lanfredi (2009)
		B. tenagophila	–/+	–/+	Maldonado et al. (2001), Maldonado and Lanfredi (2009)
	E. parvocirrus	*B. glabrata*	+/+	–/+	Keeler and Huffman (2009), Nassi and Dupouy (1988)
	E. revolutum	*B. glabrata*	–/–	–/+	Anderson and Fried (1987), Fried et al. (1997a)
	E. trivolvis	*B. glabrata*	–/+	–/+	Fried et al. (1998), Reddy et al. (2004)
Echinoparyphium	*E. elegans*	*B. glabrata*	–/–	–/+	Mouahid and Moné (1988), King and Van As (1996)
	E. ralphaudyi	*B. alexandrina*	–/–	–/+	Keeler and Huffman (2009), Lie et al. (1975)
		B. glabrata	–/–	–/+	Keeler and Huffman (2009), Lie et al. (1975)
	E. recurvatum	*B. glabrata*	–/–	–/+	McCarthy (1999)
	Echinoparyphium sp.	*B. glabrata*	–/–	–/+	Fried et al. (1998)
Stephanoprora	*S. denticulate*	*B. glabrata*	–/+	–/–	Schell (1985)
Petasiger	*P. caribbensis*	*B. glabrata*	–/+	–/–	Nassi (1980)
Fasciolidae					
Fasciola	*F. gigantica*	*B. alexandrina*	+/–	–/–	Farag and el-Sayad (1995)
	Fasciola sp.	*B. alexandrina*	+/–	–/–	El-Shazly et al. (2002), Dar et al. (2005)
Fascioloides	*F. magna*	*B. glabrata*	–/+	–/–	Laursen and Yoshino (1999)

Psilostomidae					
Ribeiroia	*R. marini*	*B. glabrata*	+/+	–/–	Johnson et al. (2004), Johnson and McKenzie (2009)
	R. m. guadeloupensis	*B. glabrata*	+/+	–/–	Johnson et al. (2004), Johnson and McKenzie (2009)
	R. ondatrae	*B. obstructa*	+/–	–/–	Johnson et al. (2004), Johnson and McKenzie (2009)
	Ribeiroia sp.	*B. prona*	+/–	–/–	Johnson et al. (2004), Johnson and McKenzie (2009)
		B. choanomphala	+/–	–/–	Johnson et al. (2004), Johnson and McKenzie (2009)
		B. sudanica	+/–	–/–	Johnson et al. (2004), Johnson and McKenzie (2009)
Paramphistomoidea					
Cladorchiidae					
Paraibatrema	*P. inesperata*	*B. glabrata*	–/+	–/–	Ueta et al. (1981)
		B. tenagophila	+/+	–/–	Ueta et al. (1981)
Zygocotylidae					
Zygocotyle	*Z. lunata*	*B. peregrina*	+/+	–/–	Ostrowski de Núñez et al. (2003), Fried et al. (2009)
		B.oligoza	–/+	–/–	Ostrowski de Núñez et al. (2003), Fried et al. (2009)
		B. orbiguyi	–/+	–/–	Ostrowski de Núñez et al. (2003), Fried et al. (2009)
		B. tenagophila t.	–/+	–/–	Ostrowski de Núñez et al. (2003), Fried et al. (2009)
Pronocephaloidea					
Notocotylidae					
Notocotylus	*N. biomphalariae*	*B. peregrina*	+/–	–/–	Flores and Brugni (2005)
Hippocrepis	*H. fuelleborni*	*B. peregrina*	+/–	–/–	Ostrowski de Núñez (1976)

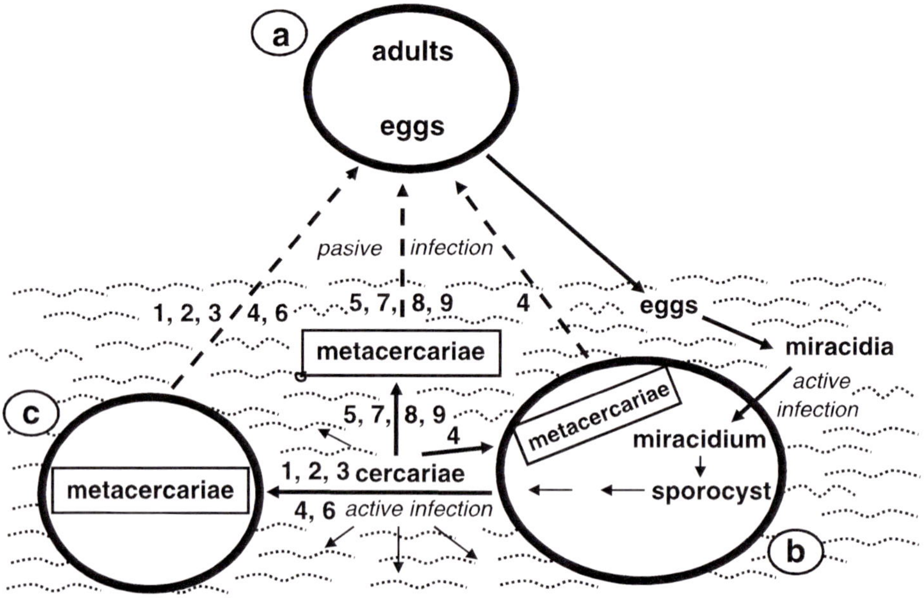

Fig. 6.1 Schematic representation of the life cycle patterns of families of digenetic trematodes (1. Clinostomidae; 2. Diplostomidae; 3. Strigeidae; 4. Echinostomatidae; 5. Fasciolidae; 6. Psilostomidae; 7. Cladorchiidae; 8. Zygocotylidae; 9. Notocotylidae) that use freshwater pulmonate snails of the genus *Biomphalaria* as intermediate hosts. (**a**) Definitive host; (**b**) first intermediate host; (**c**) second intermediate host

aspects of biology, ecology, life cycle, and pathogenesis of these parasites (Johnson et al. 2004; Johnson and McKenzie 2009). *Ribeiroia* infection can cause severe pathology in snails and amphibians. In the latter, the infection has been implicated as a widespread cause of limb malformation in several species of frogs, toads, and salamanders. Malformed amphibians rarely survive to maturity, suffering from an inability to escape predators or obtain food. It is likely that such malformations are actively induced by the parasite via production of a chemically active signaling compound.

The most extensive body of literature focuses on experimental studies using *Biomphalaria* spp. as first intermediate host of trematodes of the Echinostomatidae family, and specifically on the genus *Echinostoma* belonging to the 37-collar-spined group (Fried and Graczyk 2000, 2004). Within this group, the literature on the *Echinostoma caproni*/*Biomphalaria glabrata* system is the most extensive, since this echinostome can be easily cycled in the laboratory which makes this trematode species a suitable experimental model to provide information on the host–parasite relationships in intestinal helminth infections (Fried and Huffman 1996; Esteban and Muñoz-Antolí 2009).

Miracidial infectivy in the *Echinostoma*/*Biomphalaria* system has been the subject of several studies (Jeyarasasingam et al. 1972; Kuris 1980; Fried et al. 1987; Huffman and Fried 1990). Infectivity of *E. caproni* miracidia to *B. glabrata* snails increased markedly when eggs were incubated for 10–24 days compared with snails

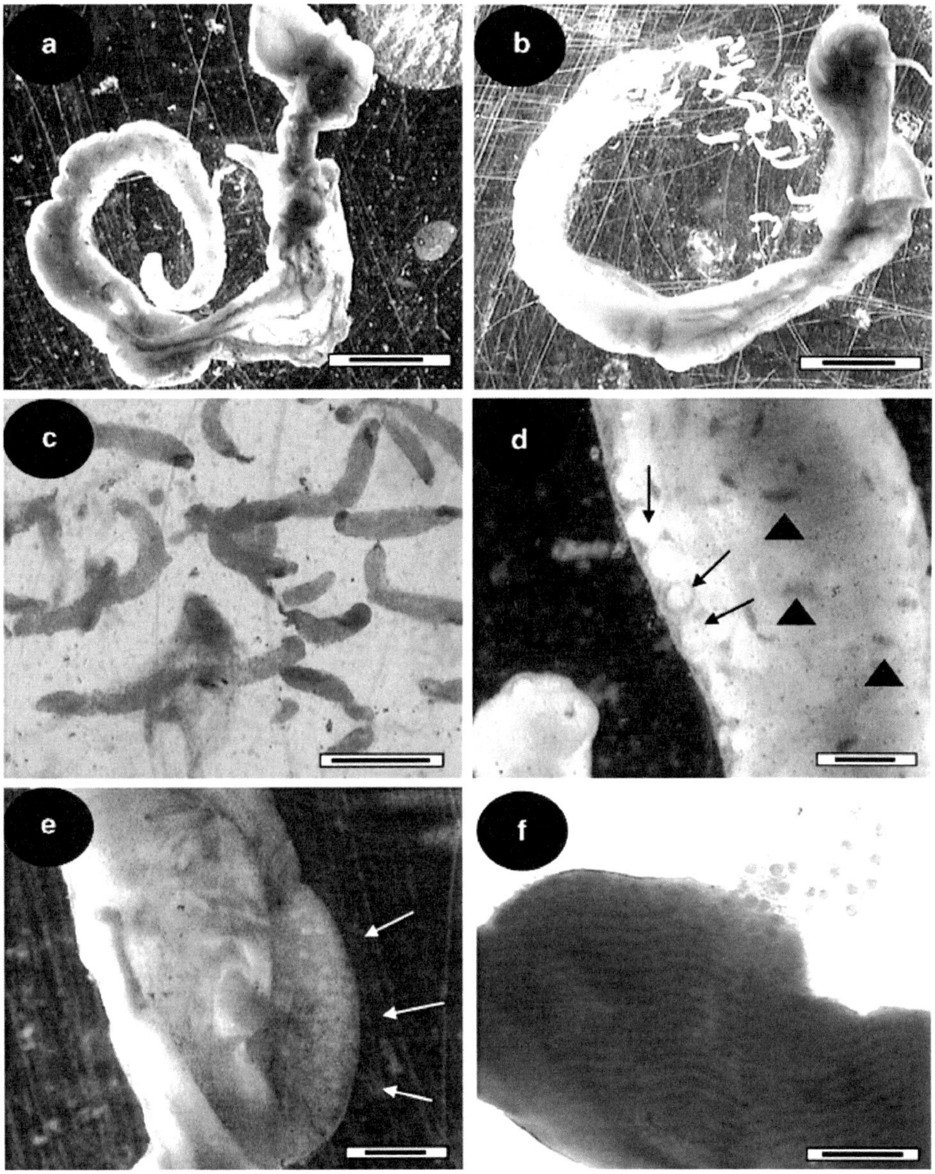

Fig. 6.2 Photomicrographs of some living cycle stages of *Echinostoma friedi* from an experimentally infected *Biomphalaria glabrata* snail (Guadaloupe): (**a**) general view of a dissected specimen of *B. glabrata* with the shell removed; (**b**) rediae of *E. friedi en masse* in the dissected hepatopancreas of an infected *B. glabrata* snail; (**c**) isolated rediae; (**d**) Hepatopancreas harboring metacercariae (*black arrows*) encysted and rediae (*black triangles*) of the same *Biomphalaria* snail specimen; (**e**) Detail of pericardial cavity showing encysted metacercariae; (**f**) some free metacercariae obtained through the dissection of pericardial cavity. (**a**, **b**, **d** and **e**): dark field; (**c**) and (**f**): bright field. Scale bars: (**a**) and (**b**): 1,000 μm; (**c**): 500 μm; (**d**): 300 μm; (**e**): 800 μm; (**f**): 600 μm

infected with miracidia derived from eggs incubated for 30–42 days. A surprisingly high infectivity rate of *B. glabrata* snails was reached (54–66%) when 25 snails were exposed to 125 miracidia in 1 l of water (Kuris 1980). Exposure of 1–2 mm *B. glabrata* snails to *E. caproni* miracidia affected the survival and longevity of the snails whether they became infected or not. Detrimental effects of exposure or infection with miracidia of *E. caproni* in 4–6 mm shell diameter *B. glabrata* were not as apparent (Kuris 1980).

Biomphalaria specimens play a major role in epidemiological studies aiming at the elucidation of miracidial host specificity of Echinostomatidae species. In the 37-collar-spined species of the "*revolutum* group", Muñoz-Antoli et al. (2006) analyzed the infectivity of *Echinostoma friedi* miracidia under experimental conditions toward a range of laboratory-reared snail species belonging to different families and from different geographical origins. The maximum infection rate obtained was found in *Radix peregra* (36.7%), *Lymnaea fuscus* (23.3%), and *Gyraulus chinensis* (26.7%), all of which are sympatric with the echinostome, followed by the allopatric snail *B. glabrata* (20.0%). In turn, no individual of *Biomphalaria alexandrina* was infected. *Echinostoma paraensei* also developed in the sympatric *B. glabrata* snail with different infection rates depending on miracidial doses: 8% for one miracidium and 28% for five miracidia (Maldonado et al. 2001).

Muñoz-Antolí et al. (2007) working with the *E. friedi/B. glabrata* system observed that the longevity of infected snails is shorter than that of controls, which agrees with other *Echinostoma/Biomphalaria* models (Kuris 1980; Schmidt et al. 1998). In the *E. friedi/B. glabrata* model, the death process in infected snails starts late. The growth of the infected snails in the patent period (during cercarial emission) exceeds that of control snails, in the form of gigantism (Muñoz-Antolí et al. 2007). In infected *B. glabrata* snails with five miracidia of *E. friedi*, the death process is prolonged, that is, the snails reach their maximum life span, just as in the *E. paraensei/B. glabrata* model (Maldonado et al. 2001).

The age of the snail intermediate host may be a factor in some echinostome infectivity studies. Young *B. glabrata* are more susceptible to the miracidia of *E. caproni* than are older snails (Jeyarasasingam et al. 1972). This situation differs in miracidial infections of other echinostome/snails systems (*Echinostoma trivolvis/Helisoma trivolvis*) (Fried et al. 1987).

Echinostoma/Biomphalaria systems have been used to cast light on the development of the different intramolluscan larval stages within the digestive gland gonad complex of infected snails (Esteban and Muñoz-Antolí 2009). The metamorphosis, migration, development, and life span of mother sporocysts of *E. caproni* and *E. paraensei* in *Biomphalaria feifferi*, *B. alexandrina*, *B. glabrata*, and *B. glabrata* embryonic cell lines have been the subject of several studies (Lie and Basch 1967; Jeyarasasingam et al. 1972; Mohamed 1992; Beers et al. 1995; Ataev et al. 1997, 1998, 2001, 2005, 2006).

B. glabrata and *B. alexandrina* have been used in studies to shed light on different stages of behavior of some echinostomes, mainly *E. caproni*, *E. trivolvis*, and *E. paraensei*. In relation to miracidial behavior, different factors, such as light and snail-conditioned water from *B. glabrata* snails, have been observed to affect the hatching of *E. caproni* miracidia (Fried and Reddy 1999; Haas et al. 1995; Haas 2000). Significantly greater hatching was obtained when snails were maintained in intact or perforated dialysis sacs in multiwell chambers as compared with sacs without snails. Substances in snail-conditioned-water significantly increased the hatching of *E. caproni* miracidia. The labeling with radioselenium did not affect miracidial behavior or infectivity of *E. caproni* miracidia toward *B. alexandrina*, and this snail species infected with such miracidia retained the label (Christensen et al. 1980).

Miracidial responses to environmental stimuli in *E. caproni* have also been studied, detecting that the miracidium shows a negative geotaxis that is dominated by a positive phototaxis (Behrens and Nollen 1992). However, *E. paraensei* miracidia show no distinct geo-orientation and only weak photo-orientation (Meece and Nollen 1996). The miracidial chemo-orientation in the *E. caproni*/*B. glabrata* system showed an increased rate of change of direction in increasing and a turnback response in decreasing concentration gradients of the excretory–secretory products emitted by *B. glabrata* (Behrens and Nollen 1992). This result is contradictory to that obtained by Haberl et al. (2000), who noted that *E. caproni* miracidia responded exclusively to high molecular weight glycoproteins of *B. glabrata* conditioned water.

Only one study has focused on the behavior of intramolluscan stages of echinostomes in *Biomphalaria* specimens. Experimental studies have demonstrated that *E. caproni* rediae moved toward individuals of the other species, but also toward conspecifics (Reddy and Fried 1996).

The laboratory experiments on the *E. caproni*/*B. glabrata* system have revealed some relevant data on cercarial behavior. The cercariae of this echinostome species show a diurnal emergence rhythm (Christensen et al. 1980). Fried et al. (2002) described the pattern of emergence of cercariae of *E. caproni* under laboratory conditions. Among the numerous conditions tested, the addition of lettuce, the use of snail-conditioned water from *B. glabrata*, and a temperature of 35°C significantly increased the shedding of *E. caproni* cercariae. A temperature of 12°C caused a significant decrease in *E. caproni* cercariae shedding. Increased snail activity associated with feeding behavior was probably responsible for the enhanced *E. caproni* cercarial sheds detected (Fried et al. 2002). Fried and King (1989) demonstrated chemoattraction of *E. trivolvis* cercariae toward *B. glabrata* dialysate in a petri dish bioassay. Fried et al. (1997b) studied chemoattraction and penetration of *E. caproni* and *E. trivolvis* cercariae in the presence of snail dialysate from different snails, among them *B. glabrata*. They reported that lipophilic extracts of dialysates placed on agar plugs in a petri dish bioassay attracted cercariae, whereas hydrophilic extracts were either nonattractive or repulsive. Working with *E. caproni* cercariae obtained from infected *B. glabrata* snails, Reddy et al. (1997) found that these cercariae were significantly attracted to an array of amino acids and to some carbohydrates, i.e., alanine, arginine, aspartic acid, cystine, glutamic acid, glycine, lysine, proline, serine, valine, and maltose. However, cercariae were significantly repelled by dextrose and butyric acid. Moreover, this study suggested that there are specific differences in cercarial chemoattractants depending on the echinostome species considered (Reddy et al. 1997). In this sense, Haberl et al. (2000) noted that *E. caproni* cercariae responded nonspecifically to organic and hydrophilic, low molecular weight components of conditioned water, with *B. glabrata* showing their typical turning response. Moreover, these authors observed that miracidia and cercariae of *E. caproni* may use different signals to identify the same *B. glabrata* snail indicating an independent evolution of the host-finding mechanism in the two larval stages. Haas et al. (2008) studied the distribution of *E. caproni* cercariae obtained from *B. glabrata* in 18-cm-sized cuvettes under different lighting conditions, in Plexiglas cylinders (80 cm high) vertically placed in a pond, and when swimming freely in a pond. In vertical cuvettes, *E. caproni* cercariae accumulated near the water surface when they were exposed to bright light and they shifted from a negative to a positive photo-orientation during their lifetime. This could mean that they search for hosts that prefer light microhabitats, as well as hosts that reside in shady habitats near the water surface (Haas et al. 2008). Platt et al. (2009) investigated the effect of light and gravity on the transmission of cercariae in *E. caproni*/*B. glabrata* model and suggested that these cercariae exhibit negative photo- and geotaxis when searching for a second intermediate host. Stereotypical releaser responses to environmental trigger cues (light and gravity) allow *E. caproni* cercariae to exploit alternative strategies to complete their life cycle, consistent with the broad range second intermediate host used by *E. caproni* cercariae.

Pechenik and Fried (1995) investigated the infectivity of *E. trivolvis* cercariae of different ages to determine if a loss of infectivy occurs prior to cercarial death. *B. glabrata* snails were exposed to cercariae of different ages and cyst counts were taken at 1–2 days p.i. Fifty percent of the survival time of the cercariae depended on temperature, supporting the hypothesis that cercarial life spans are limited by the rates at which energy stores are used.

Fried et al. (1995) studied infectivty in *B. glabrata* snails (0.7–1.0 mm shell diameter) exposed to *E. trivolvis* cercariae. Forty snails were exposed en masse to 1,000 cercariae for 24 h, and all the snails were positive at necropsy with 1–29 cysts per snail. The snails exposed individually to 25 cercariae showed high mortality within a few days, probably being related to cercarial penetration and encystment in the snail tissue.

To ascertain the dynamics of cercarial transmission of *E. caproni*, Meyrowitsch et al. (1991) examined the effects of temperature and host density on the snail-finding capacity of this echinostome cercariae. They described the effect of temperature (19–36°C) and *Biomphalaria* host density (0.014–10 snails/l) on the snail-finding capacity of this cercaria. As expected, the initial swimming speed of cercariae increased, whereas the length of the infective period decreased with increasing temperature. The *E. caproni* cercarial snail-finding capacity was temperature independent in the 19–36°C range at a snail density of 0.014 snails/l. These authors suggested that a relatively low and biologically realistic snail host density should be used in experimental studies if realistic estimates of the dynamics of cercarial transmission were to be obtained.

Relatively little information is available on the pathobiochemistry of *Biomphalaria* specimens infected with echinostomes. Various chromatography techniques (thin-layer and gas-liquid chromatographies) have been used to study the effects of *E. caproni* infection on different biochemical components of the digestive gland–gonad complex of *B. glabrata*, and the results are different depending on the biological substance studied. Fried et al. (1989, 1993) and Shetty et al. (1992) noted significant reduction in the concentration of various lipids, and Perez et al. (1994) and Wagner et al. (2001) reported similar reductions in several carbohydrates. However, Perez et al. (1995) showed significant phospholipid elevation in infected snails and Beers et al. (1995) observed that the digestive gland–gonad of infected snails maintained on a yolk diet showed a significant increase in free sterols and a significant decrease in triacylglycerols compared with uninfected snails maintained on a lettuce-Tetramin diet. Marsit et al. (2000) detected that the most abundant lipid fraction in the cercariae and rediae of *E. caproni* was free sterol, with lesser amounts of triacylglycerols and steryl esters. Moreover, Layman et al. (1996) did not find significant differences in metallic-ion concentrations between the infected vs. uninfected *B. glabrata* with *E. caproni*, and Ponder et al. (2004) obtained similar results on the study on free-pool amino acids in the same trematode snail system. Bandstra et al. (2006) noted that the major neutral lipids present in the whole bodies and digestive gland–gonad complexes in *B. glabrata* specimens patently infected with the daughter rediae of *E. caproni* were free sterols, free fatty acids, and triacylglycerols, and the major polar lipids were phosphatidylcholine and phosphatidylethanolamine. Quantitative analysis showed no significant differences in the concentrations of these lipids in whole bodies of infected snails vs. the controls, but the concentration of triacylglycerols in the infected digestive gland–gonad complex was significantly lower than that of the uninfected. No quantitative differences in neutral lipids and phospholipids in shell or plasma samples were found between infected vs. uninfected snails (Bandstra et al. 2006). Using ion-exchange chromatography, the calcium carbonate concentrations of the shells of *B. glabrata* snails experimentally infected with *E. caproni* miracidia were determined. The infection of *B. glabrata* was correlated with reduced calcium content in the shell, but also with increased size of the snail (White et al. 2005).

Reddy et al. (2004) studied the effects of copper toxicity on cercariae of *E. caproni*, as well as on the survival of *B. glabrata* snails, and noted that concentrations of copper (>0.001%) sufficient to eliminate juvenile snails are also sufficient to kill the cercariae of *E. caproni*.

6.2.2 As Second Intermediate Host

The low specificity toward the second intermediate host shown by trematodes of a triheteroxenous life cycle is well known. The cultivation and maintenance of *Biomphalaria* spp. has turned them into suitable secondary intermediate hosts, containing the infective stage of metacercariae, of many trematodes species (see Table 6.1) able to use freshwater pulmonate snails as second intermediate hosts. Moreover, this has made it possible to cast light on various aspects of trematode biology in general, and on that of representatives of the Echinostomatidae family in particular. In this sense, although cercariae of echinostomes encyst in a broad range of second intermediate hosts, there are differences in the susceptibility of second intermediate hosts for cercarial penetration. Specifically, cercarial preference toward *B. glabrata* as second intermediate hosts has been the subject of some studies (Jeyarasasingam et al. 1972; Christensen et al. 1980; Fried et al. 1987; Anderson and Fried 1987; Huffman and Fried 1990; McCarthy 1999; Fried 2001). Furthermore, *Echinostoma/Biomphalaria* is a useful experimental host for studies on cercarial encystment (Frazer and Fried 1998).

Cercariae begin to emerge from infected *Biomphalaria* snails 4–6 weeks postinfection and have a predilection for the kidney-pericardial region of the same or other snails (Jeyarasasingam et al. 1972; Christensen et al. 1980). Cysts may remain viable within *B. glabrata* for months and metacercariae of some echinostomes have also been found in *B. glabrata* harboring rediae and cercariae and serving as the first intermediate host; some metacercariae encyst within the rediae of the same echinostome species (Esteban and Muñoz-Antolí 2009).

Frazer and Fried (1998) characterized the differences between single-species infections of different pulmonate snails, among them *B. glabrata*, with *E. caproni* and concurrent infections of pulmonate snails with *E. caproni* and *E. trivolvis* cercariae. The percentage encystment of *E. caproni* was significantly greater in *B. glabrata* than in the other snail species. Moreover, in concurrent infections, *E. trivolvis* localized only in the saccular kidney, whereas *E. caproni* occurred in both the saccular kidney and the pericardicum.

Snail size is another relevant factor in determining the mortality associated with echinostome cercarial penetration and encystment in the second intermediate host. In general, neonatal snails are more susceptible to infection but show a significant decrease in survival. Sullivan (1985) examined metacercarial infection of *E. caproni* in neonatal *B. alexandrina* exposed to an undetermined number of cercariae. He found that neonatals (0.8–1.4 mm in shell diameter) were infected with 2–13 *E. caproni* cysts within 48 h postinfection. Kuris and Warren (1980) noted that continuous exposure of *B. glabrata*, 3–8 mm in shell diameter, to about 150 *E. caproni* cercariae per day caused high snail mortality after 4–6 days. Fried et al. (1995) detected a high mortality in the *E. trivolvis*-infected *B. glabrata* juveniles (neonatal), probably related to cercarial penetration and encystment in the snail tissue. Schneck and Fried (2004) studied encystment of *E. caproni* in neonatal (0.7–1.1 mm shell diameter) and juvenile (2–3 mm shell diameter) *B. glabrata* (NMRI strain) and concluded that snail size was a major factor in determining cyst burden in *B. glabrata* and in mortality associated with echinostome cercarial penetration and encystment. That neonatal survival was affected detrimentally by cercarial infection suggests that echinostomes may play a role in the biological control of *Biomphalaria* spp.

Reddy et al. (2004) noted that a concentration >0.001% $CuSO_4$ increased *B. glabrata* mortality, as well as that of the excysted metarcercariae of *E. caproni*, but not the encysted metacercariae, which may be protected by their cyst walls.

In genera other than *Echinostoma*, *Biomphalaria* specimens have also been used as second intermediate hosts. Cercariae of *E. trivolvis* and *Echinoparyphium* encysted in *B. glabrata* snails in single and concurrent infections, in the saccular kidney of the snail at 24 h postinfection (Fried et al. 1998). *Echinoparyphium elegans* cercariae released from the first intermediate host encysted in the pericardial sac and kidney of different snails, among them *B. glabrata* (Mouahid and Moné 1988). McCarthy (1999) examined the potential influence of the second intermediate host species on the infectivity of metacercarial cysts of *Echinoparyphium recurvatum*. The metacercarial cysts were obtained 14 days postcercarial infection from different second intermediate hosts, among them *B. glabrata*. The author concluded that the species of second intermediate host utilized does not influence the infectivity of the metacercarial cyst of *E. recurvatum* or the subsequent establishment and reproductive status of the parasite in the definitive host.

In the absence of a second intermediate host, in vitro encystment of echinostome cercariae can occur. Concretly, *E. paraensei* and *E. caproni* have been reported to encyst in cultures with *B. glabrata* embryonic cells (Stein and Basch 1977; Loker et al. 1999).

6.3 Effects of Nonschistosome Larval Trematodes on *Biomphalaria* spp.

Little information is available on the histopathological effects of nonschistosome larval trematodes in *Biomphalaria* spp. However, it is well known that several species of trematodes cause premature death, growth alterations, and castration in the snails.

Over a period of 26 days, *B. glabrata* infected with *Ribeiroia marini* exhibited 30% higher mortality than uninfected control snails (Huizinga 1973). Similarly, Nassi (1978) showed that infection of *B. glabrata* with three miracidia of *R. marini* induced 25% mortality and only one miracidium developed on average in surviving snails. Exposure to >5 miracidia was associated with extremely high snail mortality (Nassi 1978; Pointier and Jourdane 2000).

Although *Plagiorchis elegans* is unable to establish patent infections in *B. glabrata*, the parasite nevertheless is able to penetrate and kill or castrate this host. Zakikhani and Rau (1998) showed that almost 20% of *B. glabrata* exposed to eggs of *P. elegans* died in the first week and none survived to the age of 25 weeks. In contrast, no mortality was observed among uninfected snails until the age of 20 weeks (Zakikhani and Rau 1998).

An increase in snail mortality and growth alterations have been reported in several *Echinostoma–Biomphalaria* combinations (Christensen et al. 1980; Kuris 1980; Schmidt et al. 1998; Muñoz-Antolí et al. 2007). Christensen et al. (1980) noted that various species of *Biomphalaria* infected with miracidia of an Egyptian strain of *E. caproni* (referred as *E. liei* in their study) showed an increase in death rate compared with the uninfected snails. Kuris (1980) observed that an Ethiopian strain of *E. caproni* (referred as *E. liei* in his study) decreased the growth rate and survival of 4–6 mm juvenile *B. glabrata*. Infection of *B. glabrata* with *E. friedi* miracidia also has a significant effect on snail longevity and growth (Muñoz-Antolí et al. 2007). The *B. glabrata* snails infected with one miracidium had a maximum life span of 14 weeks, whereas 63% of uninfected control snails were still alive at 25 weeks. Furthermore, the growth of infected snails significantly exceeded that of uninfected snails during the patent period (Muñoz-Antolí et al. 2007).

The development of larval digeneans in *Biomphalaria* spp. may affect the normal reproduction of the snail, inducing castration. This fact is based on the ability of the intramolluscan stages of these parasites to suppress the reproductive functions of their hosts. Reproductive damage may be direct, as in instances where rediae actively consume gonadal tissues, or indirect, most probably by way of modulating the host's neuroendrocrinological system (Nassi 1979; Hurd 1990). Parasitic castration of the snails by digenean sporocysts is thought to be of the indirect type (De Jong-Brink 1995).

Ribeiroia spp. cause castration in infected *Biomphalaria* (Huizinga 1973). Castration occurred within 11–19 days after infection, possibly as a result of mechanical damage to the brain. Nassi (1979) found that localization of rediae of *R. marini* near to the brain region of *B. glabrata* coincided closely with the loss of reproductive function.

Zakikhani and Rau (1998) showed that infections with mother sporocysts of *P. elegans* had a significant effect on the reproductive output of *B. glabrata* exposed as juvenile or adults. The total number of eggs was reduced to approximately 7 and 13% of their respective controls. In contrast to that observed in immature *B. glabrata* infected with *Schistosoma mansoni*, in which effects on host reproduction are associated with the beginning of cercarial production (Thornhill et al. 1986), *P. elegans* elicited prepatent parasitic castration in both adult and juvenile *B. glabrata*. Furthermore, unlike the findings by Sturrock and Sturrock (1970) from *B. glabrata* infected with *S. mansoni*, there was no evidence of a limited resumption of egg production in old snails (Zakikhani and Rau 1998). Snails infected with *P. elegans* that retained even a limited level of reproductive activity survived almost 4 weeks longer than nonreproductive snails. Zakikhani and Rau (1998) suggested that the intensity of the infection may affect the longevity and reproductive output; the most heavily infected snails may be the most reproductively impaired and among the first to die.

Regarding the effect of echinostomes on the fecundity of *Biomphalaria* spp., several situations have been described: total castration in *E. friedi–B. glabrata* (Muñoz-Antolí et al. 2007), *E. caproni–B. glabrata* (Langand and Morand 1998; Sandland et al. 2007), and *Echinostoma togoensis–Biomphalaria pfeifferi* (Jourdane 1983) combinations or slight reductions of oviposition in *E. caproni–B. glabrata* combination (Schmidt et al. 1998). In general, the effect of echinostomes appears to be similar than that of schistosomes. The level of fecundity inhibition is related to the sexual maturation of the snail host when infection occurs. A total inhibition is produced when juvenile snails are infected, whereas in adult snails an initial increase in oviposition occurs followed by a reduction without reaching total castration. This phenomenon is called fecundity compensation and could serve to compensate for future reproductive losses due to parasitism at the beginning of the infection (Minchella and LoVerde 1981; Minchella 1985; Fryer et al. 1990). However, the occurrence of this phenomenon in *Biomphalaria* spp. infected with echinostomes has not been confirmed.

6.4 Antagonism Between Nonschistosome Trematodes and Schistosomes in *Biomphalaria* spp.

Digeneans also interact antagonistically within their mollusk intermediate host (Combes 1992). Rediae primarily manifest direct antagonism in which they actively prey the larval stages of other species (Lie 1973). Indirect antagonism, in contrast, acts at a distance, mediated by the internal defense mechanisms of the snail host (i.e., interference; see Sect. 6.5.2 of this chapter). Sporocysts are thought to be capable of only indirect antagonism and are generally eliminated by concurrent dominant redial infections (Lim and Heyneman 1972). Moreover, recent studies suggest that coinfection can influence the evolution of hosts and parasites by modulating host life history, parasite transmission, and the virulence of the interaction (Escribano et al. 2001; Paul et al. 2002; Zakikhani et al. 2003; Sandland et al. 2007).

Larval stages belonging to the Echinostomatidae appear to be particularly effective in terms of antagonistic effects. Interactions between echinostomes and schistosomes in *Biomphalaria* spp. often result in echinostome dominance. The absence of *S. mansoni* infection in *B. glabrata* coexposed to *E. paraensei* has been cited and may be the result of direct consumption of schistosomes by echinostome rediae (Basch and DiConza 1975). Furthermore, interspecific competition between *E. caproni* and *S. mansoni* in *B. glabrata* was observed (Jourdane and Mounkassa 1986). This resulted in shifting of the mother sporocysts of *S. mansoni* out of their usual sites into deeper snail tissues, probably to protect them from predation (Jourdane and Mounkassa 1986).

Sandland et al. (2007) assessed the coevolutionary consequences of coexposure of *B. glabrata* with *E. caproni* and *S. mansoni* in the parasite establishment, host life history, parasite reproduction, and virulence. This study confirmed the dominance of *E. caproni* vs. *S. mansoni* in concomitant infections in *B. glabrata*. The prevalence of *S. mansoni* was significantly reduced compared with hosts exposed to *S. mansoni* alone. Moreover, coexposure appeared to enhance virulence when *E. caproni* established successfully. Under these circumstances, snails died earlier, produced fewer eggs, and released more larvae than snails that were exposed to *E. caproni* alone. Echinostomes may have increased their larval output because of perceived competition with coinfecting parasites (Sandland et al. 2007). In contrast, snails coexposed but infected with *S. mansoni* exhibited higher survival, higher egg output, and lower cercarial release than do snails exposed exclusively to *S. mansoni*. This suggests that coexposure to *E. caproni* reduces *S. mansoni* fitness even in the absence of echinostome establishment (Sandland et al. 2007).

Similar to that which occurs with echinostomes, antagonism between *S. mansoni* and *R. marini* and *P. elegans* in *B. glabrata* has also been observed. Huizinga (1973) showed that *B. glabrata* infected with *S. mansoni* or *R. marini* were only partly susceptible to infection with the other parasite. In simultaneous exposures, some snails developed double infections with *R. marini* dominant and *S. mansoni* sporocysts and cercariae reduced.

Zakikhani et al. (2003) demonstrated that the presence of an incompatible trematode species (*P. elegans*) that failed to reach patency in *B. glabrata*, still dramatically compromised *S. mansoni* reproduction, growth, and survival. The above effects were attributable to the establishment and persistence of the incompatible snail host. The indirect antagonistic interaction with the larval stages of *S. mansoni* may be mediated in part, through their long-term stimulation of the host's internal defense mechanisms.

All the above studies have suggested that nonschistosome larval digeneans can be used as a component of integrated schistosomiasis control programes. Three main ideas have been considered and studied in this sense: (1) the interspecific larval trematode antagonism within the snail hosts; (2) the parasitic castration of the snail hosts; and (3) the pathogenic or lethal action of the larval trematodes (Pointier and Jourdane 2000). In this context, the presence of the echinostome and *R. marini* or *P. elegans* infections in *B. glabrata* have raised the possibility of control of *S. mansoni* transmission using these parasites. Several field studies have been performed to evaluate the potential role of these trematodes on schistosome transmission (Nassi et al. 1979; Pointier and Jourdane 2000). This topic is discussed in further detail in Chap. 10.

6.5 Immunology of the Infections

6.5.1 *Immunological Interactions Between* Biomphalaria *spp. and Nonschistosomes Larval Trematodes*

In recent years, numerous studies have been done to understand the basis of compatibility and incompatibility between mollusks and larval trematodes. The success of a trematode infection depends largely on the immunobiological interactions between the snail and the parasite. *Biomphalaria* spp. are not passive hosts for trematodes, since they possess an internal system that can recognize, manage, and eliminate pathogens (Adema et al. 1994a, b, 2000; Loker and Adema 1995; Ataev and Coustau 1999; Coustau et al. 2009). Information obtained on this topic may serve for a better understanding of several aspects on *Schistosoma* spp. transmission. In this context, studies on the relationships between *Biomphalaria* spp. and non-schistosomatid trematodes have been very useful. Among the different trematodes that use *Biomphalaria* spp. as the first intermediate host, echinostomes are the most commonly used to study the immune response of snails against trematodes. Several aspects of echinostome infections in *Biomphalaria* spp. are more obvious for observations relative to schistosome and other trematode infections. Comparative studies of echinostome infections in *Biomphalaria* spp. facilitate functional interpretation of factors that characterize digenean–snail interactions. This section will focus mainly on two echinostome species, *E. paraensei* and *E. caproni*, that have been extensively studied with respect to their interactions with *B. glabrata*. The first intermediate host is crucial in the life cycle of echinostomes, since an extensive asexual parasite reproduction occurs in the circulatory system and tissues of the snail host. An understanding of the echinostome survival strategies in the snail host can be applied to the control of other digeneans.

The first indication that snails are not passive hosts comes from the fact that echinostomes are restricted in their snail host range. *E. caproni* can complete its development in *B. pfeifferi* and *B. glabrata* (Ataev et al. 1997, 1998). *E. paraensei* can develop in *B. glabrata* and *Physa rivalis*, but infection of *Biomphalaria straminea* was unsuccessful (Lie and Basch 1967). Additionally, Sullivan (1988) lists *Biomphalaria obstructa*, *Phibalapteryx virgata*, and *Helisoma trivolvis* as unsuitable hosts for *E. paraensei*. Although several efforts have been made, the factors determining whether a given echinostome can or cannot utilize a certain snail species remain unknown. A particular species of the genus *Biomphalaria* may be nonsusceptible by providing an unsuitable environment. However, even if a snail species offers the right environment, its defence system can recognize and eliminate the parasite (Van der Knaap and Loker 1990; Lardans and Dissous 1998; Adema et al. 2000; Coustau et al. 2009). The success of an echinostome infection in *B. glabrata* depends largely on the immunobiological interactions between the snail and the parasite.

The immune system of *B. glabrata* relies on both humoral and cellular factors that cooperate in the recognition and elimination of pathogens, including echinostomes (Loker and Bayne 1986; Yoshino and Vasta 1996; Coustau et al. 2009). Initial experiments revealed that response of *B. glabrata* against *E. paraensei* comprises a complex mixture of plasma proteins with a wide range of molecular weights (Loker and Hertel 1987). Infection with schistosomes or bacteria did not result in changes in the composition of polypeptides present in the cell-free hemolymph of *B. glabrata*. In contrast, infection with *E. paraensei* induced changes in plasma of several strains of *B. glabrata* (Loker and Hertel 1987; Hertel et al. 2005; Zhang et al. 2008a, b). A potential nonself recognition function was further indicated by showing that these polypeptides have lectin-like properties and bind and precipitate parasite antigens. Although little is generally known about the invertebrate lectins, some of the most extensively studied snail lectins were discovered using the *E. paraensei–B. glabrata* model (Monroy et al. 1992; Monroy and Loker 1993; Hertel et al. 2000; Adema et al. 2000). Analysis of the snail-derived components of this precipitate showed that they fall into three categories on the basis of their molecular weight: 200, 80–120, and 65 kDa (Adema et al. 1997a, b). The study of the 65-kDa snail plasma lectins produced in response to echinostome infections and reactive with the parasite antigens showed that they posses a unique molecular structure, having one or two immunoglobulin superfamily domains (IgSFs) at the N terminus and a fibrinogen domain at the C terminus (Léonard et al. 2001; Zhang et al. 2001). Fibrinogen-related proteins (FREPs) are encoded by a large gene family first documented in the snail *B. glabrata* and have also been found in other genera of snails (Adema et al. 1997b). Further analyses of these proteins have shown that they are capable of binding to the surface of miracidia, sporocysts, and rediae of *E. paraensei*. These proteins exhibit extensive variation in the IgSFs regions. These domains are diversified at the genomic level at high rates and the sequence variants are derived from a small set of source sequences by point mutation and recombinatorial processes (Zhang et al. 2004). In the absence of adaptative immunity, the diversification of innate immune factors may play an important role. In practice, a continually diversifying repertoire of FREP sequences might change the immune-recognition capabilities and possibly with that the compatibility of individual *B. glabrata* for particular digeneans as a function of time. However, no mechanisms have been described from *B. glabrata* to explain somatic mutation, though this feature has been described in a variety of animals that do not have lymphocyte-based capabilities of acquired immunity (Coustau et al. 2009).

FREPs are thought to function in the immune response of *B. glabrata* because some family members are up-regulated following exposure to *E. paraensei* (Loker and Hertel 1987; Hertel et al. 2005; Zhang et al. 2008a, b), have lectin-like capabilities to bind carbohydrates (Monroy et al. 1992; Monroy and Loker 1993), precipitate parasite antigens (Adema et al. 1997a, b), and are capable of binding to the surface of miracidia, sporocysts, and rediae of *E. paraensei* (Adema et al. 1997b). Recently, it has been shown that FREP proteins are able to bind *E. paraensei* sporocysts and their excretory/secretory products (ESPs) and a variety of microbes (Zhang et al. 2008b). Furthermore, this binding capability showed evidence of specificity with respect to pathogen type; for example, 65–75-kDa FREPs (mainly FREP4) bind to *E. paraensei* sporocysts and their ESPs, whereas 95- and 125-kDa FREPs bind the microbes assayed. These results suggest that FREPs can recognize a wide range of pathogens, from prokaryotes to eukaryotes, and different categories of FREPs seem to exhibit functional specialization with respect to the pathogen encountered (Zhang et al. 2008a). Although considerable information on the structure, diversity, and expression of FREPs have been obtained recently (Léonard et al. 2001; Zhang et al. 2001, 2004; Zhang and Loker 2003; Hertel et al. 2005; Zhang et al. 2008a, b), the precise role of FREPs, and probably, other related functions in snail physiology need further research.

Regarding cellular factors, hemocytes are of great importance. Studies focused on *E. paraensei* and *E. caproni* suggest that echinostome immunoevasion relies on the suppression of host cellular functions, and the ESP released by the parasite appear to play a major role. Hemocytes collected from *E. paraensei*-infected *B. glabrata* showed significantly less adhesion and spreading and phagocytic capabilities than do hemocytes from control snails (Noda and Loker 1989a, b). Adema et al. (1994a, b) reported that hemocytes in close proximity to *E. paraensei* sporocysts and rediae were more affected when closer than more distant, suggesting that the parasite releases active factors to generate a gradient of interference around it. DeGaffe and Loker (1998) showed that susceptibility of *B. glabrata* to infection with *E. paraensei* is correlated with the ability of ESP to interfere with the spreading behavior of host hemocytes. Although the mechanism by which ESP products inhibit hemocyte function remains to be resolved, interference may be effected though interaction with signal transduction pathways of the snail (Walker 2006).

The in vitro exposure of *B. glabrata* hemocytes to *E. paraensei* sporocysts or their ESP yields a calcium wave in the cytoplasm of adherent hemocytes (indicative of activation of signaling pathways that employ Ca ions as second messenger) prior to the rounding up of these cells. Thus, *E. paraensei* may affect hemocyte function in different ways (Hertel et al. 2000). Similarly, ESP collected from in vitro transformed *E. caproni* sporocysts induced a total loss of *B. glabrata* hemocyte defense functions encompassing adhesion, spreading, and phagocytosis (Humbert and Coustau 2001).

Interestingly, this immune suppressive effect of *E. paraensei* and *E. caproni* ESP products may significantly contribute to the specificity of host–parasite compatibility, as the potency of ESP to affect hemocytes correlated with echinostome infectivity (DeGaffe and Loker 1998), the degree to which hemocyte function was interfered was less in snails exposed to but not infected with echinostomes (Noda and Loker 1989b), and ESP failed to affect hemocytes from nonhost species (Adema et al. (1994a, b)) or from genetically selected resistant strains of host species (Humbert and Coustau 2001).

6.5.2 *Echinostome-Mediated Interference of* B. glabrata *Immune Defense System*

For survival, digeneans must prevent an effective defense response from the snail host. Two strategies have been proposed by which digenean parasites may escape from the host defense system. Some digeneans may prevent recognition as nonself by the host by either absorbing snail host factors (molecular masking) or through expressing gene products that mimic host factors (molecular mimicry). A second survival strategy consists in the interference of the snail immune defense system by the digenean parasite. The combination of *E. paraensei–B. glabrata* has served to gain further insight into digenean-mediated interference strategy.

Lie and coworkers (1976) showed that hemocytes of *B. glabrata* snails that harbored echinostome infections lost their ability to adhere to nonself objects. Furthermore, hemocytes accumulated in irrelevant locations in the snail host. Infection with echinostomes caused diminution of resistance to *S. mansoni* in *B. glabrata* strains selected for their resistance to the schistosome (Lie et al. 1977). Sporocysts that were normally encapsulated by hemocytes from resistant *B. glabrata* strains were nonencapsulated in double infections in the proximity of *E. paraensei* sporocysts (Lie 1982). Based on these observations, digenean larvae, and echinostomes in particular, were postulated to interfere directly with the internal defenses of *B. glabrata*, affecting functional aspects of the immune defense system mostly by targeting snail hemocytes (Lie 1982; Adema et al. 2000). The term interference was originally coined to describe the inhibitory effects exerted by echinostome larvae on the snail host immune defense system, though interference can also be observed in other digenean–snail combinations such as *Diplostomum spathaceum–Lymnaea stagnalis* (Riley and Chappell 1992). However, most of the studies on this topic are focused on the *Echinostoma* spp.–*B. glabrata* system.

Additional observations support the notion that echinostomes aggressively interfere with the *B. glabrata* immune system. The in vitro killing of *S. mansoni* sporocysts by hemocytes from uninfected resistant *B. glabrata* was significantly reduced when *E. paraensei* larvae were included in the assay (Loker and Bayne 1986). The proportion of hemocytes not spreading or only minimally spreading on plastic slides was increased in *E. paraensei*-infected *B. glabrata* (Noda and Loker 1989a), showing rounded cells, reduced adherence, and diminished phagocytic activity (Noda and Loker 1989b). Living *E. paraensei* larvae strongly affect hemocyte functions: hemocytes from unexposed *B. glabrata* do not adhere to the surface of living *E. paraensei* sporocysts, but are able to encapsulate dead *E. paraensei* sporocysts (Loker et al. 1989). This suggested that interference is mediated through release of soluble factors.

Loker et al. (1992) detected components of the ESP of *E. paraensei* of more than 100 kDa that mediate the rounding effect of *B. glabrata* hemocytes. Other components (<100 kDa) of *E. paraensei* ESP inhibit *B. glabrata* hemocyte motility (Loker et al. 1992). DeGaffe and Loker (1998) showed that the potency of a particular ESP sample to cause *B. glabrata* hemocyte rounding in vitro was correlated to the infection success of *E. paraensei* miracidia from the same batch of eggs. Furthermore, ESP had a greater rounding effect on hemocytes from hemolymph samples obtained from juvenile snails compared with those obtained from adult snails.

It is known that interference relies on more components of ESP. However, little is known about the identity of these components (Adema et al. 2000). In this context, proteomic approaches are of great interest.

6.6 Genomics and Proteomics of the Infections

The genomic and proteomic aspects of *Biomphalaria* spp. and infections with non-schistosomatid trematodes have been studied in relation to the snail resistance to infections. The finding of a *B. glabrata* strain resistant to *E. caproni* infection (Langand and Morand 1998), together with the application of molecular approaches, has provided new tools to gain further insights on the molecular basis of the snail resistance to trematode infections. These studies mainly focused on hemocytes and humoral factors, which have been shown to play key roles in the susceptibility/resistance process (Ataev and Coustau 1999).

Regarding humoral factors, Vergote et al. (2005) compared the plasma protein profiles of plasma collected from *B. glabrata* snails susceptible and resistant to *E. caproni* infection. This proteomic approach revealed five proteins differentially represented in the plasma of resistant and susceptible snails. They correspond to two isoforms of a glycolytic enzyme (Bg endo-1, 4-β-mannanase), two isoforms of a calcium-binding protein of the EF-hand type (Bg CaBP-1 and -2), and an inhibitor of cysteine proteases (Bg type-2-cystatin). All these proteins were expressed in the albumen gland of the snail (Vergote et al. 2005). The higher expression of the endo-1, 4-β-mannanase, and the type-2-cystatin in the susceptible snails was suggested to have an immunosuppressive effect facilitating parasite establishment. Glycolytic enzymes can modify the structure of the parasite molecules, and consequently the interactions with host factors and the inhibitors of cysteine proteases may down regulate immunity (Vergote et al. 2005). In contrast, the higher expression of CaBPs in resistant snails could make parasite establishment difficult, since they may have an antiparasitic effect or affect adhesion processes (Vergote et al. 2005).

Regarding the cellular factors involved in susceptibility/resistance to *E. caproni* infection, studies have focused on hemocytes, since previous studies have shown that they play a key role in defense against *E. caproni*. Excretory–secretory products of *E. caproni* from in vitro transformed sporocysts inhibited defense functions of susceptible host hemocytes, whereas hemocytes from resistant snails remained unaffected (Humbert and Coustau 2001). In this sense, two molecular approaches based on the comparison of transcripts and protein expression between both snail strains have been performed (Bouchut et al. 2006a, b).

The transcriptomic approach by Bouchut et al. (2006a) focused on genes involved in adhesion processes and revealed four transcripts that were differentially represented between hemocytes from resistant and susceptible snails. These genes encode two dermatopontin-like proteins, a matrilin-like protein and a cadherin-like protein. Bouchut et al. (2006a) suggested that these are of importance in the processes of capsule formation and the coagulation system, particularly those enconding dermatopontin-like proteins and the matrilin-like protein. The transcript content of genes of dermatopontin-like proteins was higher in hemocytes from resistant snails, and the expression was greatly enhanced 48 h after exposure to *E. caproni*, coinciding with the capsule formation. Because of the similarities of the Bg matrilin with matrilin from vertebrates, Bouchut et al. (2006a) hypothesized that this protein participates in an induced filamentous network in the hemolymph of *B. glabrata* exposed to *E. caproni*. Since the transcription of Bg matrilin was higher in naïve susceptible snails, these authors suggested that this *B. glabrata* strain possesses a more potent hemolymph coagulation system that may prevent hemocyte migration to the parasite larvae, therefore facilitating parasite establishment.

Another approach identified a total of five proteins differentially represented in hemocytes of the two strains of *B. glabrata* (Bouchut et al. 2006b). Two of them (Bg aldolase and Bg ribosomal protein P1) were better represented in hemocytes from susceptible snails, whereas the remainder three proteins (Bg intermediate filament protein, Bg cytidine deaminase, and Bg histone H4) were more abundant in those from resistant snails. Of particular interest appears to be the Bg histone H4 protein, since it could be subjected to major posttraslational modifications reflecting a differential chromatin structure in hemocytes of susceptible and resistant snails responsible for different patterns of gene expression. In addition, a differential regulation of the corresponding gene between strains was enhanced following parasite exposure. While no variation was observed in resistant snails, a high increase of Bg histone H4 transcript was observed in susceptible snails.

Bouchut et al. (2007) developed a comparative transcriptomic approach by constructing substractive libraries from resistant and susceptible snails to identify genes differentially expressed. This study was not limited to factors present in plasma or hemocytes, but in contrast, the approach was performed on entire snails in order to investigate the potential involvement of genes in other snail tissues. This work revealed some candidate transcripts already identified in previous studies

(CaBPs and glycolytic enzymes) but also new candidates belonging to novel functional groups. Regarding these novel groups, the most promising candidates belonged to the immunity class. Within this group, several proteases (Bg cathepsin L-like and Bg serine protease beta-like 1) and cell adhesion proteins (Bg DEC 1-likc and Bg ependymin-like) showed higher expression in susceptible snails, which could facilitate parasite invasion and establishment. Several other novel clusters belonging to other functional groups like transport (Bg haemoglobin and Bg fatty acid binding protein) cholesterol homeostasis proteins displayed a transcript content that was superior in susceptible snails. However, their putative role in parasite establishment was difficult to evaluate. Regarding genes identified in the library from resistant *B. glabrata*, several genes involved in immune processes were also identified and they may have antiparasitic effects (Bg aplysianin-A-like protein), are involved in pathogen recognition (Bg C-type lectin-like protein), and inhibit proteases (Bg serpin-like 1 and 2 proteins).

All these approaches have allowed identifying a number of transcript and proteins differentially expressed in susceptible and resistant strains of *B. glabrata*, suggesting that they may be involved in the success of larval trematode infections in snails. However, future studies are needed to identify other factors involved in this issue and to determine more exactly the role of these factors in the resistance to trematode infections.

6.7 Molecular Detection of Trematodes in *Biomphalaria* spp.

Without understanding the ecologically important questions, especially on the transmission and distribution of trematode larvae and the antagonistic effects of other organisms under field conditions, precise alternative methods for the control of schistosomes cannot be developed. However, detailed studies on the ecology, transmission, and/or distribution of nonschistosome larval digeneans that parasite *Biomphalaria* spp. have not been performed. One of the reasons for the absence of studies is the difficulty entailed in the specific identification of the larval digeneans. Similarity between larval stages of trematodes, echinostomes in particular, often results in improper identification. Additionally, trematode larvae are very fragile and are often damaged during the sampling procedures, making their identification difficult. In this context, DNA-based molecular biology tools for differentiation between digeneans parasitizing *Biomphalaria* spp. may be very useful and of help in field studies. As occurs with other topics, these kind of studies have been focused on echinostomes.

Molecular studies on echinostomes are very preliminary. Only 474 DNA sequences are available in the databases (Marcilla 2009), from which 358 correspond to the EST project performed with *E. paraensei* sporocysts (Nowak and Loker 2005). The remaining sequences mainly represent molecules used in phylogenetic studies, such as ribosomal (rDNA) or mitochondrial molecules (mtDNA) (Morgan and Blair 1995, 1998a, b; Sorensen et al. 1998; Kostadinova et al. 2003).

Approaches to develop methods that may be useful to detect larval trematodes in the field are scarce. There exist PCR primers for the differentiation between *E. caproni* and *E. trivolvis* collected from *B. glabrata* (Fujino et al. 1995, 1997). These authors developed a random polymorphic DNA analysis (RAPD-PCR) showing some intense bands characteristic of each species. Petrie et al. (1996) used the same RAPD-PCR to differentiate *E. paraensei, E. caproni*, and *S. mansoni* (parasites of *B. glabrata*), using primers originally developed for schistosomes by Dias Neto et al. (1993). These species produced distinct banding patterns, suggesting that this method can be useful for the identification of digenean parasites of *B. glabrata*.

Hertel et al. (2003) cloned, sequenced, and analyzed a 192-bp tandem repeated DNA sequence of *E. caproni* that was used to detect this parasite in *B. glabrata* snails and plankton samples. PCR primers against this sequence detected <10 fg of *E. caproni* DNA, two miracidia in *B. glabrata* snails 1 day postinfection, or one metacercaria in 50 mg of snail tissue with high levels of specificity.

These tools might be employed to obtain field data on host specificity, ecology, and transmission of trematode larvae. Tandem repeated sequences have proven to be highly specific and to have some potential use in quantification. It is also important to investigate the ecology and the effects of antagonism. However, some problems should be solved before the application of these methods. These methods may not be specific on field samples, which may contain a lot of different digenean species. Furthermore, ecological studies may require determining if *Biomphalaria* spp. is acting as first or second intermediate host, which is difficult using these methods.

6.8 Concluding Remarks

As shown in this chapter, the interest in *Biomphalaria* snails in parasitology is not limited to their role as intermediate host of schistosomes. The interactions between these snails and the nonschistosome larval digeneans have important implications. Apart from the fact that transmission of these trematodes may have importance in wildlife, the *Biomphalaria*–nonschistosome larval digeneans systems are relevant in many other areas of biological research. These host–parasite systems are in fact valuable experimental models for the study of general parasitological and immunological questions. Insights derived from the analysis of the interactions on these models may guide and direct studies aimed at further understanding of transmission of digenean parasites by snails. There are several examples illustrating this fact. The discovery of FREPs as antidigenean response factors is a direct result of experimental use of the *Biomphalaria*–nonschistosome larval digenean systems. Moreover, several proteins of importance in the snail–larval digenean interactions have been identified using these experimental models. Further work using these models may provide much needed information on the development and consequences of larval digenean parasitism in snails.

Acknowledgments This work was supported by Project CGL2005-02321/BOS of Ministerio de Educación y Ciencia (Spain) and FEDER (Europea Union) and by Project PROMETEO 2009/081 of Conselleria d'Educació of Generalitat Valenciana (Valencia, Spain).

References

Adema CM, Arguello DF, Striker SA, Loker ES (1994a) A time-lapse study of interactions between *Echinostoma paraensei* intramolluscan larval stages and adherent hemocytes from *Biomphalaria glabrata* and *Helix aspersa*. J Parasitol 80:719–727

Adema CM, Hertel LA, Loker ES (1997a) Infection with *Echinostoma paraensei* (Digenea) induces parasite-reactive polypeptides in the hemolymph of the gastropod host *Biomphalaria glabrata*. In: Beckage N (ed) Parasite effects on host physiology and behavior. Chapman, New York, pp 77–99

Adema CM, Hertel LA, Miller RD, Loker ES (1997b) A family of fibrinogen-related proteins that precipitates parasite-derived molecules is produced by an invertebrate after infection. Proc Natl Acad Sci USA 94:8691–8696

Adema CM, Sapp KK, Hertel LA, Loker ES (2000) Immunobiology of the relationship of echinostomes with snail intermediate hosts. In: Fried B, Gracyzk TK (eds) Echinostomes as experimental models for biological research. Kluwer, Dordrecht, pp 149–173

Adema CM, Van Deutekom-Mulder EC, Van der Knaap PW, Sminia T (1994b) Schistosomicidal activities of *Lymnaea stagnalis* hemocytes: the role of oxygen radicals. Parasitology 109:479–485

Anderson JW, Fried B (1987) Experimental infection of *Physa heterostropha*, *Helisoma trivolvis*, and *Biomphalaria glabrata* (Gastropoda) with *Echinostoma revolutum* (Trematoda) cercariae. J Parasitol 73:49–54

Ataev G, Coustau C (1999) Cellular response to *Echinostoma caproni* infection in *Biomphalaria glabrata* strains selected for susceptibility/resistance. Dev Comp Immunol 23:187–198

Ataev GL, Dobrovol'skii AA, Avanessian AV, Loker ES (2001) Germinal elements and their development in *Echinostoma caproni* and *Echinostoma paraensei* (Trematoda) miracidia. J Parasitol 87:1160–1164

Ataev G, Dobrovol'skii AA, Fournier A, Jourdane J (1997) Migration and development of mother sporocysts of *Echinostoma caproni* (Digenea, Echinostomatidae). J Parasitol 83:444–453

Ataev GL, Dobrovol'skii AA, Isakova NP (2005) Development of the partenitae infrapopulation of *Echinostoma caproni* (Trematoda: Echinostomatidae). Parazitologiia 39:124–136, In Russian

Ataev G, Fournier A, Coustau C (1998) Comparison of *Echinostoma caproni* mother sporocyst development in vivo and in vitro using *Biomphalaria glabrata* and *B. glabrata* embryonic cell line. J Parasitol 84:227–235

Ataev GL, Isakova NP, Dobrovol'skii AA (2006) Development of mother sporocysts of *Echinostoma caproni* (Trematoda: Echinostomatidae). Parazitologiia 40:47–56, In Russian

Bandstra SR, Fried B, Sherma J (2006) High-performance thin-layer chromatographic analysis of neutral lipids and phospholipids in *Biomphalaria glabrata* patently infected with *Echinostoma caproni*. Parasitol Res 99:414–418

Basch PF, DiConza JJ (1975) Predation by Echinostome rediae upon Schistosome sporocysts in vitro. J Parasitol 61:1044–1047

Beers K, Fried B, Fujino T, Sherma J (1995) Effects of diet on the lipid composition of the digestive gland-gonad complex of *Biomphalaria glabrata* (Gastropoda) infected with larval *Echinostoma caproni* (Trematoda). Comp Biochem Physiol 110B:729–737

Behrens AC, Nollen PM (1992) Responses of *Echinostoma caproni* miracidia to gravity, light, and chemicals. Int J Parasitol 22:673–675

Bouchut A, Coustau C, Gourbal B, Mitta G (2007) Compatibility in the *Biomphalaria glabrata/Echinostoma caproni* model: new candidate genes evidenced by a suppressive subtractive hybridization approach. Parasitology 134:575–588

Bouchut A, Roger E, Coustau C, Gourbal B, Mitta G (2006a) Compatibility in the *Biomphalaria glabrata/Echinostoma caproni* model: potential involvement of adhesion genes. Int J Parasitol 36:175–184

Bouchut A, Sautiere PE, Coustau C, Mitta G (2006b) Compatibility in the *Biomphalaria glabrata/ Echinostoma caproni* model: Potential involvement of proteins from hemocytes revealed by a proteomic approach. Acta Trop 98:234–246

Carney WP (1991) Echinostomiasis – a snail-borne intestinal trematode zoonosis. Southeast Asian J Trop Med Public Health 22:206–211

Christensen NO, Frandsen F, Roushdy MZ (1980) The influence of environmental conditions and parasite-intermediate host-related factors on the transmission of *Echinostoma liei*. Z Parasitenkd 63:47–63

Combes C (1992) Trematodes: Antagonism between species and sterilizing effects on snails in biological control. Parasitology 84:151–175

Coustau C, Gourbal B, Mitta G, Adema C (2009) Echinostomes and snails: exploring complex interactions. In: Fried B, Toledo R (eds) The biology of echinostomes: from the molecule to the community. Springer, New York, pp 35–59

Crompton DWT (1999) How much human helminthiasis is there in the world? J Parasitol 85:397–403

Dar YD, Rondelaud D, Dreyfuss G (2005) Update of fasciolosis-transmitting snails in Egypt (review and comment). J Egypt Soc Parasitol 35:477–490

De Jong-Brink M (1995) How schistosomes profit from the stress responses they elicit in their hosts. Adv Parasitol 35:177–256

DeGaffe G, Loker ES (1998) Susceptibility of *Biomphalaria glabrata* to infection with *Echinostoma paraensei*: correlation with the effect of parasite secretory-excretory products on host hemocyte spreading. J Invertebr Pathol 71:64–72

Dias MLGG, Eiras JC, Machado MH, Souza GTR, Pavanelli GC (2003) The life cycle of *Clinostomum complanatum* Rudolphi, 1814 (Digenea, Clinostomidae) on the floodplain of the high Paraná river, Brazil. Parasitol Res 89:506–508

Dias Neto E, de Souza CP, Rollinson D, Katz N, Pena SD, Simpson AJ (1993) The random amplification of polymorphic DNA allows the identification of strains and species of schistosome. Mol Biochem Parasitol 57:83–88

Dubois G, Nassi H (1977) [Redescription of d'*Apatemon (Apatemon) graciliformis* Szidat, 1928 (Trematoda: Strigeidae) with a preliminary note on its life cycle]. Ann Parasitol Hum Comp 52:507–510

El-Shazly AM, Helmy MM, Haridy FM, el-Sharkawy EM, Morsy TA (2002) *Fasciola* immature stages sought in *Lymnaea* species and *Biomphalaria* species in the water bodies of Dakahlia Governorate. J Egypt Soc Parasitol 32:109–118

Escribano A, Williams T, Goulsson D, Cave RD, Chapman JW, Caballero P (2001) Consequences of interspecific competition on the virulence and genetic competition of nucleopolyhedrovirus in *Spodoptera frugiperla* larvae parasitized by *Chelonus insularis*. Biol Sci Technol 11:649–662

Esteban JG, Muñoz-Antolí C (2009) Echinostomes: systematic and life cycles. In: Fried B, Toledo R (eds) The biology of echinostomes. From the molecule to the community. Springer, New York, pp 1–34

Farag HF, el-Sayad MH (1995) *Biomphalaria alexandrina* naturally infected with *Fasciola gigantica* in Egypt. Trans R Soc Trop Med Hyg 89:36

Flores V, Brugni N (2005) *Notocotylus biomphalariae* n. sp. (Digenea: Notocotylidae) from *Biomphalaria peregrina* (Gastropoda: Pulmonata) in Patagonia, Argentina. Syst Parasitol 61:207–214

Frazer BA, Fried B (1998) Single-species infections of *Echinostoma caproni* cercariae in pulmonate snails and concurrent infections of *E. caproni* and *Echinostoma trivolvis* cercariae in *Biomphalaria glabrata*. Int J Parasitol 28:595–597

Fried B (2001) Biology of echinostomes except *Echinostoma*. Adv Parasitol 49:163–210

Fried B, Graczyk TK (eds) (2000) Echinostomes as experimental models for biological research. Kluwer, Dordrecht, The Netherlands, p 273

Fried B, Graczyk TK (2004) Recent advances in the biology of *Echinostoma* species in the "*revolutum*" group. Adv Parasitol 58:139–195

Fried B, Huffman JE (1996) The biology of the intestinal trematode *Echinostoma caproni*. Adv Parasitol 38:311–368
Fried B, King W (1989) Attraction of *Echinostoma revolutum* cercariae to *Biomphalaria glabrata* dialysate. J Parasitol 75:55–57
Fried B, Reddy A (1999) Effects of snail-conditioned water from *Biomphalaria glabrata* on hatching of *Echinostoma caproni* miracidia. Parasitol Res 85:155–157
Fried B, Frazer BA, Kanev I (1998) Comparative observations on cercariae and metacercariae of *Echinostoma trivolvis* and *Echinoparyphium* sp. J Parasitol 84:623–626
Fried B, Frazer BA, Reddy A (1997a) Chemoattraction and penetration of *Echinostoma trivolvis* and *E. caproni* cercariae in the presence of *Biomphalaria glabrata, Helisoma trivolvis*, and *Lymnaea elodes* dialysate. Parasitol Res 83:193–197
Fried B, Huffman JE, Keeler S, Peoples RC (2009) The biology of the caecal trematode *Zygocotyle lunata*. Adv Parasitol 69:1–40.
Fried B, Idris N, Ohsawa T (1995) Experimental infection of juvenile *Biomphalaria glabrata* with cercariae of *Echinostoma trivolvis*. J Parasitol 81:308–310
Fried B, LaTerra R, Kim Y (2002) Emergence of cercariae of *Echinostoma caproni* and *Schistosoma mansoni* from *Biomphalaria glabrata* under different laboratory conditions. J Helminthol 76:369–371
Fried B, Schafer S, Kim S (1989) Effects of *Echinostoma caproni* infection on the lipid composition of *Biomphalaria glabrata*. Int J Parasitol 19:353–354
Fried B, Scheuermarn S, Jr M (1987) Infectivity of *Echinostoma revolutum* miracidia for laboratory-raised pulmonate snails. J Parasitol 73:1047–1048
Fried B, Schmidt KA, Sorensen RE (1997b) In vivo and ectopic encystment of *Echinostoma revolutum* and chemical excystation of the metacercariae. J Parasitol 83:251–254
Fried B, Sherma J, Rao KS, Ackman RG (1993) Fatty acid composition of *Biomphalaria glabrata* (Gastropoda: Planorbidae) experimentally infected with the intramolluscan stages of *Echinostoma caproni* (Trematoda). Comp Biochem Physiol 104B:595–598
Fryer SE, Oswald RC, Probert AJ, Runham NW (1990) The effect of *Schistosoma haematobium* infection on the growth and fecundity of three sympatric species of bulinid snails. J Parasitol 76:557–563
Fujino T, Takahashi Y, Fried B (1995) A comparison of *Echinostoma trivolvis* and *E. caproni* using random amplified polymorphic DNA analysis. J Helminthol 69:263–264
Fujino T, Wu Z, Nagano I, Takahashi Y, Fried B (1997) Specific primers for the detection of genomic DNA of *Echinostoma trivolvis* and *E. caproni* (Trematoda:Echinostomatidae). Mol Cell Probes 11:77–80
Goodman JD, Panesar TS (1976) *Parasymphylodora manteri* n.sp., and adult trematode from fresh-water pulmonate snail, *Biomphalaria choanomphala choanomphala*, from Lake Victoria, Uganda. Trans Am Microsc Soc 95:204–209
Haas W (2000) The behavioral biology of echinostomes. In: Fried B, Gracyzk TK (eds) Echinostomes as experimental models for biological research. Kluwer, Dordrecht, pp 175–197
Haas W, Beran B, Loy C (2008) Selection of the host's habitat by cercariae: from laboratory experiments to the field. J Parasitol 94:1233–1238
Haas W, Haberl B, Kalbe M, Körner M (1995) Snail-host finding by miracidia and cercariae: chemical host cues. Parasitol Today 11:468–472
Haberl B, Körner M, Spengler Y, Hertel J, Kalbe M, Haas W (2000) Host-finding in *Echinostoma caproni*: miracidia and cercariae use different signals to identify the same snail species. Parasitology 120:470–496
Hertel LA, Adema CM, Loker ES (2005) Differential expression of FREP genes in two strains of *Biomphalaria glabrata* following exposure to the digenetic trematodes *Schistosoma mansoni* and *Echinostoma paraensei*. Dev Comp Immunol 29:295–303
Hertel J, Haberl B, Hamburger J, Haas W (2003) Description of a tandem repeated DNA sequence of *Echinostoma caproni* and methods for its detection in snail and plankton samples. Parasitology 126:443–449

Hertel LA, Stricker SA, Loker ES (2000) Calcium dynamics of hemocytes of the gastropod *Biomphalaria glabrata*: effects of digenetic trematodes and selected bioactive compounds. Invertebr Biol 119:27–37

Huffman JE, Fried B (1990) *Echinostoma* and echinostomiasis. Adv Parasitol 29:215–269

Huizinga HW (1973) *Ribeiroia marini*: Pathogenesis and larval trematode antagonism in the snail, *Biomphalaria glabrata*. Exp Parasitol 33:350–364

Humbert E, Coustau C (2001) Refractoriness of host *hemocytes* to parasite immunosuppressive factors as a putative resistance mechanism in the *Biomphalaria glabrata* - *Echinostoma caproni* system. Parasitology 122:651–660

Hurd H (1990) Physiological and behavioural interactions between parasites and invertebrate hosts. Adv Parasitol 29:272–317

Jeyarasasingam U, Heyneman D, Lim HK, Mansour N (1972) Life cycle of a new echinostome from Egypt, *Echinostoma liei* sp.nov. (Trematoda: Echinostomatidae). Parasitology 65:203–222

Johnson PTJ, McKenzie VJ (2009) Effects of environmental change on helminth infections in amphibians: exploring the emergence of *Ribeiroia* and *Echinostoma* infections in North America. In: Fried B, Toledo R (eds) The biology of echinostomes. From the molecule to the community. Springer, New York, pp 249–280

Johnson PTJ, Sutherland DR, Kinsella JM, Lunde KB (2004) Review of the trematode genus *Ribeiroia* (Psilostomidae): ecology, life history and pathogenesis with special emphasis on the amphibian malformation problem. Adv Parasitol 57:191–253

Jourdane J (1983) Effects of hyperinfestations by *Echinostoma togoensis* Jourdane and Kulo, 1981 on growth and survival of *Biomphalaria pfeifferi* snails. Ann Parasitol Hum Comp 58:103–108

Jourdane J, Mounkassa JB (1986) Topographic shifting of primary sporocysts of *Schistosoma mansoni* in *Biomphalaria pfeifferi* as a result of coinfection with *Echinostoma caproni*. J Invertebr Pathol 48:269–274

Keeler SP, Huffman JE (2009) Echinostomes in the second intermediate host. In: Fried B, Toledo R (eds) The biology of echinostomes: from the molecule to the community. Springer, New York, pp 61–87

Kechemir N, Jourdane J, Mas-Coma S (2002) Life cycle of a new African echinostome species reproducing by parthenogenesis. J Nat Hist 36:1777–1784

King PH, Van As JG (1996) A description of the life stages of *Echinoparyphium elegans* (Trematoda: Echinostomatidae). S Afr J Zool 31:145–153

Kostadinova A, Gibson DI (2000) The systematic of the Echinostomes. In: Fried B, Graczyk TK (eds) Echinostomes as experimental models for biological research. Kluwer, Dordrecht, The Netherlands, pp 31–57

Kostadinova A, Herniou EA, Barret J, Littlewood DTJ (2003) Phylogenetic relationships of *Echinostoma* Rudolphi, 1809 (Digenea: Echinostomatidae) and related genera re-assessed via DNA and morphological analyses. Syst Parasitol 54:159–176

Kuris AM (1980) Effect of exposure to *Echinostoma liei* miracidia on growth and survival of young *Biomphalaria glabrata* snails. Int J Parasitol 10:303–308

Kuris AM, Warren J (1980) Echinostome cercarial penetration and metacercarial encystment as mortality factors for a second intermediate host, *Biomphalaria glabrata*. J Parasitol 66:630–635

Langand J, Morand S (1998) Heritable non-susceptibility in an allopatric host-parasite system: *Biomphalaria glabrata* (Mollusca)-*Echinostoma caproni* (Plathelminth Digenea). J Parasitol 84:739–742

Lardans V, Dissous C (1998) Snail control strategies for reduction of schistosomiasis transmission. Parasitol Today 14:413–417

Laursen JR, Yoshino TP (1999) *Biomphalaria glabrata* embryonic (Bge) cell line supports in vitro miracidial transformation and early larval development of the deer liver fluke, *Fascioloides magna*. Parasitology 118:187–194

Layman LR, Dory AC, Koehnlein KM, Fried B, Sherma J (1996) Nmeasurement of metallic ions in *Biomphalaria glabrata* (Gastropoda) infected with *Echinostoma caproni* (Trematoda) and in uninfected sanils. J Helminthol Soc Wash 63:256–258

Léonard PM, Adema CM, Zhang S-M, Loker ES (2001) Structure of two FREP genes that encode IgSF and fibrinogen domains, with comments on diversity of the FREP gene family in the snail *Biomphalaria glabrata*. Gene 269:155–165

Lie KJ (1973) Larval trematode antagonism: principles and possible application as a control method. Exp Parasitol 33:343–349

Lie KJ (1982) Survival of *Schistosoma mansoni* and other trematode larvae in the snail *Biomphalaria glabrata*. A discussion of the interference hypothesis. Trop Geogr Med 34:111–122

Lie KJ, Basch PF (1966) Life history of *Echinostoma barbosai* sp. n. (Trematoda: Echinosto matidae). J Parasitol 52:1052–1057

Lie KJ, Basch PF (1967) The life history of *Echinostoma paraensei* sp. n. (Trematoda: Echinosto matidae). J Parasitol 53:1192–1199

Lie KJ, Heyneman D (1976) Studies on resistance in snails. 3. Tissue reactions to *Echinostoma lindoense* sporocysts in sensitized and resensitized *Biomphalaria glabrata*. J Parasitol 62:51–58

Lie KJ, Basch PF, Hoffman MA (1967) Antagonism between *Paryphostomum segregatum* and *Echinostoma barbosai* in the snail *Biomphalaria straminea*. J Parasitol 53:1205–1209

Lie KJ, Heyneman D, Jeyarasasingam V, Mansour N, Lee HF, Lee HA, Kostanian N (1975) The life cycle of *Echinoparyphium ralphaudyi* sp. n. (Trematoda: Echinostomatidae). J Parasitol 61:59–65

Lie KJ, Heyneman D, Richards CS (1977) Studies on resistance in snails: interference by nonirradiated echinostome larvae with natural resistance to *Schistosoma mansoni* in *Biomphalaria glabrata*. J Invertebr Pathol 29:118–125

Lim HK, Heyneman D (1972) Intramolluscan inter-trematode antagonism: a review of factors influencing the host-parasite system and its possible role in biological control. Adv Parasitol 10:191–268

Lo CT (1973) Experimental second intermediate hosts of *Echinostoma malayanum* Leiper, 1911. J Parasitol 59:746–747

Loker ES, Adema CM (1995) Schistosomes, echinostomes and snails: comparative immunobiology. Parasitol Today 11:120–124

Loker ES, Bayne CJ (1986) Immunity to trematode larvae in the snail *Biomphalaria*. Zoolog Symp 56:199–220

Loker ES, Hertel LA (1987) Alterations in *Biomphalaria glabrata* plasma induced by infection with the digenetic trematode *Echinostoma paraensei*. J Parasitol 73:503–513

Loker ES, Boston ME, Bayne CJ (1989) Differential adherence of M-line *Biomphalaria glabrata* hemocytes to *Schistosoma mansoni* and *Echinostoma paraensei* larvae, and experimental manipulation of hemocyte binding. J Invertebr Pathol 54:260–268

Loker ES, Cimino DF, Hertel LA (1992) Excretory-secretory products of *Echinostoma paraensei* sporocyst mediate interference with *Biomphalaria glabrata* hemocyte functions. J Parasitol 78:104–115

Loker ES, Coustau C, Ataev GL, Jourdane J (1999) In vitro culture of rediae of *Echinostoma caproni*. Parasite 6:169–174

Maldonado A, Lanfredi R (2009) Echinostomes in the wild. In: Fried B, Toledo R (eds) The biology of echinostomes: from the molecule to the community. Springer, New York, pp 129–145

Maldonado A Jr, Vieira GO, Garcia JS, Rey L, Lanfredi RM (2001) Biological aspects of a new isolate of *Echinostoma paraensei* (Trematoda: Echinostomatidae): susceptibility of sympatric snails and the natural vertebrate host. Parasitol Res 87:853–859

Maldonado A Jr, Vieira GO, Lanfredi RM (2003) *Echinostoma luisreyi* n. sp. (Platyhelminthes: Digenea) by light and scanning electron microscopy. J Parasitol 89:800–808

Marcilla A (2009) Echinostomes: genomics and proteomics. In: Fried B, Toledo R (eds) The biology of echinostomes: from the molecule to the community. Springer, New York, pp 207–228

Marsit CJ, Fried B, Sherma J (2000) Neutral lipids in cercariae, encysted metacercariae, and rediae of *Echinostoma caproni*. J Helminthol 74:365–367

McCarthy AM (1999) The influence of second intermediate host species on the infectivity of metacercarial cysts of *Echinoparyphium recurvatum*. J Helminthol 73:143–145

Meece JK, Nollen PM (1996) A comparison of the adult and miracidial stages of *Echinostoma paraensei* and *E. caproni*. Int J Parasitol 26:37–43

Meyrowitsch D, Christensen NO, Hindsbo O (1991) Effects of temperature and host density on the snail-finding capacity of cercariae of *Echinostoma caproni* (Digenea: Echinostomatidae). Parasitology 102:391–396

Minchella DJ (1985) Host life history variation in response to parasitism. Parasitology 90:205–216

Minchella DJ, LoVerde PT (1981) A cost of increased early reproductive effort in the snail *Biomphalaria glabrata*. Am Nat 118:876–881

Mohamed SH (1992) Ultrastructure aspects of intramolluscan developing cercariae of *Echinostoma liei*. J Egypt Soc Parasitol 22:479–485

Monroy FP, Loker ES (1993) Production of heterogeneous carbohydrate-binding proteins by the host snail *Biomphalaria glabrata* following exposure to *Echinostoma paraensei* and *Schistosoma mansoni*. J Parasitol 79:416–423

Monroy F, Hertel LA, Loker ES (1992) Carbohydrate-binding plasma proteins from the gastropod *Biomphalaria glabrata*: strain specificity and the effects of trematode infection. Dev Comp Immunol 16:355–366

Morgan JAT, Blair D (1995) Nuclear rDNA ITS sequence variation in the trematode genus *Echinostoma*: an aid to establishing relationships within the 37-collar-spine group. Parasitology 111:609–615

Morgan JAT, Blair D (1998a) Relative merits of nuclear ribosomal internal transcribed spacers and mitochondrial CO1 and ND1 genes for distinguishing among *Echinostoma* species (Trematoda). Parasitology 116:289–297

Morgan JAT, Blair D (1998b) Mitochondrial ND1 gene sequences used to identify echinostome isolates from Australia and New Zealand. Int J Parasitol 28:493–502

Mouahid A, Moné H (1988) *Echinoparyphium elegans* (Looss, 1899) (Digenea: Echinostomatidae): the life cycle and redescription of the adult with a revision of the 43-spined members of the genus *Echinoparyphium*. Syst Parasitol 12:149–157

Muñoz-Antolí C, Marin A, Toledo R, Esteban JG (2007) Effect of *Echinostoma friedi* (Trematoda: Echinostomatidae) experimental infection on longevity, growth and fecundity of juvenile *Radix peregra* (Gastropoda: Lymnaeidae) and *Biomphalaria glabrata* (Gastropoda: Planorbidae) snails. Parasitol Res 101:1663–1670

Muñoz-Antoli C, Trelis M, Toledo R, Esteban JG (2006) Infectivity of *Echinostoma friedi* miracidia to different snail species under experimental conditions. J Helminthol 80:323–325

Nassi H (1978) Données sur le cycle biologique de *Ribeiroia marini guadeloupensis* n. ssp., trematoda stérilisant *Biomphalaria glabrata* en Guadaloupe. Acta Trop 35:41–56

Nassi H (1979) Coïncidence entre le blocage précoce de la ponte de *Biomphalaria glabrata* (Gasteropoda: Pulmonata) et la localisation cérébrale des jeunes rédies mères de *Ribeiroia marini guadeloupensis* (Trematoda: Cathaemasiidae). C R Acad Sci III 289:165–168

Nassi H (1980) Données experimentales sur le cycle biologique de *Petasiger caribbensis* n. sp. (Trematoda: Echinostomatidae) parasite larvaire de *Biomphalaria glabrata* en Guadeloupe. Ann Parasitol Hum Comp 55:41–55

Nassi H, Bayssade-Dufour C (1980) Life cycle of *Clinostomum golvani n. sp.* (Trematoda: Clinostomidae) a larval parasite of *Biomphalaria glabrata*, the snail vector of *Schistosoma mansoni* in Guadeloupe. Ann Parasitol Hum Comp 55:527–540, In French

Nassi H, Dupouy J (1988) Étude experimentale du cicle biologique d´*Echinostoma parvocirrus* n. sp. (Trematoda: Echinostomatidae), parasite larvaire de *Biomphalaria glabrata* en Guadeloupe. Ann Parasitol Hum Comp 63:103–118

Nassi H, Pointier JP, Golvan YJ (1979) Bilan d'un essai de contrôle de *Biomphalaria glabrata* en Guadaloupe â l'aide d'un trématode stérilisant. Ann Parasitol Hum Comp 54:185–192

Noda S, Loker ES (1989a) Effect of infection with *Echinostoma paraensei* on the circulating hemocyte population of the host snail *Biomphalaria glabrata*. Parasitology 98:35–41

Noda S, Loker ES (1989b) Phagocytic activity of hemocytes of M-line *Biomphalaria glabrata* snails: effect of exposure to the trematode *Echinostoma paraensei*. J Parasitol 75:261–269

Nowak TS, Loker ES (2005) *Echinostoma paraensei*: differential gene transcription in the sporocyst stage. 109:94–105

Ostrowski de Núñez M (1976) Sobre el ciclo biológico de *Hippocrepis fuelleborni* Travassos y Vogelsang 1930 (Trematoda, Notocotylidae). Physis, Sección B 35:125–130

Ostrowski de Núñez M, Spatz L, González Cappa SM (2003) New intermediate hosts in the life cycle of *Zygocotyle lunata* in South America. J Parasitol 89:193–194

Paul RE, Un VA, Krettli AU, Brey PT (2002) Interspecific competition during transmission of two sympatric malaria parasite species to the mosquito vector. Proc Biol Sci 269:2551–2557

Pechenik JA, Fried B (1995) Effect of temperature on survival and infectivity of *Echinostoma trivolvis* cercariae: a test of the energy limitation hypothesis. Parasitology 111:373–378

Perez MK, Fried B, Sherma J (1994) High-performance thin-layer chromatographic analysis of sugars in *Biomphalaria glabrata* (Gastropoda) infected with *Echinostoma caproni* (Trematoda). J Parasitol 80:336–338

Perez MK, Fried B, Sherma J (1995) Comparison of mobile phases HPTLC qualitative and quantitative analysis, on preadsorbent silica gel plates, of phospholipids in *Biomphalaria glabrata* (Gastropoda) infected with *Echinostoma caproni* (Trematoda). J Planar Chromatogr Mod TLC 7:340–343

Petrie JL, Burg EF 3rd, Cain GD (1996) Molecular characterization of *Echinostoma caproni* and *E. paraensei* by random amplification of polymorphic DNA (RAPD) analysis. J Parasitol 82:360–362

Platt TR, Burnside L, Bush E (2009) The role of light and gravity in the experimental transmission of *Echinostoma caproni* (Digenea: Echinostomatidae) cercariae to the second intermediate host, *Biomphalaria glabrata* (Gastropoda: Pulmonata). J Parasitol 95:512–516

Pointier JP, Jourdane J (2000) Biological control of the snail hosts of schistosomiasis in areas of low transmission: the example of the Caribbean area. Acta Trop 77:53–60

Ponder EL, Fried B, Sherma J (2004) Free-pool amino acids in *Biomphalaria glabrata* infecetd with *Echinostoma caproni* as determined by thin-layer chromatography. J Parasitol 90:665–666

Reddy A, Fried B (1996) In vitro studies on intraspecific and interspecific chemical attraction in daughter rediae on *Echinostoma trivolvis* and *E. caproni*. Int J Parasitol 26:1081–1085

Reddy A, Frazer BA, Fried B, Sherma J (1997) Chemoattraction of *Echinostoma trivolvis* (Trematoda) rediae to lipophilic excretory-secretory products and thin layer-chromatographic analysis of redial lipids. Parasite 4:37–40

Reddy A, Ponder EL, Fried B (2004) Effects of copper sulfate toxicity on cercariae and metacercariae of *Echinostoma caproni* and *Echinostoma trivolvis* and on the survival of *Biomphalaria glabrata* snails. J Parasitol 90:1332–1337

Richard J, Brygoo ER (1978) Life cycle of the trematode *Echinostoma caproni* Richard, 1964 (Echinostomatoidea). Ann Parasitol Hum Comp 53:265–275

Riley EM, Chappell LH (1992) Effect of infection with *Diplostomum spathaceum* on the intestinal defense system of *Lymnaea stagnalis*. J Invertebr Pathol 59:190–196

Sandland GJ, Rodgers JK, Minchella DJ (2007) Interspecific antagonism and virulence in hosts exposed to two parasite species. J Invertebr Pathol 96:43–47

Schell SC (1985) Trematodes of North America, North of Mexico. University Press of Idaho, Moscow, Idaho, p 263

Schmidt KA, Fried B, Haseeb MA (1998) Effect of *Echinostoma caproni* infection on survival, growth, and fecundity of *Biomphalaria glabrata*. J Helminthol Soc Wash 65:95–98

Schneck JL, Fried B (2004) Effects of snail size on encystment of *Echinostoma caproni* in juvenile *Biomphalaria glabrata* (NMRI strain) and observations on survival of infected snails. J Helminthol 78:277–279

Shetty PH, Fried B, Sherma J (1992) Effects of patent *Echinostoma caproni* infection on the sterol composition of the digestive gland-gonad complex of *Biomphalaria glabrata* as determined by gas-liquid chromatography. J Helminthol 66:68–71

Sorensen RE, Curtis J, Minchella DJ (1998) Intraspecific variation in the rDNA ITS loci of 37-collar-spined echinostomes from North America: implications for sequence-based diagnoses and phylogenetics. J Parasitol 84:992–997

Stein PC, Basch PF (1977) Metacercarial cyst formation in vitro of *Echinostoma paraensei*. J Parasitol 63:1031–1040

Sturrock BM, Sturrock RF (1970) Laboratory studies of the host-parasite relationship of *Schistosoma mansoni* and *Biomphalaria glabrata* from St Lucia, West Indies. Ann Trop Med Parasitol 64:357–363

Sullivan JJ (1985) Juvenile snails as hosts for echinostome metacercariae. Southeast Asian J Trop Med Public Health 16:343–344

Sullivan JT (1988) Hematopoiesis in three species of gastropods following infection with *Echinostoma paraensei* (Trematoda: Echinostomatidae). Trans Am Microsc Soc 107:355–361

Thornhill JA, Jones JT, Kusel JR (1986) Increased oviposition and growth in immature *Biomphalaria glabrata* after exposure to *Schistosoma mansoni*. Parasitology 93:443–450

Ueta MT, Deberaldini ER, Da Silva N, De Toledo P (1981) Ciclo biológico de *Paraibatrema inesperata* n. g., n. sp. (Trematoda, Paramphistomidae) a partir de metacercárias desenvolvidas en *Biomphalaria tenagophila* (D'Orbigny, 1835) (Mollusca, Planorbidae). Mem Inst Oswaldo Cruz 76:15–21

Van der Knaap WP, Loker ES (1990) Immune mechanisms in trematode-snail interactions. Parasitol Today 6:175–182

Vergote D, Bouchut A, Sautière PE, Roger E, Galinier R, Rognon A, Coustau C, Salzet M, Mitta G (2005) Characterisation of proteins differentially present in the plasma of *Biomphalaria glabrata* susceptible or resistant to *Echinostoma caproni*. Int J Parasitol 35:215–224

Violante-González J, García-Varela M, Rojas-Herrera A, Gil Guerrero S (2009) Diplostomiasis in cultured and wild tilapia Oreochromis niloticus in Guerrero State, Mexico. Parasitol Res 105:803–807

Wagner SD, Kim Y, Fried B, Sherma J (2001) Thin-layer chromatographic analysis glucose, maltose, and raffinose in *Biomphalaria glabrata* (Gastropoda) infecetd with *Echinostoma caproni* (Trematoda). J Planar Chromatogr Mod TLC 14:459–461

Walker AJ (2006) Do trematode parasites disrupt defence-cell signalling in their snail hosts? Trends Parasitol 22:154–159

White MM, Chejlava M, Fried B, Sherma J (2005) Effects of various larval digeneans on the calcium carbonate content of the shells of *Helisoma trivolvis*, *Biomphalaria glabrata*, and *Physa* sp. Parasitol Res 95:252–255

Yoshino TP, Vasta GR (1996) Parasite-invertebrate host immune interaction. In: Cooper EL (ed) Advances in comparative and environmental physiology, vol 24. Springer, Berlin, pp 125–167

Yousif F, Haroun N (1986) Intramolluscan development of *Echinostoma liei* (Trematoda: Echinostomatidae) in its snail host *Biomphalaria alexandrina*. J Egypt Soc Parasitol 16:127–140

Zakikhani M, Rau ME (1998) Effects of *Plagiorchis elegans* (Digenea: Plagiorchiidae) infection on the reproduction of *Biomphalaria glabrata* (Pulmonata: Planorbidae). J Parasitol 84:927–930

Zakikhani M, Smith JM, Rau ME (2003) Effects of *Plagiorchis elegans* (Digenea: Plagiorchiidae) infection of *Biomphalaria glabrata* (Pulmonata: Planorbidae) on a challenge infection with *Schistosoma mansoni* (Digenea: Schistosomatidae). J Parasitol 89:70–75

Zhang S-M, Loker ES (2003) The FREP gene family in the snail *Biomphalaria glabrata*: additional members, and evidence consistent with alternative splicing and FREP retrosequences. Fibrinogen-related proteins. Dev Comp Immunol 27:175–187

Zhang S-M, Adema CM, Kepler TB, Loker ES (2004) Diversification of Ig superfamily genes in an invertebrate. Science 305:251–254

Zhang S-M, Léonard PM, Adema CM, Loker ES (2001) Parasite responsive IgSF members in the snail Biomphalaria glabrata: characterization of novel genes with tandemly arranged IgSF domains and a fibrinogen domain. Immunogenetics 53:684–694

Zhang S-M, Nian H, Zeng Y, Dejon RJ (2008a) Fibrinogen-bearing protein genes in the snail *Biomphalaria glabrata*: characterization of two novel genes and expression studies during ontogenesis and trematode infection. Dev Comp Immunol 32:1119–1130

Zhang S-M, Zeng Y, Loker ES (2008b) Expression profiling and binding porperties of fibrinogen-related proteins (FREPs), plasma proteins from the schistosome snail host *Biomphalaria glabrata*. Innate Immunol 14:175–189

Chapter 7
Immunobiology of *Biomphalaria*–Trematode Interactions

Timothy P. Yoshino and Christine Coustau

Abstract As the exclusive intermediate host of the human blood fluke *Schistosoma mansoni*, species of the snail *Biomphalaria*, especially *Biomphalaria glabrata*, have been the subjects of numerous studies focused on the immunobiology of parasite–host interactions. With the recent applications of molecular, genomic, proteomic and glycomic approaches to the study of *Biomphalaria*'s immune response to schistosomes and other trematode species, there is now accumulating a wealth of information that is beginning to address the mechanisms underlying these complex parasite–host associations. In this chapter, we attempt to broadly review our past understanding of *Biomphalaria* immunity and integrate recent information on the cellular and molecular events surrounding initial immune responses to larval trematode infections, the consequences of immune interactions and counter-immune strategies used by these parasites.

7.1 Introduction

7.1.1 Why Study Biomphalaria?

Freshwater snails of the genus *Biomphalaria* would be considered just another "pond snail" were it not for the fact that species of this group represent the obligate intermediate host for the human blood fluke *Schistosoma mansoni*, a major causative agent of hepatosplenic schistosomiasis affecting an estimated 83 million people in 54 countries worldwide (Crompton 1999). A cursory PubMed search (http://www.ncbi.nlm.nih.gov/) of the terms "*Biomphalaria*" or "*Australorbis*" (its former taxonomic name) revealed 2,374 and 2,522 published works, respectively, dating back to the late 1940s. The first listed publication for *Biomphalaria* entitled "Susceptibility

T.P. Yoshino (✉)
Department of Pathobiological Sciences, University of Wisconsin,
Madison, WI 53706, USA
e-mail: yoshinot@svm.vetmed.wisc.edu

R. Toledo and B. Fried (eds.), Biomphalaria *Snails and Larval Trematodes*,
DOI 10.1007/978-1-4419-7028-2_7,

of the snail *Biomphalaria boissyi* to infection with certain strains of *S. mansoni*" was by late Dr. Emile A. Malek, one of the pioneers of modern-day medical malacology (Abdel-Malek 1950). For *Australorbis*, the first listed publication was by von Brand and Files (1947) entitled "Chemical and histological observations of *S. mansoni* infections in *Australorbis glabratus*". Of particular relevance to the present chapter, these earliest research topics addressed the differences among different *Biomphalaria* spp. in their susceptibility to *S. mansoni* infection, and the tissue responses evoked in this snail upon larval infection. After more than 60 years since these investigations were conducted, researchers are still continuing to study the underpinnings of schistosome-snail compatibility including the mechanisms underlying host response to infection. Thus, in large part, these earliest works have played (and continue to play) a significant role in the establishment of contemporary studies of molluscan immunobiology with *Biomphalaria* serving as a model organism.

7.1.2 **Biomphalaria/Schistosoma** ***Compatibility and Immunobiological Studies***

As a direct result of this important connection between *Biomphalaria* and its essential role in human blood fluke transmission, many early studies have also investigated the genetic basis of variation in host–parasite compatibility. They were mostly based on experimental infections of inbred laboratory stocks of *Biomphalaria glabrata* originating from allopatric or sympatric populations, and they demonstrated a substantial polymorphism in compatibility (Richards 1975; Richards and Shade 1987). Genetic studies then demonstrated that both snail susceptibility and schistosome infectivity were heritable and could be selected for in the laboratory (Richards and Shade 1987; Richards et al. 1992; Webster et al. 2004). However, subsequent efforts in elucidating the genetic determinism of resistance revealed a variety of underlying mechanisms ranging from a single major locus to a potentially high number of loci (Richards 1975; Richards et al. 1992; Webster and Davies 2001). In addition, snail strains that were naturally incompatible or selected for high incompatibility with one particular schistosome strain were often highly compatible with other schistosome strains (Richards and Shade 1987; Webster and Woolhouse 1998; Webster et al. 2004). These complex results support the idea that *B. glabrata/S. mansoni* compatibility relies on complex genotype-by-genotype interactions (Basch 1975; Theron et al. 2008), where the success or failure of infection of an individual snail by an individual parasite larvae would depend on the concordance or discordance of potentially numerous host and parasite genes (and their products). The random loss or fixation of alleles at loci that participate in this genotype-by-genotype interaction may largely explain the diversity in compatibility status or compatibility determinism observed in the laboratory (Theron et al. 2008).

Although compatibility in the *B. glabrata/S. mansoni* system appears to be a joint trait of the host and parasite, the terms "susceptibility/resistance" can be used when focusing on variation in compatibility among different hosts toward the same strain of parasite. Similarly, the term "infectivity" refers to parasite phenotypes measured against the same strain of host. Host and parasite laboratory strains differing in their resistance/susceptibility or infectivity status, respectively, have been particularly useful for functional approaches investigating snail immunity and trematode immune evasion processes. A vast body of literature has accumulated on the fundamental mechanisms of the immune defense system in this snail that represents an important comparative model for other lophotrochozoan invertebrates (Loker et al. 2004; Raghavan and Knight 2006). In the following review, we attempt to summarize broadly our current knowledge of the immune defense system of *Biomphalaria* spp., specifically its role in regulating interactions with larval trematodes. Although overlap with previous reviews of this topic (Loker and Bayne 1986; Bayne and Yoshino 1989; Yoshino and Vasta 1996; Adema and Loker 1997; Bayne 2009; Coustau et al. 2009) is unavoidable, we attempt to integrate recent information on the molecular events surrounding initial establishment of snail–trematode relationships and immune–counterimmune interactions involved in continued maintenance of larval infections.

7.2 Back to Basics: Components of *Biomphalaria* Innate Immunity

Like other gastropod molluscs, the internal defense or immune system of *Biomphalaria* is composed of both cellular and humoral components, which may act either alone or in concert with each other to recognize and respond to invading microbes or parasites (Yoshino et al. 2001; Lockyer et al. 2004a; Loker et al. 2004; Bayne 2009). Circulating phagocytic cells, termed hemocytes or amebocytes, are found in the hemolymph of the snail and represent the primary effector cells involved in host reactions to invading pathogens, including larval trematodes. The soluble portion of hemolymph, termed plasma (Bayne et al. 1980a), serves as the hemocyte "transportation" system and contains humoral immune factors functioning to facilitate/enhance cell-mediated responses or alone, as immune effectors directly interacting with and eliminating pathogens.

In the last decade, considerable efforts have been made to start characterizing genes or gene products involved in *B. glabrata* immune responses, thanks to the development of methodologies allowing identification and/or comparisons of expressed genes in nonmodel organisms. Gene discovery studies have incorporated various technical approaches including differential display reverse transcriptase, suppression subtractive hybridization, random sequencing of ESTs or ORESTES, and gene expression profiling through microarray studies. Regardless of the methodology used, theses studies have focused on identifying transcripts from circulating hemocytes (Mitta et al. 2005), transcripts differentially expressed before and after an immune challenge (Miller et al. 2001; Raghavan et al. 2003; Lockyer et al. 2000, 2004b, 2007; Nowak et al. 2004; Guillou et al. 2007a; Hanelt et al. 2008), or differentially expressed transcripts in snails differing in their trematode susceptibilities (Schneider and Zelck 2001; Bouchut et al. 2006a, 2007; Lockyer et al. 2008; Adema et al. 2010; Ittiprasert et al. 2010). Proteomic studies also have been conducted with the aim of characterizing *B. glabrata* proteins that were differentially represented in the plasma or hemocytes from parasite-susceptible and -resistant *B. glabrata* (Vergote et al. 2005; Bouchut et al. 2006b). Taken together, these studies have yielded a multitude of immune-relevant candidates belonging to several major functional categories (see Bayne 2009 ; Coustau et al. 2009 for review) such as genes involved in nonself recognition, cell–cell or cell–matrix adhesion, proteases and protease inhibitors, antimicrobial proteins, immune regulators, antioxidants, as well as components of the three major signaling pathways involved in immune responses (Toll, Imd, MAPK). Thus, although it is clear that the genetic machinery exists in *Biomphalaria* for "building" an effective system of internal defense, the challenge now is to incorporate this vast genetic information into a unified picture of the mechanisms underlying cellular immune recognition and reactivity to invading trematode larvae and other pathogens.

7.2.1 Cellular Components

Hemocytes represent the primary effector cells comprising the snail's internal defense system (Fig. 7.1). Based on the morphology of adherent hemocytes under in vitro conditions, *B. glabrata* hemocytes were initially reported as having two

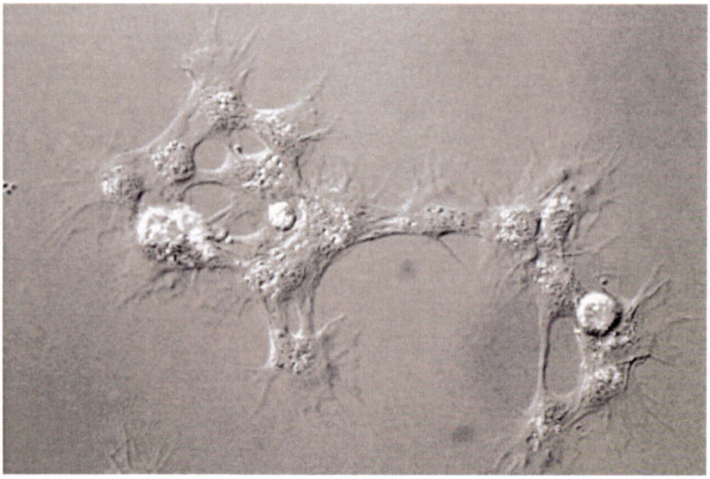

Fig. 7.1 Photomicrograph of live hemocytes of *Biomphalaria glabrata* adhering to and spreading over a glass coverslip surface. Circulating granular hemocytes (granulocytes) represent the primary effector cells involved in the snail's internal defense including parasitic encapsulation reactions. (Nomarski DIC)

distinctive cell types; granulocytes and nongranular hyalinocytes (Cheng 1975; Cheng and Auld 1977), although more recently, variations in the degree of granularity (Martins-Souza et al. 2009), cell size (Matricon-Gondran and Letocart 1999; Martins-Souza et al. 2009), ultrastructural features (Jeong et al. 1983; Matricon-Gondran and Letocart 1999), and cell surface/biochemical markers (Granath and Yoshino 1983; Yoshino and Granath 1985; Martins-Souza et al. 2006) have prompted modifications of the snail's cell classification scheme. Regardless, it is clear that *Biomphalaria* hemocytes represent a morphologically and functionally heterogeneous population(s) whose cellular composition can change with snail age, external factors (e.g., infection), and assay methodology.

Production of hemocytes in *B. glabrata*, and presumably other *Biomphalaria* spp., appears to take place primarily in an "amebocyte-producing organ" (APO; Lie et al. 1975; Jeong et al. 1983) located in the anterior pericardial wall separating the pericardial sac from the mantle cavity, although other sites of blood cell formation also have been proposed (Souza and Andrade 2006). The APO is responsible for maintaining steady-state populations of circulating cells and, importantly, is responsive to introduction of foreign invaders or substances. For example, hematopoeitic activity has been shown to be specifically stimulated by infection with selected trematode species/strains (Lie et al. 1976) or by injection of various foreign materials including schistosome larval extracts (Sullivan et al. 2004; Sullivan 2007) or cell signal-activating agents (Salamat and Sullivan 2009) (Fig. 7.2).

7.2.2 Immune Recognition

Discrimination of "self" from "nonself" by circulating hemocytes of *Biomphalaria* is believed to be accomplished through pattern recognition receptors (PRRs; Janeway and Medzhitov 2002), in which "invariant" cell-associated or soluble receptors bind to fixed foreign chemical structures, referred to as pathogen-associated molecular patterns or PAMPs (Janeway 1989). However, recent discovery of a family of highly diversified pathogen-binding proteins, the fibrinogen-related proteins or Freps (Adema et al. 1997; Zhang et al. 2004), now provides new evidence that *Biomphalaria,* and other snail species (Zhang et al. 2009), possess a structurally and functionally related group of divergent molecules with potential wide-spread pathogen-binding activities (Zhang et al. 2008). Together, with similar findings in arthropod and echinoderm species (e.g., Brites et al. 2008; Buckley et al. 2008; Dong and Dimopoulos 2009; Schmucker and Chen 2009), the notion of invertebrate PRRs as an "invariant" system of immunorecognition is certainly being challenged (see Sect. 7.2.3). Whether "invariant" or "highly diversified", PRR binding to their corresponding PAMPs at the hemocyte surface is presumed to generate intracellular molecular signals that result in a variety of hemocyte responses. Chief among these responses are (a) phagocytosis – internalization and destruction of small particles (e.g., microbes, protozoan parasites), (b) encapsulation – multicellular

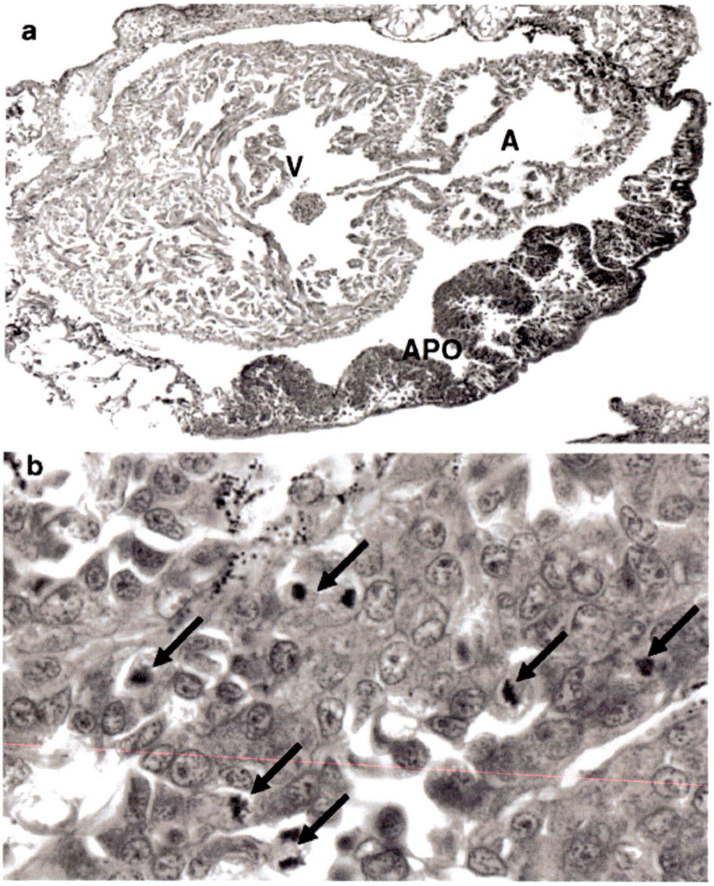

Fig. 7.2 Histological sections of the amebocyte-producing organ (APO), a major source of circulating hemocytes in *Biomphalaria glabrata*. This organ is located in the anterior pericardial wall that forms part of the pericardial sac surrounding the heart (*V* ventricle; *A* atrium) (**a**). Stimulation by specific larval trematode infection or injection with the parasite extracts may result in hyperplastic expansion of this tissue due to increased mitotic activity of hemocyte progenitor cells (**b**, *arrows*). Photomicrographs kindly provided by Dr. John T. Sullivan (Univ. of San Francisco)

reactions in response to foreign bodies or organisms too large to be phagocytosed, and (c) cytotoxicity – extracellular killing of foreign cells or tissue grafts by contact cytolysis. Although phagocytosis represents an extremely important hemocyte function for eliminating microbial infections in *Biomphalaria* and other gastropods (Michelson 1975; Bayne and Fryer 1994; Yoshino and Granath 1985; Martins-Souza et al. 2009), emphasis in this chapter will be given to the encapsulation/cytotoxic responses of hemocytes due to their role in immune defense against larval trematode infections. The role of hemocytes in immune encapsulation of trematodes will be covered in detail later in this chapter.

7.2.3 Cell-Associated and Plasma-Associated Pattern Recognition Receptors

What do we currently know about pattern recognition receptors (PRRs) in *Biomphalaria*? In the context of this chapter, the designation "PRR" is functionally defined as a hemocyte molecule (usually a protein) that binds to constitutively expressed (fixed) structures displayed by the parasite, and elicits in the cell either directly (through a membrane-associated PRR) or indirectly (through soluble PRRs) immune-related responses. In *B. glabrata*, hemocyte-associated adhesion-type molecules, including α and ß integrin-like proteins, cadherins, dermatopontins, and matrilin-like proteins (Davids and Yoshino 1998, 1999; Mitta et al. 2005; Bouchut et al. 2006a), may be functioning as "self" recognition receptors promoting cellular interactions with host extracellular matrix, cytoskeletal or snail plasma proteins, and thereby influencing hemocyte behavior such as cell migration, chemotactic targeting, or cell–cell adhesion during encapsulation or wound healing responses. Other *Biomphalaria* PRRs with known or presumed microbial- or larval trematode-binding capabilities include short- and long-form peptidoglycan recognition proteins (PGRPs), a gram-negative bacterial binding protein (GNBP) (Zhang et al. 2007), LPS-binding proteins/bactericidal permeability increasing proteins (LBP/BPIs) (Mitta et al. 2005; Guillou et al. 2007a), and the large family of fibrinogen-related proteins or Freps (Adema et al. 1997; Zhang et al. 2004, 2008). Unfortunately, except in the cases of ß integrins and selected Frep members (Stout et al. 2009), these adhesion molecules in *Biomphalaria* were identified solely via transcriptomic studies, and therefore little functional data are available for these proteins in this snail, including their role(s) in mediating pathogen recognition. Structurally, however, Zhang et al. (2007) showed that while the GNBP gene appeared to encode a secreted molecule, the three long-form PGRPs lack a signal peptide and thus are presumed to be intracellular in their action. These results are consistent with *Drosophila melanogaster* PGRPs in which the *Dm*PGRP-LE gene encodes three isoforms including two truncated forms, none of which have a signal peptide. Interestingly, these *Dm*PGRP-LE can function by two distinct mechanisms, either as extracellular or intracellular receptors (Kaneko et al. 2006). Based on this information, it seems likely that *Biomphalaria* PGRPs also differ in their location and function (Zhang et al 2007), and future studies are clearly needed to determine whether these PGRPs can function as extracellular receptors.

One class of recognition receptors believed to be of particular importance in initial hemocyte interactions with larval trematodes are the nonenzymatic carbohydrate-binding proteins known as lectins (Endo et al. 2006; Vasta et al. 2007). Given that complex carbohydrates (CHOs) represent major components of the larval tegument (Lehr et al. 2007, 2008; Peterson et al. 2009) and glycoproteins released during miracidial transformation (Wu et al. 2009), it has been hypothesized that a CHO-based recognition system may be pivotal in determining hemocyte responsiveness and ultimately the compatibility phenotype in *B. glabrata*/*S. mansoni* systems (Yoshino and Vasta 1996; Adema and Loker 1997; Loker et al. 2004; Bayne 2009).

This notion is supported by functional data demonstrating involvement of larval CHO in hemocyte-mediated cytotoxicity and phagocytosis (Boswell and Bayne 1985; Fryer et al. 1989), induction of reactive oxygen species (ROS) production in hemocytes (Hahn et al. 2000; Lacchini et al. 2006; Zelck et al. 2007; Humphries and Yoshino 2008), and in vitro adherence of host plasma or hemocyte proteins to the *S. mansoni* sporocyst tegument or released proteins (Johnston and Yoshino 1996, 2001; Castillo and Yoshino 2002; Castillo et al. 2007).

Prominent among lectin-based PRR proteins in *Biomphalaria* are the previously mentioned Freps. This is a highly diversified molecular family comprised of 314 unique gene sequences encoding >200 different proteins (Stout et al. 2009). The rapid and selective upregulation of specific Frep transcripts in response to parasite challenge in resistant snails and the ability of Freps to precipitate larval secretory proteins may be functionally tied to the snail's antiparasite response. This response appears to be quite selective, as only certain Frep members are stimulated/upregulated in reaction to a given trigger (e.g., differing responses to schistosome vs. echinostome infections; Zhang et al. 2008). However, the precise mechanisms of interaction between Freps, hemocytes, and parasite larvae, and how such interactions are related to susceptibility status of a given snail–trematode association still remain to be answered in this system (Bayne 2009).

In addition to Freps, other lectins have been identified in *B. glabrata* including selectin-like proteins (Duclermortier et al. 1999; Guillou et al. 2004), a C-type lectin related to mannose-binding protein (EST only; Bouchut et al. 2007), and a tandem-repeat galectin (Yoshino et al. 2008). The selectins, galectin, and Freps appear to be produced in hemocytes, but only the galectin has been shown to be associated with the hemocyte surface membrane. In functional studies, both recombinant galectin and certain native Freps bind to trematode antigens (CHOs), but differ in their primary sugar-binding specificities (gal/lac/galNAc for galectin vs. fuc for Freps) and divalent cation-dependencies (Adema et al. 1997; Yoshino et al. 2008). In summary, although the in vivo functions of these hemocyte-associated lectins are still a matter of conjecture, current evidence supports a central role of lectin-like PRRs in parasite recognition and initiation of anti-parasite responses: (1) they are synthesized in, and for some, expressed on the surface of hemocytes, (2) they are able to bind/recognize larval CHOs, (3) expression of some are selectively induced by pathogen-challenge, and (4) they possess significant gene homology with vertebrate lectins with known immune function.

7.2.4 Other Immune-Related Plasma Components

In addition to soluble PRRs, *Biomphalaria* plasma contains other proteins with potential immune activity. For example, hydrolytic/cytolytic enzymes such as lysozyme, acid phosphatases, or proteinases (Cheng et al. 1978; McKerrow and Doenhoff 1988; Cheng and Dougherty 1989; Bouchut et al. 2007) or putative antimicrobial proteins (Mitta et al. 2005; Guillou et al. 2007a; Hanelt et al. 2008) also may be responsible for pathogen killing. Moreover, glycanases such as identified endo-ß mannanases 1 and 2 in *B. glabrata* plasma (Vergote et al. 2005) could function to modify CHO structures at the larval surface rendering the parasite recognizable by lectin PRRs.

7.3 Immune Interactions Between Trematodes and *Biomphalaria*

7.3.1 Encapsulation and Larval Killing In Vivo and In Vitro

In snails exhibiting a resistance phenotype hemocytic encapsulation represents their primary response to invasion by parasitic helminths (Bayne et al. 2001; Bayne 2009; Carton et al. 2005). Typically this involves circulating hemocytes being attracted to and infiltrating the area surrounding the parasite, followed by the formation of a multilayered cellular capsule leading eventually to larval death (Sullivan and Richards 1981; Loker et al. 1982). As mentioned earlier, the capacity of *Biomphalaria* hemocytes to recognize and encapsulate various trematode species, most notably *S. mansoni* and several species of echinostomes, is genetically determined, which has led to the isolation and cultivation of a number of genetically inbred *B. glabrata* strains selected for their susceptibilities to a given isolate of *S. mansoni* or *Echinostoma* spp. (Langand and Morand 1998; Ataev and Coustau 1999; Lewis et al. 2001). Although such inbreeding greatly reduces the genetic variability seen in natural populations (Lockyer et al. 2004a; Theron et al. 2008), these inbred snail lines continue to serve as valuable research resources for investigating host–parasite immune mechanisms under controlled laboratory conditions.

7.3.1.1 Methods of Investigating Immune Interactions

In addition to the establishment of inbred *B. glabrata* strains differing in parasite compatibilities, several other important advancements in the study of hemocyte-mediated encapsulation responses have been made. Of particular significance is the development of in vitro methods for manipulating both the relevant early larval stages of trematode development and elements of the host's immune system. These included the axenic isolation of free-living miracidia of *S. mansoni* and their induced transformation to the first intramolluscan developmental stage, the primary or mother sporocyst (Voge and Seidel 1972), and, based on the manipulation of sporocysts in culture, the design and development of an in vitro assay for evaluating snail hemocyte–sporocyst interactions at the cellular and molecular levels (cell-mediated cytotoxicity or CMC assay; Bayne et al. 1980a, b) (Fig. 7.3). Since these initial reports, there have been many refinements in the cultivation methods and media for maintaining both trematode larvae and snail cells (Ivanchenko et al. 1999;

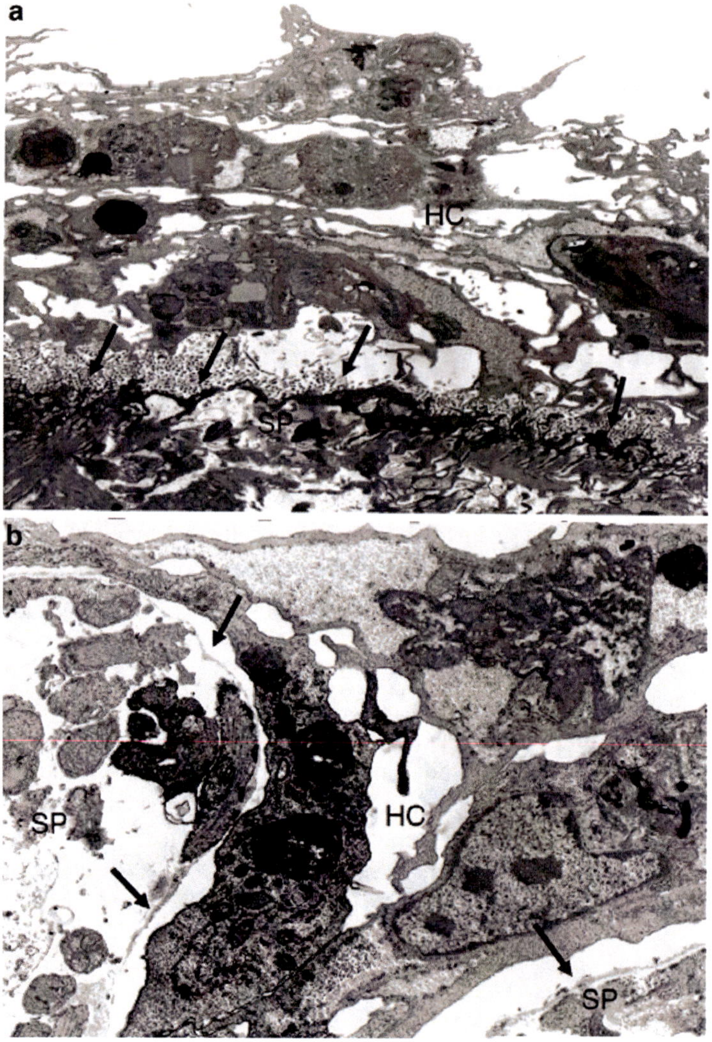

Fig. 7.3 Electron photomicrographs of hemocytic encapsulation of *Schistosoma mansoni* primary sporocysts (SP) by circulating hemocytes (HC) from a schistosome-susceptible (**a**) and -resistant (**b**) strain of *Biomphalaria glabrata*. After 24 h of in vitro hemocyte-sporocyst cocultivation, note the presence of an intact tegument (*arrows*) at the interface with susceptible snail hemocytes (**a**), compared to complete destruction of the tegument and internal damage associated with sporocysts during resistant snail encapsulation reactions (**b**). *Arrows* in panel (**b**) point to the basal lamina, normally located just proximal to the larval tegument, which is missing

Bixler et al. 2001), as well as adaptations of the CMC assay (Bayne 2009) to explore the basic mechanisms underlying *Biomphalaria* susceptibility and resistance to trematode infections.

Another important discovery significantly contributing to the schistosome-snail research field was isolation of the first, and presently only, molluscan cell line from 5-day old embryos of *B. glabrata* (Hansen 1976). The Bge (*B. glabrata* embryonic)

cell line has been an invaluable tool in investigations of snail–trematode interactions by significantly promoting in vitro *S. mansoni* larval development from the miracidium through multiple daughter sporocyst (Yoshino and Laursen 1995) and cercarial (Ivanchenko et al. 1999) stages in coculture experiments. Due to its unique capacity to "nonlethally" encapsulate *S. mansoni* sporocysts Bge cells have been used to investigate adhesion receptors (Duclermortier et al. 1999; Castillo and Yoshino 2002; Castillo et al. 2007), cell signaling pathways (Humphries et al. 2001; Humphries and Yoshino 2006), influence of parasite ESP on snail cell gene expression (Coustau et al. 2003) and snail cell influences on larval *S. mansoni* gene expression (Coppin et al. 2003; Vermeire et al. 2004). Although not ontologically related to hemocytes, the behavioral and molecular attributes shared between these cell types have provided important insights into potential mechanisms involved in larval immune responses (Yoshino et al. 1999; Lockyer et al. 2004a).

7.3.2 Recognition and Initial Triggering of Encapsulation Responses

As alluded to earlier, miracidial penetration of the host snail is accompanied by protein secretion from the lateral and apical penetration glands and release of a barrage of larval proteins during the subsequent transformation to the primary sporocyst stage. It is during this time that the chemical signals "announcing" the parasite's presence is conveyed to the host, and to which the host response is initiated. Although previously referred to as excretory-secretory proteins/products (ESP; Lodes and Yoshino 1989; Guillou et al. 2007b), the term larval transformation proteins or LTPs has been suggested to more accurately reflect this complex group of proteins. This was based on recent findings that the majority of identified *S. mansoni* LTPs lacked signal peptides or characteristics of nonclassical secretion, and that a significant source of released proteins appeared to originate from degenerating ciliated epidermal plates (Wu et al. 2009; Peterson et al. 2009). However, regardless of the origins of ESP/LTPs, the fact that many of these proteins, as well as those expressed at the sporocyst tegument, are heavily glycosylated (Nyame et al. 2002; Lehr et al. 2008; Peterson et al. 2009) suggests that terminal CHOs may play a prominent role in determining hemocytic responses to larval infection. Since ESP continues to be used in the literature, for the purpose of this review the terms ESP and LTP are used synonymously.

Recent analyses of specific glycotopes expressed at the tegumental surface of *S. mansoni* sporocysts and in ESP/LTPs indicate an array of fucosylated CHO structures built around the terminal lacdiNAc or LDN [GalNAcß1-4GlcNAcß1] backbone of expressed schistosome oligosaccharides (Peterson et al. 2009). These include F-LDN, LDN-F, F-LDN-F and LDN-DF. What is not known is whether or not these prominent sugar determinants actually serve as ligands for snail hemocyte or plasma PRRs. Using a cell-adhesion assay modified from Bayne et al. (1984), Castillo and Yoshino (2002) found that the binding of Bge cells to *S. mansoni* sporocysts was inhibited by fucoidan, a sulfated poly-fucose, and monoclonal antibodies to LDN-F, suggesting the involvement of fucosylated sugars in Bge cell–sporocyst binding. Fucoidan also inhibited hemocyte–sporocyst binding interactions (unpublished observation), suggesting a functional association between fucosylated sugar–snail cell-binding interactions. The observation that precipitation reactions involving Frep-*Echinostoma* ESP were inhibited by L-fucose further suggests that Freps may represent candidate PRRs responsible for mediating initial encapsulation reactions involving hemocytes in resistant (R) *B. glabrata* snails, or blocking of hemocyte reactivity in susceptible snails by interaction with selected fucosylated proteins (Adema et al. 1997; see Sect. 7.4.1). Although other candidate molecules as either host PRRs or potential parasite ligands may be functioning in this system, current evidence favors a lectin-CHO-based recognition system operating to regulate parasite–host immune interactions.

7.3.2.1 Immune Signaling Pathways

In order for hemocytes to mount antiparasite responses, receptor–ligand binding must produce an initial stimulus that triggers the transduction of intracellular signals resulting in the eventual activation of target genes and/or proteins responsible for generating hemocytic effector functions. In *Biomphalaria*, as pointed out earlier, CHOs represent an important class of "ligand" molecules with demonstrated capacity to induce a spectrum of cell-mediated reactions including substrate adhesion, phagocytosis, encapsulation, ROS production and the like. However, until only recently have studies been initiated to determine putative signaling pathways linking together receptor–ligand binding and cell effector function. The involvement of Erk-like mitogen-activated protein kinase (Erk-MAPK) and protein kinase C-like (PKC) signaling proteins in regulating in vitro cell adherence and spreading (Humphries et al. 2001) and phagocytosis (Humphries and Yoshino 2003) was first demonstrated in *Biomphalaria* using the Bge cell line as a model system. That hemocytes of *B. glabrata* and a related snail, *Lymnaea stagnalis*, also rely, at least in part, on MAPK/PKC signaling mechanisms during phagocytosis (Humphries and Yoshino 2003; Plows et al. 2005; Zelck et al. 2007) indicates a functional connection between Bge cells and hemocytes regarding the commonality of their signaling pathways. In addition, studies on *Lymnaea* hemocytes have revealed important aspects of intracellular signaling that relate to *Biomphalaria*: (1) signaling pathways (especially PKC/Erk) can be modulated by specific CHOs (Plows et al. 2005; Lacchini et al. 2006) and (2) PKC signaling may represent a primary pathway functionally linking CHO recognition and the generation of hemocyte ROS (Lacchini et al. 2006). Since treatment of *L. stagnalis* hemocytes with schistosome-related CHOs (neoglycoproteins fuc-BSA and gal-BSA) resulted in downregulation of PKC/Erk signaling (Plows et al. 2005), this set the stage for subsequent studies investigating signaling pathways involved in regulating CHO–ROS interactions (Humphries and Yoshino 2008) and the role of larval ESP/LTPs in modulating this system of intracellular signals (see Sect. 7.4.2.4 for further discussion). To date, although still a relatively new area of investigation, evidence favors the involvement of the PKC/MAPK (Humphries and Yoshino 2003, 2006, 2008;

Zelck et al. 2007; Zahoor et al. 2008) and the phosphatidylinositol 3-kinase (PI3-K) (Zelck et al. 2007; Humphries and Yoshino 2008; Salamat and Sullivan 2009) pathways of signal transduction in hemocytes of *B. glabrata*.

Probably due to the fact that little attention has been given to *Biomphalaria* antimicrobial response, information regarding the well-known Toll and Imd immune pathways are scarce. A Genbank database search reveals the existence of a transcript predicted to correspond to NF-kappa B (Genbank accession #FJ711166), a key element in both pathways (Tanji et al. 2007), but to our knowledge this result has not yet been published. Intriguing also is the fact that PGRPs, known to serve as receptor complexes in the IMD pathways (Maillet et al. 2008), have been identified in *B. glabrata* (see Sect. 7.2.3). Although it is likely that Toll- and Imd-related pathways exist in *B. glabrata*, there is no evidence to date suggesting that they may be involved in regulating parasite infection.

7.3.2.2 Consequences of Parasite Recognition: Effector Mechanisms

Depending on the specific intracellular signaling pathway(s) triggered by receptor–ligand binding, hemocytes may respond by either upregulating or downregulating its effector cell activity. As discussed below, "purposeful" downregulating of hemocyte function may represent an important anti-immune mechanism by some trematode species or strains. However, in *Biomphalaria* snails that efficiently fight against a trematode infection, cellular activation usually leads to larval killing. How is this accomplished? Using an in vitro CMC assay, Bayne and colleagues (2001) have demonstrated that ROS, especially H_2O_2, and reactive nitrogen species (RNS) elaborated from hemocytes of *S. mansoni*-resistant (R) strains of *B. glabrata* are responsible for the killing of encapsulated sporocysts. In a series of follow-up studies, summarized in a recent review by Bayne (2009), they have implicated a Cu/Zn-superoxide dismutase (SOD1) as a key enzyme in the oxidative killing activity of R hemocytes based on their experimental findings: (a) SOD1 gene expression and enzyme activity were higher in R hemocytes compared to those of a susceptible (S) *B. glabrata* strain (Goodall et al. 2004), and this correlated directly with greater H_2O_2 productivity in the R strain (Bender et al. 2005); (b) their finding that *B. glabrata* SOD is comprised of three alleles, of which one (SOD1 B allele) was significantly associated with the R snail phenotype (Goodall et al. 2006), and (c) data indicating that SOD1 B allelic expression was greater in R hemocytes and significantly correlated with snail resistance (Bender et al. 2007). Based on their findings, it was suggested that snail differences in SOD1 hemocyte expression is causally linked to the observed susceptibility and resistance in their host–parasite strains system. Clearly the connection between elevated SOD enzyme expression/activity and cytotoxic potential involves the augmentation of H_2O_2 formation from superoxide during the respiratory burst, and eloquently explains the ultimate outcome of R phenotype expression (Bayne 2009).

Whether other effector mechanisms exist in *Biomphalaria* to deal with trematode infections is currently not known. Recent transcriptome analyses of R versus S strain snails (whole body or hemocytes) reveal a variety of proteases and protease inhibitors that may play a role by directly distrupting tegumental membrane proteins resulting in fatal breaching of the parasite surface, or indirectly by modifying surface proteins or glycoproteins uncovering hidden molecular patterns that are recognized by host hemocytes or plasma PRRs (e.g., Lockyer et al. 2007; Hamlet et al. 2008). Other hydrolytic enzymes are described both within hemocytes and in plasma, but their role in antitrematode responses is unclear (Cheng 1975). Their activity more likely is focused on destruction of microbial invaders that are more sensitive to these lysosomal-type enzymes (Cheng et al. 1978).

7.4 Anti-immune Counter-Measures by Parasites

7.4.1 Mimicry or Immunological Smokescreen

Together with antigenic variation and immune suppression or modulation, molecular mimicry is considered as one of the major process of parasite immune evasion (Damian 1989, 1997). The idea that schistosomes avoid immune recognition partly through molecular mimicry processes originates from early studies showing cross-reactivity between polyclonal antibodies raised against *Biomphalaria* hemolymph and schistosome miracidia and/or sporocysts (Yoshino and Cheng 1978; Yoshino and Bayne 1983). In these studies, both miracidia and in vitro-transformed mother sporocysts were derived in media devoid of snail host material, preventing the possibility that host antigens may have been acquired by the parasite during intra-host development. These results therefore showed that miracidial and sporocysts stages of *S. mansoni* constitutively harbor snail-like antigens (Yoshino and Cheng 1978; Yoshino and Bayne 1983). A reciprocal study confirmed the existence of mimicked antigens, as it showed that antibodies raised against schistosome miracidia cross-reacted with hemocytes or soluble plasma from *B. glabrata* (Bayne and Stephens 1983).

Although the nature of these shared antigens was unknown, previous results pointed out the possible importance of carbohydrates as shared moieties. Investigations using carbohydrate-binding lectins showed that some lectins were binding both parasite larval stages (miracidia and mother sporocysts) and soluble plasma components (Stanislawski et al. 1976; Yoshino et al. 1977). Using a monoclonal antibody to a schistosome oligosaccharide epitope, Dissous et al. (1986) also detected cross-reactivity between glycoproteins on miracidia surface and a 39-kDa glycoprotein from *B. glabrata*. More recently, studies that focused on the structural characterization of *N*-glycans from *S. mansoni* soluble egg antigens confirmed that cross-reacting species represent about 5% of the total glycans (Lehr et al. 2007), further supporting the potential importance of carbohydrate moieties as shared epitopes participating in molecular mimicry in this host–parasite system (Lehr et al. 2008; Peterson et al. 2009). As pointed out by Damian (1989), the concept of molecular mimicry refers both to the fact that parasites or pathogens constitutively express shared moieties with their host (probably through convergent evolution), and that this molecular sharing benefits the parasite by preventing their recognition as nonself. However, evidence for such a beneficial effect is scarce in the *Biomphalaria-Schistosoma* system, and the question of molecular mimicry as a major schistosome immune evasion process continues to be debated (Yoshino and Boswell 1986; Yoshino and Vasta 1996; Adema and Loker 1997; Bayne 2009).

Another possible mechanism involving glycoproteins participating in schistosome immune protection has been reported recently. As an approach to gain insights into the molecular determinant of *B. glabrata/S. mansoni* compatibility, Roger et al. (2008a) compared the proteomes of two strains of *S. mansoni* that were compatible and incompatible with a particular strain of *B. glabrata.* In their initial investigation, several mucin-like molecules were identified among differentially expressed proteins. Subsequent characterization of this set of proteins revealed a high degree of molecular polymorphism (Roger et al. 2008b) resulting from their being encoded by an extensive multigenic family whose individual members frequently recombine and are transcribed as multiple splice variants (Roger et al. 2008c). Because these polymorphic mucins, designated *Sm*PoMuc, are only expressed in the larval stages that interact with the snail host and are released during miracidial transformation as ESP, it has been suggested that these highly polymorphic and glycosylated proteins potentially may bind snail PRRs, such as snail Freps, thus serving as an immunological smokescreen functioning to overwhelming the snail immune system (Roger et al. 2008b, c). Assuming that CHOs represent the ligands (PAMPs) for lectin-like freps and other lectin PRRs, the diversity of sugar structures associated with larval N- and O-linked glycans (Peterson et al. 2009) provides an even greater repertoire of potential divergent structures for each SmPoMuc. Current efforts to begin identifying the glycan structures associated with larval mucins and other glycoproteins, and their reactive lectin counterparts are now in progress.

7.4.2 *Direct Interference with Immune Effector Cell Function*

7.4.2.1 Echinostome Versus Schistosome: Mechanisms of Survival

In addition to the possible avoidance of recognition through molecular mimicry or by "overwhelming" the immune system via the release of polymorphic glycoproteins such as the *Sm*PoMuc, larval trematodes also can influence the host cellular response by directly modulating or interfering with hemocyte activity/function. This is most evident in the case of echinostome infections of *B. glabrata* in which it has been shown that hemocytes from infected snails lose adherence and phagocytic capabilities in vitro, as well as their larval encapsulation responsiveness (Loker 1994; Adema and Loker 1997; Humbert and Coustau 2001; Coustau et al. 2009). Early larval elaboration of ESP/LTP appears to be the source of hemocyte-modulating molecules (Loker et al. 1992), providing additional evidence for the importance of these parasite products in the initial establishment echinostome infections. By comparison, *S. mansoni* also appears to modulate hemocyte defense functions, although *S. mansoni* disruption of *B. glabrata* hemoocyte responses is much less aggressive than interference observed in other trematodes such as echinostomes (Guillou et al. 2007b; Bayne 2009; Coustau et al. 2009).

Past in vitro studies showed that *S. mansoni* sporocysts or their ESP could affect hemocyte motility (Lodes and Yoshino 1990), intracellular protein synthesis (Lodes et al. 1991) and phagocytosis (Fryer and Bayne 1990). At the time of these early studies, the molecular characterization of active ESP factors had not developed beyond determination of the molecular weight of protein fractions, and therefore the identity of putative immunomodulatory factors remained unknown. However, with recent proteomic studies of *S. mansoni* and *E. caproni* ESP/LTP, protein candidates with possible immunomodulating activities are now being identified (Guillou et al. 2007b; Wu et al. 2009). For example, a calreticulin (CRT) has been identified in the ESP proteome of *S. mansoni*. CRTs are well-conserved proteins that control Ca^{2+} homeostasis and are therefore involved in many cellular functions, including the Ca-dependent cellular adhesion and spreading processes. As in a parasitoid wasp, in which CRT has been shown to inhibit host hemocyte spreading, Guillou et al. (2007b) hypothesized that secretion of the *S. mansoni* CRT could modify the extracellular concentration of Ca^{2+} and prevent the hemocyte encapsulation process. In another proteomic study (Wu et al. 2009), several other Ca^{2+}-binding proteins with potential immune modulatory activity were identified. One of particular interest was a calcium-binding dynein light chain sharing significant homology with the fish (*Apinephelus*) and human PIN or protein inhibitor of nitric oxide (NO) synthase, an essential enzyme involved in NO formation and NO-mediated larval killing by *B. glabrata* hemocytes (Hahn et al. 2001b). Other candidate proteins included a phospholipid-binding protein and an annexin, both of which are capable of membrane binding and cell entry, and affecting various cell signaling activities (Wu et al. 2009).

7.4.2.2 Antioxidant Molecules in ESP/LTP Associated with Larval Protection

Parasite survival relies, at least in part, on their ability to detoxify the ROS present within their normal host environment or produced as a consequence of their host's immune response. Trematodes are exposed to ROS throughout their life cycle, including their intramolluscan stages. As mentioned earlier, hydrogen peroxide (H_2O_2) and NO are known to be important in parasite killing (Hahn et al. 2001a, b), and for this reason, a number of complementary studies have investigated the presence or the activity of antioxidant molecules in larval *S. mansoni*. Early investigations have shown that *S. mansoni* ES products/proteins (=LTPs; Wu et al. 2009) released during the miracidium-to-mother sporocyst transformation could inhibit superoxide production in snail hemocytes (Connors and Yoshino 1990). Molecular separation of crude ESP/LTP resulted in the isolation of a 108-kDa polypeptide fraction capable of scavenging both exogenously produced and *B. glabrata* hemocyte-derived superoxide (O_2^-) anions (Connors et al. 1991). In another study, Zelck and von Janowsky (2004) targeted three key antioxidant enzymes, SOD, glutathione peroxidase (GPx) and glutathione-*S*-transferase (GST), and showed that their transcripts were expressed in the sporocyst stages of *S. mansoni*. Although the potential secretion of these enzymes was not studied, their increase in expression upon in vitro exposure to ROS or to snail hemocytes strongly suggested that they were involved in ROS detoxification (Zelck and von Janowsky 2004). Using a similar approach, Vermeire and Yoshino (2007) targeted three peroxiredoxin genes (Prx1, Prx2, Prx3) that are key enzymes in the glutathione (GSH)-thioredoxin redox pathway (Sayed and Williams 2004). They showed that Prx1 and Prx2 transcripts were upregulated during early *S. mansoni* sporocyst development, and that enzyme proteins were expressed within the apical papillae of miracidia and the sporocyst tegument, as well as in secretions released during in vitro larval transformation. Further evidence that these enzymes may function in the protection of *S. mansoni* sporocysts was provided by the fact that removal of Prx1 and 2 from larval ES products by immunoabsoption significantly reduced its ability to breakdown exogeneous H_2O_2 (Vermeire and Yoshino 2007).

Another series of studies suggesting the importance of antioxidant molecules consisted of proteomic analyses of ESP/LTP released during in vitro transformation of *S. mansoni* miracidia to mother sporocysts (Guillou et al. 2007b; Wu et al. 2009). Guillou and coworkers (2007b) identified three major groups of proteins in the ESP of *S. mansoni* sporocysts: antioxidant enzymes, glycolytic enzymes and calcium-binding proteins. The secreted antioxidant enzymes were Cu/Zn SOD and GST. The Cu–Zn SOD catalyzes the dismutation of superoxide anion (O_2^-) into H_2O_2, and as a result of its peroxidative activity, SOD is able to utilize its own dismutation product, H_2O_2, as a substrate catalyzing further breakdown of H_2O_2 to H_2O and O_2 (Kim and Kang 1997). Thus, it appears that *S. mansoni* Cu/Zn SOD could represent a first line of cellular antioxidant defense through its peroxidative activity, resulting in H_2O_2 inactivation and production of either H_2O/O_2 or •OH (via the Fenton rxn). •OH has been shown to be considerably less toxic than H_2O_2 for sporocysts (Hahn et al. 2001a). A second group of antioxidant enzymes identified in proteomic studies was the GSTs including GST26, GST28 and GSTomega (Guillou et al. 2007b; Wu et al. 2009). GSTs neutralize cytotoxic by-products of lipid peroxidation arising from ROS acting on the cell membrane (Tew and Ronai 1999).

In addition to antioxidant enzymes, *S. mansoni* ESP/LTPs were found to contain a number of glycolytic enzymes, for example, triosephosphate isomerase (TPI), a glyceraldehyde-3-phosphate dehydrogenase (GAPDH) and an NAD-dependent malate dehydrogenase, possibly acting as reducing agents. Although the presence of glycolytic enzymes may be surprising in an extracellular environment, these enzymes have been reported in ES products from cercarial or adult schistosomes (Knudsen et al. 2005; Perez-Sanchez et al. 2006). TPI and GAPDH are involved in two successive reactions of the glycolysis involving the reduction of a molecule of NAD^+ to NADH. The NADH generated can be used as a reducing agent to reduce oxidized glutathione (GSSG) accumulating due to the GST activity, thus restoring intracellular pools of reduced glutathione (GSH). These results strongly support the view that these enzymes of glycolysis may be involved in antioxidant stress response by contributing to the synthesis of NADH involved in the glutathione redox cycle

(Kum-Tatt et al. 1975; Guillou et al. 2007b). In addition to the aforementioned antioxidant proteins, Wu et al. (2009) also identified several additional redox/antioxidant molecules in their shotgun proteomics screen including a thioredoxin, H_2O_2-scavenger Prx, and hydroxyacylglutathione hydrolase. This study therefore largely confirmed the existence of the antioxidant proteins reported in previous studies and completed the identification of this parasite's antioxidant arsenal.

7.4.2.3 Role of Endogenous Antioxidant Enzyme Systems Against External Oxidative Stress

Although elaboration of potentially protective antioxidant proteins during the first hours of larval development following snail infection may be achieved by ESP/LTPs release during larval transformation, maintaining an intact and functional system throughout the development within the snail also would be of critical importance. This notion was recently explored in a study by Mourão et al. (2009a) in which RNA interference (RNAi) was used to knockdown transcript and protein expression of several antioxidants in *S. mansoni* sporocysts to determine whether endogenous larval antioxidants functioned to protect parasites from external oxidative stress. Results showed that sporocysts treated with double-stranded (ds)RNA for GST26, GST28, Prx and GPx exhibited higher sensitivity to the toxic effects of sublethal H_2O_2 treatment compared to larvae treated with control green fluorescent protein (GFP) dsRNA. Similarly, antioxidant dsRNA-treated sporocysts were more readily killed by susceptible NMRI hemocytes than GFP dsRNA-treated larvae in in vitro CMC assays, demonstrating for the first time a sporocyst-protective function of endogenous antioxidants against external oxidative stress, including that produced during hemocytic encapsulation reactions.

7.4.2.4 Interference with Signaling Pathways

As discussed previously, larval trematode entry into its snail host may result either in stimulating the immune system (serving as "danger" signals; Bianchi 2007) or suppressing immunity through various hypothetical mechanisms including molecular mimicry (Damian 1989), compatibility polymorphism (Roger et al. 2008c) or by direct modulation of hemocyte reactivity. Although these general hypotheses of parasite–host immune interactions have been recognized, the specific molecular mechanism(s) by which either immune stimulation or suppression occurs is still poorly understood. One hypothesis for parasite-mediated suppression suggests that molecules (mainly glycoproteins) comprising ESP/LTPs released during early intramolluscan development can affect hemocyte function by modulation of cell signaling pathways (Walker 2006). This may be accomplished, in theory, by one of the following two mechanisms: (1) binding to a specific receptor that transmits a negative (downregulating) signal in cells or (2) molecules directly interferring with receptors involved in the cell's signaling network. These notions are based on earlier findings that ESP from echinostome larvae can directly interfere with *B. glabrata* hemocyte functions including phagocytosis and encapsulation reactions (Loker 1994; Loker et al. 1992; Coustau et al. 2009), and in *S. mansoni*, the inhibition of hemocyte motility and intracellular superoxide production (Lodes and Yoshino 1990; Connors and Yoshino 1990) by in vitro ESP treatment.

More recently, this hypothesis has taken on broader support as a result of experiments designed to assess the effect of ESP/LTPs on various signaling pathway molecules involved in the regulation of ROS and RNS production in hemocytes. For example, Erk and p38 MAPK/PKC-associated H_2O_2 release from hemocytes stimulated by PMA or BSA-gal was significantly reduced in the presence of ESP indicating that parasite products were either nonstimulatory, were preventing peroxide production/release or were metabolizing/scavenging H_2O_2 as it was being released from hemocytes (Humphries and Yoshino 2008). An earlier finding that larval ESP contained the H_2O_2-scavenging proteins Prx 1 and 2 (Vermeire and Yoshino 2007) implies the latter explanation. However, given that secreted schistosome proteins are highly glycosylated (Lodes and Yoshino 1989; Wu et al. 2009), direct binding of CHOs associated with larval ESP to hemocyte receptors also could be transducing a negative or inhibitory signal thus blocking specific biochemical or cellular activities (Walker 2006). Recently, Zahoor et al. (2008) demonstrated that *S. mansoni* ESP and fixed whole sporocysts significantly impaired Erk signaling in susceptible (S) *B. glabrata* hemocytes, but not in those of resistant (R) snails, implying a direct differential effect of ESP/sporocyst binding on the Erk MAPK signaling pathway in cells of R and S strain snails. In a recent follow-up study, production of NO in *B. glabrata* hemocytes was shown to be regulated through the Erk MAPK pathways, and that NO production could be induced in R strain hemocytes, but not those of S snails (Zahoor et al. 2009). Since *S. mansoni* sporocysts have been shown to be sensitive to NO-mediated killing in vitro (Hahn et al. 2001b), the data suggest that component(s) in ESP may be differentially regulating NO-based cytotoxicity through manipulation of the Erk signaling pathway. Current evidence supports the involvement of the PKC-MAPK pathway in signaling hemocyte–larval interactions and its modulation by larval CHOs. However, given the considerable contributions made by investigators to this parasite–host system, a cohesive picture of the linkages between specific CHO recognition by hemocytes, the signals propagated as a consequence of "recognition," and the effector cell response (ROS generation) has yet to be created.

7.5 Conclusions and Future Directions

7.5.1 General Conclusions

Although our current understanding of *Biomphalaria/Schistosoma* interactions remains fragmentary, the data accumulated so far provide evidence for highly complex interactions. When considering the extraordinary diversification of both putative parasite-binding molecules of the host, such as the Freps, and parasite-secreted *Sm*PoMuc proteins, one can only envision the diverse and complex molecular mechanisms underlying the success or failure of an infection. Adding to the complexity of trematode–snail encounters, the importance of posttranslational modifications such as protein glycosylation and phosphorylation, regulatory enzymes such as protease/protease inhibitors and signal transduction networks also must be recognized as essential parts of this host–parasite molecular dialogue. This is particularly true of signaling pathways involved in snail immunity. In contrast to other invertebrates where the importance of the Toll and Imd pathways has clearly been demonstrated, in studies to date on *Biomphalaria,* strongest evidence mainly points to the involvement of the MAPK- and PKC-related pathways in the immune response. Taken together, results of recent studies support the idea that the innate immune system of *Biomphalaria* may differ substantially from what is known in other invertebrate models (e.g., *Drosophila*, *Caenorabditis*, *Manduca*) belonging to the Ecdysozoa phylum. In addition, because most invertebrate immune studies have focused on antimicrobial responses, relatively little is known about antiparasitic responses, especially those underlying regulatory aspects. It is anticipated, however, that continued efforts to elucidate snail–trematode immunobiological interactions will surely provide us with novel and important information on invertebrate immunity and host–pathogen interactions.

7.5.2 Critical Research Areas for Future Progress

With the dawning of the genomic and now postgenomic eras, new technological approaches in molecular biology, cell biology and biochemistry are now providing increased opportunities to address previously intractable questions associated with trematodes and their snail–host interactions. That being the case, however, there are still many methodological approaches and experimental tools that have yet to be developed to ensure continued future progress in this important research field. Here are some examples.

7.5.2.1 Achievements of Genome Sequencing Efforts

The complete *S. mansoni* genome has now been sequenced and partially annotated, and the first comprehensive analysis recently has been published (Berriman et al. 2009). Together with the sequencing of the *Schistosoma japonicum* genome (The Schistosoma japonicum Genome Sequencing and Functional Analysis Consortium et al. 2009), these represent the first completely sequenced genomes from lophotrochozoan species. Although the initial genomic analysis of *S. mansoni* was primarily orientated toward predicting novel drug targets (Berriman et al. 2009), the availability of this comprehensive dataset now paves the way for future postgenomic studies investigating other biological aspects such as host–parasite relationships.

Regarding *Biomphalaria* genome sequencing, efforts were initiated in 2001 by an international consortium (http://biology.unm.edu/biomphalaria-genome/index.html; Raghavan and Knight 2006). With funding support from the National Human Genome Research Institute, a *B. glabrata* (BBO2 Brazilian strain) BAC library was generated (Adema et al. 2006), and the sequencing of the entire *B. glabrata* genome was initiated through a joint UNM-BRI-TIGR effort currently coordinated by the Washington University Genome Sequencing Center (http://genome.wustl.edu/genomes/view/biomphalaria_glabrata/) (Raghavan and Knight 2006).

To our knowledge, genome sequencing at Washington University is still in progress and an anticipated completion date is not known at this time. One of the major difficulties is the large size of *B. glabrata* genome (930 Mbases), and it is anticipated that annotation will require substantial additional time and efforts. However, such data are crucial to future research developments if we are to begin to fully understand the interactions of *Biomphalaria* spp. and *S. mansoni* at the molecular level.

7.5.2.2 Investigate Host–Parasite Interactomes

Because of methodological constraints, investigations have so far focused on *Biomphalaria* immune response or *Schistosoma* immune evasion strategies. One possibly fruitful line of research is now to directly investigate *Biomphalaria* /*Schistosoma* interactomes. Current proteomic facilities are now able to analyze molecular complexes, thereby identifying interactions of specific proteins in complex mixtures (Holzmuller et al. 2008). In addition, genomic, transcriptomic and proteomic data on *Biomphalaria* and *Schistosoma* should soon be exhaustive enough to permit computational searches for predicted protein–protein or protein-CHO interaction networks between host and parasite (Cui et al. 2009) and facilitate interpretation or integration of proteomic results. Such approaches should greatly facilitate the identification of proteins playing a key role in host–parasite molecular dialogue.

7.5.2.3 Development of Tools for Functional Studies

Currently there exists only one molluscan cell line, the Bge cell line from *B. glabrata*. As pointed out earlier, this cell line has been a useful model for studying genes involved in basic cellular mechanisms such as adhesion, CHO-binding interactions, larval encapsulation and the like. Generation of additional snail cell lines, especially those that are ontologically related to hemocytes, would be a valued resource for future studies on snail internal defense mechanisms. In addition, currently there are no cell lines available for any parasitic platyhelminth, including the schistosome species. This lack of a schistosome cell line has greatly hindered development of transgenic/protein expression systems, as well as approaches to address functionality of parasite genes in their interactions with the snail or mammalian host.

Since validation of gene function is a primary aim of molecular/genomic studies, another major drawback for the *Biomphalaria/Schistosoma* research community has long been the unavailability of gene manipulation techniques (Yoshino et al. 2010). The development of RNA interference technology, although promising, still requires further refinement to improve on its sensitivity, knockdown efficacy and reproducibility. RNAi has been successfully used in both *Biomphalaria* (Zhang et al. 2005; Jiang et al 2006) and larval *Schistosoma* (Boyle et al. 2003; Dinguirard and Yoshino 2006; Mourão et al. 2009a). However, as shown in *S. mansoni*, the success of RNAi gene expression knockdown is highly variable and target gene- or DNA sequence-dependent, and accurate interpretation of results requires particular attention to details of experimental design and execution (Mourão et al. 2009b). A significant future achievement would be the ability to create stable transgenic lines of *Biomphalaria* snails carrying specific "knock-in" or "knock-out" gene characteristics. Although the general technology to reach this goal is available, development of the specific methods and approaches applicable to snails are still lacking. Perhaps, incorporating the Bge cell line in the initial development and testing of transgenic approaches (Yoshino et al. 1998) for eventual application to whole animals should be considered.

7.5.2.4 Address the Question of Interaction Polymorphism

One of the major challenges in deciphering the molecular events underlying host–parasite compatibility in natural populations will be to address the question of host and parasite polymorphism. As mentioned in the Introduction section of this chapter, interactions between *Biomphalaria* spp. and *S. mansoni* populations are extremely variable and are likely determined by complex genotype-by-genotype interactions. Future elucidation of molecular interactions underlying compatibility/incompatibility in laboratory strains should, in the future, be validated in natural snail populations exhibiting a diversity of genotypes. Because improvements and refinements in genomic, transcriptomic and proteomic analysis techniques are advancing at an extremely rapid pace, it seems highly probable that, in the near future, host-parasite interactions may be analyzed at the individual level. Previous identification of key protein families using laboratory strains should facilitate targeted investigations exploring their diversity or variability in host and parasite individuals from natural populations.

References

Abdel-Malek E (1950) Susceptibility of the snail *Biomphalaria boissyi* to infection with certain strains of *Schistosoma mansoni*. Am J Trop Med Hyg 30:887–894

Adema CM, Loker ES (1997) Specificity of digenean-mollusc association. In: Fried B, Graczyk TK (eds) Advances in trematode biology. CRC Press, Boca Raton, pp 229–263

Adema CM, Hertel LA, Miller RD, Loker ES (1997) A family of fibrinogen-related proteins that precipitates parasite-derived molecules is produced by an invertebrate after infection. Proc Natl Acad Sci U S A 94:8691–8696

Adema CM, Luo MZ, Hanelt B, Hertel LA, Marshall JJ, Zhang SM, DeJong RJ, Kim HR, Kudrna D, Wing RA, Soderlund C, Knight M, Lewis FA, Caldeira RL, Jannotti-Passos LK, Carvalho Odos S, Loker ES (2006) A bacterial artificial chromosome library for *Biomphalaria glabrata*, intermediate snail host of *Schistosoma mansoni*. Mem Inst Oswaldo Cruz 101 (Suppl 1):167–177

Adema CM, Hanington PC, Lun CM, Rosenberg GH, Aragon AD, Stout BA, Lennard Richard ML, Gross PS, Loker ES (2010) Differential transcriptomic responses of Biomphalaria glabrata (Gastropoda, Mollusca) to bacteria and metazoan parasites, *Schistosoma mansoni* and Echinostoma paraensei (Digenea, Platyhelminthes). Mol Immunol 47:849–860

Ataev GL, Coustau C (1999) Cellular response to *Echinostoma caproni* infection in *Biomphalaria glabrata* strains selected for susceptibility/resistance. Dev Comp Immunol 23:187–198

Basch PF (1975) An interpretation of snail-trematode infection rates: specificity based on concordance of compatible phenotypes. Int J Parasitol 5:449–452

Bayne CJ (2009) Successful parasitism of vector snail *Biomphalaria glabrata* by the human blood fluke trematode *Schistosoma mansoni*: a 2009 assessment. Mol Biochem Parasitol 165:8–18

Bayne CJ, Fryer SE (1994) Phagocytosis and invertebrate opsonins in relation to parasitism. Ann NY Acad Sci 712:162–177

Bayne CJ, Stephens JA (1983) *Schistosoma mansoni* and *Biomphalaria glabrata* share epitopes: antibodies to sporocysts bind host snail hemocytes. J Invertebr Pathol 42:221–223

Bayne CJ, Yoshino TP (1989) Determinants of compatibility in mollusc-trematode parasitism. Am Zool 29:399–407

Bayne CJ, Buckley PM, DeWan PC (1980a) Macrophage-like hemocytes of resistant *Biomphalaria glabrata* are cytotoxic for sporocysts of *Schistosoma mansoni* in vitro. J Parasitol 66:413–419

Bayne CJ, Buckley PM, DeWan PC (1980b) *Schistosoma mansoni*: cytotoxicity of hemocytes from susceptible snail hosts for sporocysts in plasma from resistant *Biomphalaria glabrata*. Exp Parasitol 50:409–416

Bayne CJ, Loker ES, Yui MA, Stephens JA (1984) Immune-recognition of *Schistosoma mansoni* primary sporocysts may require specific receptors on *Biomphalaria glabrata* hemocytes. Parasite Immunol 6:519–528

Bayne CJ, Hahn UK, Bender RC (2001) Mechanisms of molluscan host resistance and of parasite strategies for survival. Parasitology 123:S59–S67

Bender RC, Broderick EJ, Goodall CP, Bayne CJ (2005) Respiratory burst of *Biomphalaria glabrata* hemocytes: *Schistosoma mansoni*-resistant snails produce more extracellular H_2O_2 than susceptible snails. J Parasitol 91:275–279

Bender RC, Goodall CP, Blouin MS, Bayne CJ (2007) Variation in expression of *Biomphalaria glabrata* SOD1: a potential controlling factor in susceptibility/resistance to *Schistosoma mansoni*. Dev Comp Immunol 3:874–878

Berriman M, Haas BJ, LoVerde PT, Wilson RA, Dillon GP, Cerqueira GC, Mashiyama ST, Al-Lazikani B, Andrade LF, Ashton PD, Aslett MA, Bartholomeu DC, Blandin G, Caffrey CR, Coghlan A, Coulson R, Day TA, Delcher A, DeMarco R, Djikeng A, Eyre T, Gamble JA, Ghedin E, Gu Y, Hertz-Fowler C, Hirai H, Hirai Y, Houston R, Ivens A, Johnston DA, Lacerda D, Macedo CD, McVeigh P, Ning Z, Oliveira G, Overington JP, Parkhill J, Pertea M, Pierce RJ, Protasio AV, Quail MA, Rajandream MA, Rogers J, Sajid M, Salzberg SL, Stanke M, Tivey AR, White O, Williams DL, Wortman J, Wu W, Zamanian M, Zerlotini A, Fraser-Liggett CM, Barrell BG, El-Sayed NM (2009) The genome of the blood fluke *Schistosoma mansoni*. Nature 7253:352–358

Bianchi M (2007) DAMPs, PAMPs and alarmins: all we need to know about danger. J Leuk Biol 81:1–5

Bixler LM, Lerner JP, Ivanchenko M, McCormick RS, Barnes DW, Bayne CJ (2001) Axenic culture of *Schistosoma mansoni* sporocysts in low O_2 environments. J Parasitol 87:1167–1168

Boswell CA, Bayne CJ (1985) *Schistosoma mansoni*: lectin-dependent cytotoxicity of hemocytes from susceptible host snails, *Biomphalaria glabrata*. Exp Parasitol 60:133–138

Bouchut A, Roger E, Coustau C, Gourbal B, Mitta G (2006a) Compatibility in the *Biomphalaria glabrata/Echinostoma caproni* model: potential involvement of adhesion genes. Int J Parasitol 36:175–184

Bouchut A, Roger E, Coustau C, Gourbal B, Mitta G (2006b) Compatibility in the *Biomphalaria glabrata/Echinostoma caproni* model: potential involvement of proteins from hemocytes revealed by a proteomic approach. Acta Trop 98:234–246

Bouchut A, Coustau C, Gourbal B, Mitta G (2007) Compatibility in the *Biomphalaria glabrata/ Echinostoma caproni* model: new candidate genes evidenced by a suppressive subtractive hybridization approach. Parasitology 134:575–588

Boyle JP, Wu XJ, Shoemaker CB, Yoshino TP (2003) Using RNA interference to manipulate endogenous gene expression in *Schistosoma mansoni* sporocysts. Mol Biochem Parasitol 128:205–215

Brites D, McTaggart S, Morris K, Anderson J, Thomas K, Colson I, Fabbro T, Little TJ, Ebert D, Du Pasquier L (2008) The Dscam homologue of the crustacean *Daphnia* is diversified by alternative splicing like in insects. Mol Biol Evol 25:1429–1439

Buckley KM, Terwilliger DP, Smith LC (2008) Sequence variations in 185/333 messages from the purple sea urchin suggest posttranslational modifications to increase immune diversity. J Immunol 181:8585–8594

Carton Y, Nappi AJ, Poirie M (2005) Genetics of anti-parasite resistance in invertebrates. Dev Comp Immunol 29:9–32

Castillo MG, Yoshino TP (2002) Carbohydrate inhibition of *Biomphalaria glabrata* embryonic Bge cell adhesion to primary sporocysts of *Schistosoma mansoni*. Parasitology 125:513–525

Castillo MG, Wu XJ, Dinguirard N, Nyame AK, Cummings RD, Yoshino TP (2007) Surface membrane proteins of *Biomphalaria glabrata* embryonic cells bind fucosyl determinants on the tegumental surface of *Schistosoma mansoni* sporocysts. J Parasitol 93:832–840

Cheng TC (1975) Functional morphology and biochemistry of molluscan phagocytes. Ann NY Acad Sci 266:343–379

Cheng TC, Auld KR (1977) Hemocytes of the pulmonate gastropod *Biomphalaria glabrata*. J Invertebr Pathol 30:119–122

Cheng TC, Dougherty WJ (1989) Ultrastructural evidence for the destruction of *Schistosoma mansoni* sporocysts associated with elevated lysosomal enzyme levels in *Biomphalaria glabrata*. J Parasitol 75:928–941

Cheng TC, Guida VG, Gerhart PL (1978) Aminopeptidase and lysozyme activity levels and serum protein concentrations in *Biomphalaria glabrata* Mollusca challenged with bacteria. J Invertebr Pathol 32:297–302

Connors VA, Yoshino TP (1990) In vitro effect of larval *Schistosoma mansoni* excretory-secretory products on phagocytosis-stimulated superoxide production in hemocytes from *Biomphalaria glabrata*. J Parasitol 76:895–902

Connors VA, Lodes MJ, Yoshino TP (1991) Identification of *Schistosoma mansoni* sporocyst excretory-secretory anti-oxidant molecule and its effect on superoxide production by *Biomphalaria glabrata* hemocytes. J Invertebr Pathol 58:387–395

Coppin JF, Lefebvre C, Caby S, Cocquerelle C, Vicogne J, Coustau C, Dissous C (2003) Gene expression changes in *Schistosoma mansoni* sporocysts induced by *Biomphalaria glabrata* embryonic cells. Parasitol Res 89:113–119

Coustau C, Mitta G, Dissous C, Guillou F, Galinier R, Allienne JF, Modat S (2003) *Schistosoma mansoni* and *Echinostoma caproni* excretory-secretory products differentially affect gene expression in *Biomphalaria glabrata* embryonic cells. Parasitology 127:533–542

Coustau C, Gourbal B, Mitta G, Adema C (2009) Echinostomes and snails: exploring complex interactions. In: Fried B, Toledo R (eds) The biology of echinostomes, from the molecule to the community. Springer, New York, pp 35–59

Crompton DWT (1999) How much helminthiasis is there in the world? J Parasitol 85:397–403

Cui T, Zhang L, Wang XZ, He ZG (2009) Uncovering new signaling proteins and potential drug targets through the interactome analysis of *Mycobacterium tuberculosis*. BMC Genomics 10:118

Damian RT (1989) Molecular mimicry: parasite evasion and host defense. Curr Top Microbiol Immunol 145:101–115

Damian R (1997) Parasite immune evasion and exploitation: reflections and projections. Parasitology 115:S169–S175

Davids BJ, Yoshino TP (1998) Integrin-like RGD-dependent binding mechanism involved in the spreading response of circulating molluscan phagocytes. Dev Comp Immunol 22:39–53

Davids BJ, Yoshino TP (1999) Cloning of a ß integrin subunit cDNA from an embryonic cell line derived from the freshwater mollusc, *Biomphalaria glabrata*. Gene 228:213–223

Dinguirard N, Yoshino TP (2006) Potential role of a CD36-like class B scavenger receptor in the binding of modified low-density lipoprotein acLDL to the tegumental surface of *Schistosoma mansoni* sporocysts. Mol Biochem Parasitol 146:219–230

Dissous C, Grzych JM, Capron A (1986) *Schistosoma mansoni* shares a protective oligosaccharide epitope with fresh-water and marine snails. Nature 323:443–445

Dong Y, Dimopoulos G (2009) *Anopheles* fibrinogen-related proteins provide expanded pattern recognition capacity against bacteria and malaria parasites. J Biol Chem 284:9835–9844

Duclermortier P, Lardans V, Serra E, Trottein F, Dissous C (1999) *Biomphalaria glabrata* embryonic cells express a protein with a domain homologous to the lectin domain of mammalian selectins. Parasitol Res 85:481–486

Endo Y, Takahashi M, Fujita T (2006) Lectin complement system and pattern recognition. Immunobiology 211:283–293

Fryer SE, Bayne CJ (1990) *Schistosoma mansoni* modulation of phagocytosis in *Biomphalaria glabrata*. J Parasitol 76:45–52

Fryer SE, Dykes-Hoberg M, Bayne CJ (1989) Changes in plasma opsonization of yeast after isolation of *Biomphalaria glabrata* in small volumes of water. J Invertebr Pathol 54:275–276

Goodall CP, Bender RC, Broderick EJ, Bayne CJ (2004) Constitutive differences in Cu/Zn superoxide dismutase mRNA levels and activity in hemocytes of *Biomphalaria glabrata* (Mollusca) that are either susceptible or resistant to *Schistosoma mansoni* Trematoda. Mol Biochem Parasitol 137:321–328

Goodall CP, Bender RC, Brooks JK, Bayne CJ (2006) *Biomphalaria glabrata* cytosolic copper/zinc superoxide dismutase SOD1 gene: association of SOD1 alleles with resistance/susceptibility to *Schistosoma mansoni*. Mol Biochem Parasitol 147:207–210

Granath WO, Yoshino TP (1983) Characterization of molluscan phagocyte subpopulations based on lysosomal enzyme markers. J Exp Zool 226:205–210

Guillou F, Mitta G, Dissous C, Pierce R, Coustau C (2004) Use of individual polymorphism to validate potential functional markers: case of a candidate lectin BgSel differentially expressed in susceptible and resistant strains of *Biomphalaria glabrata*. Comp Biochem Physiol B 138:175–181

Guillou F, Mitta G, Galinier R, Coustau C (2007a) Identification and expression of transcripts generated during an anti-parasitic response in *Biomphalaria glabrata*. Dev Comp Immunol 31:657–671

Guillou F, Roger E, Mone Y, Rognon A, Grunau C, Theron A, Mitta G, Coustau C, Gourbal BE (2007b) Excretory–secretory proteome of larval *Schistosoma mansoni* and *Echinostoma caproni*, two parasites of *Biomphalaria glabrata*. Mol Biochem Parasitol 155:45–56

Hahn UK, Bender RC, Bayne CJ (2000) Production of reactive oxygen species by hemocytes of *Biomphalaria glabrata*: carbohydrate-specific stimulation. Dev Comp Immunol 24:531–541

Hahn UK, Bender RC, Bayne CJ (2001a) Killing of *Schistosoma mansoni* sporocysts by hemocytes from reisistant *Biomphalaria glabrata*: role of reactive oxygen species. J Parasitol 87:292–299

Hahn UK, Bender RC, Bayne CJ (2001b) Involvement of nitric oxide in killing of *Schistosoma mansoni* sporocysts by hemocytes from resistant *Biomphalaria glabrata*. J Parasitol 87:778–785

Hanelt B, Lun CM, Adema CM (2008) Comparative ORESTES-sampling of transcriptomes of immune-challenged *Biomphalaria glabrata* snails. J Invertebr Pathol 99:192–203
Hansen EL (1976) A cell line from embryos of *Biomphalaria glabrata* Pulmonata: establishment and characteristics. In: Maramorosch K (ed) Invertebrate tissue culture: research applications. Academic, New York, pp 75–99
Holzmuller P, Grebaut P, Brizard JP, Berthier D, Bossard G, Bucheton B, Vezilier F, Chuchana P, Bras-Goncalves R, Lemesre JL, Vincendeau P, Cuny G, Frutos R, Biron DG (2008) "Pathogeno-Proteomics" toward a new approach of host-vector-pathogen interactions. Ann NY Acad Sci 1149:66–70
Humbert E, Coustau C (2001) Refractoriness of host haemocytes to parasite immunosuppressive factors as a putative resistance mechanism in the *Biomphalaria glabrata* - *Echinostoma caproni* system. Parasitology 122:651–660
Humphries JE, Yoshino TP (2003) Cellular receptors and signal transduction in molluscan hemocytes: connections with the innate immune system of vertebrates. Integr Comp Biol 43:305–312
Humphries JE, Yoshino TP (2006) *Schistosoma mansoni* excretory-secretory products stimulate a p38 signalling pathway in *Biomphalaria glabrata* embryonic cells. Int J Parasitol 36:37–46
Humphries JE, Yoshino TP (2008) Regulation of hydrogen peroxide release in circulating hemocytes of the planorbid snail *Biomphalaria glabrata*. Dev Comp Immunol 32:554–562
Humphries JE, Elizondo L, Yoshino TP (2001) Protein kinase C regulation of cell spreading in the molluscan *Biomphalaria glabrata* embryonic Bg cell line. Biochim Biophys Acta 1540:243–252
Ittiprasert W, Miller A, Myers J, Nene V, El-Sayed NM, Knight M (2010) Identification of immediate response genes dominantly expressed in juvenile resistant and susceptible Biomphalaria glabrata snails upon exposure to *Schistosoma mansoni*. Mol Biochem Parasitol 169:27–39
Ivanchenko MG, Lerner RS, Toumadje B, Allen B, Fisher K, Barnes DW, Bayne CJ (1999) Continuous *in vitro* propagation and differentiation of cultures of the intra-molluscan stages of the human parasite *Schistosoma mansoni*. Proc Natl Acad Sci U S A 96:4965–4970
Janeway CA (1989) Approaching the asymptope? Evolution and revolution in immunology. Cold Spring Harb Symp Quant Biol 54:1–13
Janeway CA, Medzhitov R (2002) Innate immune recognition. Ann Rev Immunol 20:197–216
Jeong KH, Lie KJ, Heyneman D (1983) The ultrastructure of the amebocyte-producing organ in *Biomphalaria glabrata*. Dev Comp Immunol 7:217–228
Jiang YG, Loker ES, Zhang SM (2006) In vivo and in vitro knockdown of FREP2 gene expression in the snail *Biomphalaria glabrata* using RNA interference. Dev Comp Immunol 30:855–866
Johnston LA, Yoshino TP (1996) Analysis of lectin- and snail plasma-binding glycopeptides associated with the tegumental surface of the primary sporocysts of *Schistosoma mansoni*. Parasitology 112:469–479
Johnston LA, Yoshino TP (2001) Larval *Schistosoma mansoni* excretory-secretory glycoproteins ESPs bind to hemocytes of *Biomphalaria glabrata* (Gastropoda) via surface carbohydrate binding receptors. J Parasitol 87:786–793
Kaneko T, Yano T, Aggarwal K, Lim JH, Ueda K, Oshima Y, Peach C, Erturk-Hasdemir D, Goldman WE, Oh BH, Kurata S, Silverman N (2006) PGRP-LC and PGRP-LE have essential yet distinct functions in the *Drosophila* immune response to monmeric DAP-type peptidoglycan. Nat Immunol 7:715–723
Kim SM, Kang JH (1997) Peroxidative activity of human Cu, Zn-superoxide dismutase. Mol Cell 7:120–124
Knudsen GM, Medzihradszky KF, Lim KC, Hansell E, McKerrow JH (2005) Proteomic analysis of *Schistosoma mansoni* cercarial secretions. Mol Cell Proteomics 4:1862–1875
Kum-Tatt L, Tan IK, Seet AM (1975) A new colorimetric method for the determination of NADH/NADPH dependent glutathione reductase in erythrocytes and in plasma. Clin Chim Acta 58:101–108
Lacchini AH, Davies AJ, Mackintosh D, Walker AJ (2006) ß-1, 3-glucan modulates PKC signalling in Lymnaea stagnalis defence cells: a role for PKC in H2O2 production and downstream ERK activation. J Exp Biol 209:4829–4840

Langand J, Morand S (1998) Heritable non-susceptibility in an allopatric host-parasite system: *Biomphalaria glabrata* (Mollusca)-*Echinostoma caproni* (Platyhelminthes, Digenea). J Parasitol 84:739–742

Lehr T, Geyer H, Maass K, Doenhoff MJ, Geyer R (2007) Structural characterization of N-glycans from the freshwater snail *Biomphalaria glabrata* cross-reacting with *Schistosoma mansoni* glycoconjugates. Glycobiology 17:82–103

Lehr T, Beuerlein K, Doenhoff MJ, Grevelding CG, Geyer R (2008) Localization of carbohydrates common to *Biomphalaria glabrata* as well as to sporocysts and miracidia of *Schistosoma mansoni*. Parasitology 135:931–942

Lewis FA, Patterson CN, Knight M, Richards CS (2001) The relationship between *Schistosoma mansoni* and *Biomphalaria glabrata*: genetic and molecular approaches. Parasitology 123:S169–S179

Lie KJ, Heyneman D, Yau P (1975) The origin of amebocytes in *Biomphalaria glabrata*. J Parasitol 61:574–576

Lie JK, Heyneman D, Jeong KH (1976) Studies on resistance in snails. 4. Induction of ventricular capsules and changes in the amebocyte-producing organ during sensitization of *Biomphalaria glabrata*. J Parasitol 62:286–291

Lockyer AE, Jones CS, Noble LR, Rollinson D (2000) Use of differential display to detect changes in gene expression in the intermediate snail host *Biomphalaria glabrata* upon infection with *Schistosoma mansoni*. Parasitology 120:399–407

Lockyer AE, Jones CS, Noble LR, Rollinson D (2004a) Trematodes and snails: an intimate association. Can J Zool 82:251–269

Lockyer AE, Noble LR, Rollinson D, Jones CS (2004b) *Schistosoma mansoni*: resistant specific infection-induced gene expression in *Biomphalaria glabrata* identified by fluorescent-based differential display. Exp Parasitol 107:97–104

Lockyer AE, Spinks J, Noble LR, Rollinson D, Jones CS (2007) Identification of genes involved in interactions between *Biomphalaria glabrata* and *Schistosoma mansoni* by suppression subtractive hybridization. Mol Biochem Parasitol 151:18–27

Lockyer AE, Spinks J, Kane RA, Hoffmann KF, Fitzpatrick JM, Rollinson D, Noble LR, Jones CS (2008) *Biomphalaria glabrata* transcriptome: cDNA microarray profiling identifies resistant- and susceptible-specific gene expression in haemocytes from snail strains exposed to *Schistosoma mansoni*. BMC Genomics 9:634

Lodes MJ, Yoshino TP (1989) Characterization of excretory-secretory proteins synthesized in vitro by *Schistosoma mansoni* primary sporocysts. J Parasitol 75:853–862

Lodes MJ, Yoshino TP (1990) The effect of schistosome excretory-secretory products on *Biomphalaria glabrata* hemocytes motility. J Invertbr Pathol 56:75–85

Lodes MJ, Connors VA, Yoshino TP (1991) Isolation and functional characterization of snail hemocyte-modulating polypeptide from primary sporocysts of *Schistosoma mansoni*. Mol Biochem Parasitol 49:1–10

Loker ES (1994) On being a parasite in an invertebrate host: a short survival course. J Parasitol 80:728–747

Loker ES, Bayne CJ (1986) Immunity to trematode larvae in the snail *Biomphalaria*. In: Lackie AM (ed) Immune mechanisms in invertebrate vectors. Clarendon, Oxford, pp 199–220

Loker ES, Bayne CJ, Buckley PM, Kruse KT (1982) Ultrastructure of encapsulation of *Schistosoma mansoni* mother sporocysts by hemocytes of juveniles of the 10-R2 strain of *Biomphalaria glabrata*. J Parasitol 68:84–94

Loker ES, Cimino DF, Hertel LA (1992) Excretory-secretory products of *Echinostoma paraensei* sporocysts mediate interference with *Biomphalaria glabrata* hemocyte function. J Parasitol 78:104–115

Loker ES, Adema CM, Zhang SM, Kepler TB (2004) Invertebrate immune systems – not homogeneous, not simple, not well understood. Immunol Rev 198:10–24

Maillet F, Bischoff V, Vignal C, Hoffmann J, Royet J (2008) The *Drosophila* peptidoglycan recognition protein PGRP-LF blocks PGRP-LC and IMD/JNK pathway activation. Cell Host Microbe 15:293–303

Martins-Souza RL, Pereira CA, Martins-Filho OA, Coelho PMZ, Corrêa A Jr, Negrão-Corrêa D (2006) Differential lectin labelling of circulating hemocytes from *Biomphalaria glabrata* and *Biomphalaria tenagophila* resistant or susceptible to *Schistosoma mansoni* infection. Mem Inst Oswaldo Cruz 101:S185–S192

Martins-Souza RL, Pereira CA, Coelho PMZ, Martins-Filho OA, Negrão-Corrêa D (2009) Flow cytometry analysis of the circulating haemocytes from *Biomphalaria glabrata* and *Biomphalaria tenagophila* following *Schistosoma mansoni* infection. Parasitology 136:670–676

Matricon-Gondran M, Letocart M (1999) Internal defenses of the snail *Biomphalaria glabrata*. I. Characterization of hemocytes and fixed phagocytes. J Invertebr Pathol 74:224–234

McKerrow J, Doenhoff MJ (1988) Schistosome proteases. Parasitol Today 4:334–340

Michelson EH (1975) Cellular defense mechanisms and tissue alterations in gastropod molluscs. In: Maramorosch K, Shope RE (eds) Invertebrate immunity. Academic, New York, pp 181–195

Miller AN, Raghavan N, FitzGerald PC, Lewis FA, Knight M (2001) Differential gene expression in haemocytes of the snail *Biomphalaria glabrata*: effects of *Schistosoma mansoni* infection. Int J Parasitol 31:687–696

Mitta G, Galinier R, Tisseyre P, Allienne JF, Girerd-Chambaz Y, Guillou F, Bouchut A, Coustau C (2005) Gene discovery and expression analysis of immune-relevant genes from *Biomphalaria glabrata* hemocytes. Dev Comp Immunol 29:393–407

Mourão MM, Dinguirard N, Franco GR, Yoshino TP (2009a) Role of the endogenous antioxidant system in the protection of *Schistosoma mansoni* primary sporocysts against exogenous oxidative stress. PLoS Negl Trop Dis 3:e550

Mourão MM, Dinguirard N, Franco GR, Yoshino TP (2009b) Phenotypic screen of early-developing larvae of the blood fluke, *Schistosoma mansoni*, using RNA interference. PLoS Negl Trop Dis 38:e502

Nowak TS, Woodards AC, Jung Y, Adema CM, Loker ES (2004) Identification of transcripts generated during the response of resistant *Biomphalaria glabrata* to *Schistosoma mansoni* infection using suppressive subtractive hybridization. J Parasitol 90:1034–1040

Nyame AK, Yoshino TP, Cummings RD (2002) Differential expression of LacdiNAc, fucosylated LacdiNAc, and Lewis x glycan antigens in intramolluscan stages of *Schistosoma mansoni*. J Parasitol 88:890–897

Perez-Sanchez R, Ramajo-Hernandez A, Ramajo-Martin V, Oleaga A (2006) Proteomic analysis of the tegument and excretory–secretory products of adult *Schistosoma bovis* worms. Proteomics 6:S226–S236

Peterson NA, Hokke CH, Deelder AM, Yoshino TP (2009) Glycotope analysis in miracidia and primary sporocysts of *Schistosoma mansoni*: differential expression during the miracidium-to-sporocyst transformation. Int J Parasitol 39:1331–1344

Plows LD, Cook RT, Davies AJ, Walker AJ (2005) Carbohydrates that mimic schistosome surface coat components affect ERK and PKC signalling in *Lymnaea stagnalis* haemocytes. Int J Parasitol 35:293–302

Raghavan N, Knight M (2006) The snail *Biomphalaria glabrata* genome project. Trends Parasitol 22:148–151

Raghavan N, Miller AN, Gardner M, FitzGerald PC, Kerlavage AR, Johnston DA, Lewis FA, Knight M (2003) Comparative gene analysis of *Biomphalaria glabrata* hemocytes pre- and post-exposure to miracidia of *Schistosoma mansoni*. Mol Biochem Parasitol 126:181–191

Richards CS (1975) Genetic factors in susceptibility of *Biomphalaria glabrata* for different strains of *Schistosoma mansoni*. Parasitology 70:231–241

Richards CS, Shade PC (1987) The genetic variation of compatibility in *Biomphalaria glabrata* and *Schistosoma mansoni*. Parasitology 73:1146–1151

Richards CS, Knight M, Lewis FA (1992) Genetics of *Biomphalaria glabrata* an its effect on the outcome of *Schistosoma mansoni* infection. Parasitol Today 8:171–174

Roger E, Gourbal B, Grunau C, Pierce RJ, Galinier R, Mitta G (2008a) Expression analysis of highly polymorphic mucin proteins Sm PoMuc from the parasite *Schistosoma mansoni*. Mol Biochem Parasitol 157:217–227

Roger E, Mitta G, Mone Y, Bouchut A, Rognon A, Grunau C, Boissier J, Theron A, Gourbal BEF (2008b) Molecular determinants of compatibility polymorphism in the *Biomphalaria glabrata/ Schistosoma mansoni* model: new candidates identified by a global comparative proteomics approach. Mol Biochem Parasitol 157:205–216

Roger E, Grunau C, Pierce RJ, Hirai H, Gourbal B, Galinier R, Emans R, Cesari IM, Cosseau C, Mitta G (2008c) Controlled chaos of polymorphic mucins in a metazoan parasite *Schistosoma mansoni* interacting with its intermediate host *Biomphalaria glabrata*. PLoS Negl Trop Dis 211:e330

Salamat Z, Sullivan JT (2009) Involvement of protein kinase C signalling and mitogen-activated protein kinase in the amebocyte-producing organ of *Biomphalaria glabrata* (Mollusca). Dev Comp Immunol 33:725–727

Sayed AA, Williams DL (2004) Biochemical characterization of 2-Cys peroxiredoxins from *Schistosoma mansoni*. J Biol Chem 279:26159–26166

Schmucker D, Chen B (2009) Dscam and DSCAM: complex genes in simple animals, complex animals yet simple genes. Genes Dev 23:147–156

Schneider O, Zelck UE (2001) Differential display analysis of hemocytes from schistosome-resistant and schistosome-susceptible intermediate hosts. Parasitol Res 87:489–491

Souza SS, Andrade ZA (2006) On the origin of the *Biomphalaria glabrata* hemocytes. Mem Inst Oswaldo Cruz 101:S213–S218

Stanislawski E, Renwrantz LR, Becker W (1976) Soluble blood group reactive substances in the hemolymph of *Biomphalaria glabrata* (Mollusca). J Invertebr Pathol 28:301–308

Stout BA, Adema CM, Zhang SM, Loker ES (2009) Biology of FREPs: diversified lectins with fibrinogen-related domains from freshwater snail *Biomphalaria glabrata*. In: Vasta G, Ahmed H (eds) Animal lectins: a functional view. CRC Press Taylor & Francis, Boca Raton, pp 471–491

Sullivan JT (2007) Mitotic responses to injected extracts of larval and adult *Schistosoma mansoni* in *Biomphalaria glabrata*: effects of dose and colchicine treatment. J Parasitol 93:213–215

Sullivan JT, Richards CS (1981) *Schistosoma mansoni*, NIH-SM-PR-2 strain, in susceptible and nonsusceptible stocks of *Biomphalaria glabrata*: comparative histology. J Parasitol 67:702–708

Sullivan JT, Pikios SS, Alonzo AQ (2004) Mitotic responses to extracts of miracidia and cercariae of *Schistosoma mansoni* in the amebocyte-producing organ of the snail intermediate host *Biomphalaria glabrata*. J Parasitol 90:92–96

Tanji T, Hu X, Weber ANR, Ip YT (2007) Toll and IMD pathways synergistically activate an innate immune response in *Drosophila melanogaster*. Mol Cell Biol 27:4578–4588

Tew KD, Ronai Z (1999) GST function in drug and stress response. Drug Resist Updat 2:143–147

The *Schistosoma japonicum* Genome Sequencing and Functional Analysis Consortium; Zhou Y, Zheng H, Chen Y, Zhang L, Wang K, Guo J, Huang Z, Zhang B, Huang W, Jin K, Dou T, Hasegawa M, Wang L, Zhang Y, Zhou J, Tao L, Cao Z, Li Y, Vinar T, Brejova B, Brown D, Li M, Miller DJ, Blair D, Zhong Y, Chen Z; Functional genomics analysis: Liu F, Hu W, Wang ZQ, Zhang QH, Song HD, Chen S, Xu X, Xu B, Ju C, Huang Y, Brindley PJ, McManus DP, Feng Z, Han ZG; Sequencing and assembly: Lu G, Ren S, Wang Y, Gu W, Kang H, Chen J, Chen X, Chen S, Wang L, Yan J, Wang B, Li X, Jin L, Wang B, Pu S, Zhang X, Zhang W, Hu Q, Zhu G, Wang J, Yu J, Wang J, Yang H, Ning Z, Berriman M, Wei CL, Ruan Y, Zhao G, Wang S; Paper writing: Liu F, Zhou Y, Wang ZQ, Lu G, Zheng H, Brindley PJ, McManus DP, Blair D, Zhang QH, Zhong Y, Wang S, Han ZG, Chen Z; Project leaders: Wang S, Han ZG, Chen Z (2009) Genome annotation and evolution analysis: The *Schistosoma japonicum* genome reveals features of host-parasite interplay. Nature 460:345–351

Theron A, Coustau C, Rognon A, Gourbière S, Blouin MS (2008) Effects of laboratory culture on compatibility between snails and schistosomes. Parasitology 135:1179–1188

Vasta GR, Ahmed H, Tasumi S, Odom EW, Saito K (2007) Biological roles of lectins in innate immunity: molecular and structural basis for diversity in self/non-self recognition. Adv Exp Med Biol 598:389–406

Vergote D, Bouchut A, Sautiere PE, Roger E, Galinier R, Rognon A, Coustau C, Salzet M, Mitta G (2005) Characterization of proteins differentially present in the plasma of *Biomphalaria glabrata* susceptible and resistant to *Echinostoma caproni*. Int J Parasitol 35:215–224
Vermeire JJ, Yoshino TP (2007) Antioxidant gene expression and function in in vitro-developing *Schistosoma mansoni* mother sporocysts: possible role in self-protection. Parasitology 134:1369–1378
Vermeire JJ, Boyle JP, Yoshino TP (2004) Differential gene expression and the effects of *Biomphalaria glabrata* embryonic Bge cell factors during larval *Schistosoma mansoni* development. Mol Biochem Parasitol 135:153–157
Voge M, Seidel JS (1972) Transformation in vitro of miracidia of *Schistosoma mansoni* and *S. japonicum* into young sporocysts. J Parasitol 58:699–704
Von Brand T, Files VS (1947) Chemical and histological observations on the influence of *Schistosoma mansoni* infection on *Australorbis glabratus*. J Parasitol 33:476–482
Walker AJ (2006) Do trematode parasites disrupt defence-cell signaling in the snail hosts? Trends Parasitol 22:154–159
Webster JP, Davies CM (2001) Coevolution and compatibility in the snail-schistosome system. Parasitology 123:S41–S56
Webster JP, Woolhouse MEJ (1998) Selection and strain specificity of compatibility between snail intermediate hosts and their parasitic schistosomes. Evolution 52:1627–1634
Webster JP, Gower CM, Blair L (2004) Do hosts and parasites coevolve? Empirical support form the *Schistosoma* system. Am Nat 164:S33–S51
Wu XJ, Sabat G, Brown JF, Zhang M, Taft A, Peterson N, Harms A, Yoshino TP (2009) Proteomic analysis of *Schistosoma mansoni* proteins released during in vitro miracidium-to-sporocyst transformation. Mol Biochem Parasitol 164:32–44
Yoshino TP, Bayne CJ (1983) Mimicry of snail host antigens by miracidia and primary sporocysts of *Schistosoma mansoni*. Parasite Immunol 5:317–328
Yoshino TP, Boswell CA (1986) Antigen sharing between larval trematodes and their snail hosts: how real a phenomenon in immune evasion? In: Lackie AM (ed) Immune mechanisms in invertebrate vectors. Clarendon, Oxford, pp 221–238
Yoshino TP, Cheng TC (1978) Snail host-like antigens associated with the surface membranes of *Schistosoma mansoni* miracidia. J Parasitol 64:752–754
Yoshino TP, Granath WO (1985) Surface antigens of *Biomphalaria glabrata* (Gastropoda) hemocytes: functional heterogeneity in cell subpopulations recognized by a monoclonal antibody. J Invertebr Pathol 45:174–186
Yoshino TP, Laursen JR (1995) Production of *Schistosoma mansoni* daughter sporocysts from mother sporocysts maintained in synxenic culture with *Biomphalaria glabrata* embryonic Bge cells. J Parasitol 81:714–722
Yoshino TP, Vasta GR (1996) Parasite-invertebrate host immune interactions. In: Cooper EL (ed) Invertebrate immune responses: cell activities and the environment. Springer, Berlin, pp 125–167
Yoshino TP, Cheng TC, Renwrantz LR (1977) Lectin and human blood group determinants of *Schistosoma mansoni*: Alteration following in vitro transformation of miracidium to mother sporocyst. J Parasitol 63:818–824
Yoshino TP, Wu XJ, Liu HD (1998) Transfection and heat-inducible expression of molluscan promoter-luciferase reporter gene constructs in the *Biomphalaria glabrata* embryonic snail cell line. Am J Trop Med Hyg 59:414–420
Yoshino TP, Coustau C, Modat S, Castillo MG (1999) The *Biomphalaria glabrata* embryonic BGE molluscan cell line: establishment of an in vitro cellular model for the study of snail host-parasite interactions. Malacologia 41:331–343
Yoshino TP, Boyle JP, Humphries JE (2001) Receptor-ligand interactions and cellular signalling at the host-parasite interface. Parasitology 123:S143–S157
Yoshino TP, Dinguirard N, Kunert J (2008) Gene cloning and functional characterization of a tandem-repeat galectin from the freshwater snail *Biomphalaria glabrata*, intermediate host of the human blood fluke *Schistosoma mansoni*. Gene 411:46–58

Yoshino TP, Dinguirard N, Mourão MM (2010) In vitro manipulation of gene expression in larval Schistosoma: a model for postgenomic approaches in Trematoda. Parasitology 137:463–483

Zahoor Z, Davies AJ, Kirk RS, Rollinson D, Walker AJ (2008) Disruption of ERK signalling in *Biomphalaria glabrata* defense cells by *Schistosoma mansoni*: implications for parasite survival in the snail host. Dev Comp Immunol 32:1561–1571

Zahoor Z, Davies AJ, Kirk RS, Rollinson D, Walker AJ (2009) Nitric oxide production by *Biomphalaria glabrata* haemocytes: effects of *Schistosoma mansoni* ESPs and regulation through the extracellular signal-regulated kinase pathway. Parasit Vectors 2:18. doi: 10.1186/1756-3305-2-18

Zelck UE, Von Janowsky B (2004) Antioxidant enzymes in intramolluscan *Schistosoma mansoni* and ROS-induced changes in expression. Parasitology 128:493–501

Zelck UE, Gege BE, Schmid S (2007) Specific inhibitors of mitogen-activated protein kinase and PI3-K pathways impair immune responses by hemocytes of trematode intermediate host snails. Dev Comp Immunol 31:321–331

Zhang SM, Adema CM, Kepler TB, Loker ES (2004) Diversification of Ig superfamily genes in an invertebrate. Science 305:251–254

Zhang SM, Jiang YG, Loker ES (2005) Establishing RNAi to knockdown FREP2 expression in the snail, *Biomphalaria glabrata*, an intermediate host for *Schistosoma mansoni*. Am J Trop Med Hyg 73:20

Zhang SM, Zeng Y, Loker ES (2007) Characterization of immune genes from the schistosome host snail *Biomphalaria glabrata* that encode peptidoglycan recognition proteins and gram-negative bacteria binding protein. Immunogenetics 59:883–898

Zhang SM, Zeng Y, Loker ES (2008) Expression profiling and binding properties of fibrinogen-related proteins FREPs, plasma proteins from the schistosome snail host *Biomphalaria glabrata*. Innate Immun 14:175–189

Zhang H, Wang LL, Song LS, Song XY, Wang B, Mu Ch, Zhang Y (2009) A fibrinogen-related protein from bay scallop *Argopecten irradians* involved in innate immunity as pattern recognition receptor. Fish Shellfish Immunol 26:56–64

Chapter 8
Advances in the Genomics and Proteomics of the Freshwater Intermediate Snail Host of *Schistosoma mansoni, Biomphalaria glabrata*

Wannaporn Ittiprasert, Jocelyn Myers, Edwin C. Odoemelam, Nithya Raghavan, Fred Lewis, Joanna M. Bridger, and Matty Knight

Abstract Molecular events governing the interplay between the intermediate snail host, *Biomphalaria glabrata*, and its parasitic trematodes are gradually being unraveled. The last 20 years has seen an upsurge in the number of gene sequences and proteins that are expressed, differentially regulated, and diversified in this snail in relation to its role as an obligate host for an important human pathogen, *Schistosoma mansoni*, the causative agent of schistosomiasis in the Western Hemisphere. Although regarded as a good model organism for studying the complexities of host–pathogen interactions, *B. glabrata* also serves as being useful in bridging the information gap that exists between locotrophozoans and the more popular model organisms that belong to other clades (ecdyzoa and deuterostomes). By the application of a variety of molecular tools, emerging results show the significance of innate defense and stress-related genes in the snail host/parasite relationship. In this chapter, we will provide an overview of some of the recent advances that have been made in the field of genomics and proteomics of this snail, mainly in relation to schistosomes. Although information remains for the most part rudimentary, significant advances have been made in the molecular characterization of certain genes, such as FREPs and the *nimbus* mobile genetic element. Key enzymes participating in the snail's ability to either support or reject the parasite infection, such as hydrolases and oxidoreductases, have also been characterized. A significant milestone, the completion of the 931-Mb genome sequence of this snail, is also anticipated soon. Collectively, all these advances, unless interest and/or funding opportunities wane, should create a favorable research environment for attracting more investigators into the field of molecular malacology.

M. Knight (✉)
Biomedical Research Institute, 20852 Rockville, MD, USA
e-mail: mknight@afbr-bri.com

R. Toledo and B. Fried (eds.), Biomphalaria *Snails and Larval Trematodes*,
DOI 10.1007/978-1-4419-7028-2_8, © Springer Science+Business Media, LLC 2011

8.1 Introduction

The freshwater snail *Biomphalaria glabrata* serves as an important intermediate host for the parasitic trematode, *Schistosoma mansoni*, in the Western Hemisphere. Along with the other schistosome species, most commonly *S. haematobium* and *S. japonicum*, these parasitic flukes are the causative agents of schistosomiasis, a medically significant disease that is endemic in 74 countries of the world (Chitsulo et al. 2004; Gryseels et al. 2006). It is estimated that in the developing world, especially in sub-Saharan Africa, at least 650 million people remain at risk for exposure despite efforts to control transmission in human and snail populations by mass chemotherapy and molluscicides, respectively. While substantial improvements have been made in short-term transmission reduction, long-term control of schistosomiasis remains elusive. Our inability to break the chronic cycle of repeated re-infection often seen in human populations following intervention can be attributable to a variety of factors. One significant factor is the inadequate sustained surveillance after control efforts have been undertaken and the expansion of freshwater habitats (from dam construction to irrigation projects) that allows the snail-host to thrive. Also of significant importance are poor sanitation, inadequate health infrastructure, lack of access to clean water, and poverty that prevails in schistosomiasis-afflicted communities. Taken together, these conditions make schistosomiasis a major public health concern that is second only to challenges posed by other diseases (e.g., malaria, AIDS, and TB) that greatly impact developing countries of the tropics and neotropics.

It is anticipated that a molecular-based vaccine in conjunction with more sensitive diagnostics to monitor low disease prevalence in human and snail populations will one day become integrated into the existing control tools to help in a multi-pronged approach to, hopefully, eradicate schistosomiasis (Bergquist et al. 2009; Hubendick 1958; Wang et al. 2009). Another component proposed for achieving long-term transmission reduction is a better understanding of the molecular basis of the snail/schistosome relationship. Such information might lead to the development of novel intervention tools that will block parasite development at the snail stage of its life cycle. Thus far, this overall objective has served as the driving force for the past two decades in helping to identify molecular factors (DNA, RNA, and proteins) that influence the outcome of schistosome development in the snail host (Bayne 2009; Knight et al. 1999; Raghavan and Knight 2006).

While interest in the molecular aspect of the snail's role in schistosomiasis transmission has grown substantially in recent years, of equal importance is the growing recognition that by integrating "-omic" technologies, the snail can serve as a useful model organism for studies investigating the evolution/co-evolutionary dynamics in host–parasite interactions. Within this context, the aforementioned snail host, *B. glabrata*, and the sea hare *Aplysia californica* are seen as invertebrate model organisms in basic research relating to innate defense and neurobiology, respectively. Accordingly, genome-sequencing projects have recently been initiated for both these organisms along with a genome project for another mollusk, *Lottia gigantea* (Wu et al. 2007).

It is envisioned that the completion of these genome projects will help bridge the considerable information gap that exists in the genome, cell, and evolutionary biology between snails and the current, more extensively used invertebrate model organisms, such as *Drosophila melanogaster* and *Caenorhabditis elegans*. Furthermore, studying the genomics and proteomics of snails as model organisms will bring these lophotrochozoans (mollusks, annelids, and platyhelminths), organisms that include the most morphologically diverse phyla, into the modern era of molecular science and strengthen the new and growing field of comparative genomics. Also, it is hoped that these studies will lead ultimately to a more thorough understanding of evolutionary relationships between the lophotrochozoans and the more popular model organisms that belong entirely to other metazoan clades, ecdysozoa (arthropods and nematodes) and deuterostomes (amphioxus and sea urchins).

In this chapter, we will summarize recent advances that have been made in the subject of snail genomics and proteomics, with special emphasis on *B. glabrata*. While this snail is essentially the focus of this chapter, relevant molecular advances with other mollusks will be incorporated for the sake of comparison to illustrate progress, but not for the purpose of extrapolation. As mentioned above, because of the obligatory role that *B. glabrata* plays as a major invertebrate host for *S. mansoni*, most molecular studies undertaken in this snail are geared toward its relationship with the parasite. Consequently, several recent reviews have been written on this subject and we will provide only recent updates as related to the snail host proteomics and genomics. The reader is, therefore, advised to consult these earlier reviews for a more in-depth, general background on the subject of snail/schistosome molecular biology (Lockyer et al. 2007; Stothard et al. 2002). In addition, we will try to highlight some of the relatively newer studies involving the use of the only molluscan cell line available, the *B. glabrata* embryonic cell line (Bge), and the potential for this cell line in the *in vitro* parasite co-culture system to facilitate the development of new technology for investigating molecular mechanisms of the snail/parasite interaction.

8.2 Genetics of the *B. glabrata* Snail/Schistosome Interaction

Given the complex nature of the snail/parasite interaction, earlier studies of this relationship dealt mainly with the role that genetics plays in the outcome of the snail infection, particularly in response to schistosomes and echinostomes. Other schistosome-transmitting snails, in particular, snails of the genera *Bulinus* and *Oncomelania* that serve as hosts for *S. haematobium* and *S. japonicum*, respectively, have been less studied. Also understudied, at the molecular level, is the snail *Biomphalaria pfeifferi*, the *S. mansoni* intermediate snail host that is responsible for the transmission of the parasite throughout much of sub-Saharan Africa. These problems notwithstanding, the relative ease of propagating *B. glabrata* in the laboratory and the use of several phenotypic markers (e.g., pigment) have been a major advantage in helping to appreciate how the outcome of *S. mansoni* infection in this snail host has a strong genetic basis (Richards 1975a).

Using laboratory-maintained *B. glabrata* stocks bred for either parasite resistance or susceptibility, Richards showed by exposing different snail stocks to either compatible or non-compatible parasites that in adult snails, resistance to the parasite is a dominant single-gene trait that follows a simple Mendelian mode of inheritance (Richards 1975b). In juvenile snails, however, resistance to schistosomes is a complex trait and involves at least 4–5 genes, each with multiple alleles. Because *B. glabrata* snails are hemaphrodites, it is relatively straightforward to establish and propagate isogenic snail stocks for the different disease phenotypes by selfing representative snail stocks through several generations. Aided by the genetics of pigmentation (CC, cc), progeny snails resulting from a cross between a pigmented resistant snail (serving as male) and an albino susceptible snail (serving as female) can readily be identified (Lewis et al. 2003). The generation time for *B. glabrata* snails is approximately 6 weeks (from egg to egg), thus making it possible (although not as convenient as other model organisms) to propagate several generations of progeny snails from a single mating experiment within a moderate amount of time. From several crosses that used different snail stocks displaying a variety of susceptibility phenotypes to different strains of *S. mansoni*, Richards and Shade (Richards and Shade 1987) showed the inherent variation (that they categorized into four susceptibility types) within the complex snail/schistosome relationship (see review by Lewis et al. 2003; Lewis et al. 2001) These earlier studies helped establish the snail stocks that have become an invaluable resource for investigating the molecular basis of the *B. glabrata*/*S. mansoni* relationship. Thus far, prototype representative snails for parasite resistance used commonly in most laboratories consist of the laboratory bred stocks (10-R2, Richards 1973 and 13-16-R1, Goodall et al. 2006) and a wild-type snail that was isolated in Salvadore, Brazil (also known as BS-90) (Paraense and Correa 1963). Of these three representative resistant stocks, the BS-90 snail, in our laboratory, has consistently maintained its robust resistance phenotype (either as juveniles or adults) to several *S. mansoni* strains (NMRI, PR1 and 2). The other two resistant stocks, 10-R2 and 13-16-R1, although still useful have been less stable. Some are no longer resistant, while others are currently only partially (about 70%) resistant to the parasite. Of the representative susceptible stocks available, two are laboratory bred, M-line and NMRI. These snails are albino stocks and were bred for high susceptibility to the parasite (Newton 1955). While these two stocks have maintained this phenotype since the 1950s, they are known to display a high frequency of intra-strain genotypic variation. This feature, causing some of the NMRI susceptible stocks to become less susceptible over time, has been exploited to derive a partially resistant (non-susceptible) snail line (LAC) that has been a useful reference tool (Cooper et al. 1994) for verifying either the presence or absence of parasite resistance molecular markers. Another susceptible snail stock that is currently in use is a wild-type snail known as the BB02 strain, chosen for the *B. glabrata* full-coverage genome sequencing project (Adema et al. 2006). Representative *B. glabrata* snails that have been purposefully bred to display either resistance or susceptibility phenotypes for echinostomes are also available, but less extensive genetics have been undertaken for this particular trematode/

B. glabrata interaction compared to that for the *S. mansoni/B. glabrata* relationship. For this reason, this review will focus mostly on the snail's relationship with schistosomes rather than with echinostomes, even though it is acknowledged that a noticeably more intense cell-mediated defense is provoked by echinostomes than by schistosomes in these snails (Loker et al. 1986).

It has been shown from several studies that upon entering a resistant snail, miracidia are rapidly engulfed by hemocytes (the cellular component of the snail's innate defense system), encapsulated, and destroyed in anywhere from 24 to 72 h post-exposure (Loker et al. 1982). Assisted by as yet unknown factors in the snail's plasma (hemolymph), the hemocytes surrounding the parasite (by mechanisms that are still unknown) mediate the formation of a granulomatous-type response that is the hallmark sign of the resistance phenotype. Usually, this active response is histologically documented before any given snail can be classified as being genetically resistant.

In contrast, susceptible snails fail to mount a similar active response against the parasite. Consequently, the parasite develops successfully in these snails, resulting in the release of infective cercariae several weeks post-exposure. Another feature known to affect the outcome of infection is the age of the snail (Niemann and Lewis 1990; Richards 1975c; Richards and Minchella 1987; Richards and Shade 1987). Juvenile snails, for example, are known to be considerably more vulnerable to schistosomes than their adult counterparts. A genetic basis is known to govern this age-variation in snail susceptibility and occurs even within the same snail stock. A single dominant gene (Mr) that is only expressed in the adult but not juvenile snails has been shown to govern this trait (Richards and Minchella 1987). These snails are susceptible as juveniles, non-susceptible as young adults, and can revert to susceptibility as mature senescent snails after egg laying ceases. With this stock (93375), Richards and Minchella were able to demonstrate a correlation between the amoebocytic nodular accumulations in the pericardial aortic region, and the onset of the non-susceptibility phenotype. In these snails, the pericardial aortic region, the site of the snail's amoebocyte-producing organ (APO), is considerably more visible than in either susceptible juvenile or senescent adult snails. While we know little about the biological/physiological factor(s) that potentiates this region in these snails, the observations in the 93375 stock suggests the existence of a common factor(s) that maintains and controls atrial nodule accumulations and sexual maturity/reproduction in these snails.

8.3 Snail Innate Defense

As mentioned above, the cells that are most intimately involved with the snail's internal defense system against foreign invaders (including trematodes) are the hemocytes (also known as amoebocytes). These cells are produced in the APO and consist of different cell types classified as granulocytes and hyalinocytes (Lie et al. 1980). With the entry of miracidia into the snail, the APO is stimulated to produce large numbers of these cells that presumably migrate through the snail's open circulatory system to the site of the invasion. What factor(s) triggers the burst of recruitment of these cells is as yet unknown, but evidence exists that the parasite's own excretory secretory (ES) products play a significant role in the process (Guillou et al. 2007a). Once stimulated, the APO acquires a pronounced increase in size caused by the proliferation of cells. These cells in the region have a high mitotic activity and can remain viable outside of the snail's body for an extended period of time (Lie et al. 1980). Sullivan and Spence were able to demonstrate from adoptive transfer experiments that hemocytes (minus the simultaneous transfer of hemolymph) can confer resistance to susceptible snails (Sullivan and Spence 1994).

Although the molecular/biochemical basis of hemocyte cytotoxicity toward trematodes remains unknown, it has been shown that soluble factors in the hemolymph play a major role in the hemocyte-mediated killing of schistosomes in the snail (Goodall et al. 2004; Hahn et al. 2001; Raghavan et al. 2003; Sullivan et al. 1995). Accordingly, several candidate molecules have been identified. Thus far, evidence shows that lectins play an important part in this process, helping with cell-to-cell binding that allows the successful hemocyte encapsulation process to occur around the parasite. Some of the genes for *B. glabrata* lectins have now been isolated and fully sequenced. These include the genes for the fibrinogen (FBG)-bearing proteins, also known as FREPs (Hertel et al. 2005), selectin (Duclermortier et al. 1999), dermatopontin (Bouchut et al. 2006), galectin (Yoshino et al. 2008), C-type lectin (Guillou et al. 2007a; Ittiprasert et al. 2010), and mucin (Roger et al. 2008, Roger et al. 2008a, b; Ittiprasert et al. 2010). From the extensive sequence analysis of transcripts encoding FREPs, diverse forms of this molecule were recognized. Interestingly, the structure of FREPs showed several immunoglobulin superfamily (IgSF) domains at the amino terminus and a FBG domain at the carboxyl terminus (Leonard et al. 2001; Zhang et al. 2001). Studies have revealed the extensive diversification of FREPs at both RNA and DNA levels, suggestive of a primordial invertebrate antibody that primarily functions in the snail's innate defense. A developmental role for FREPs has also been described, especially for the recently described FBG-related gene that also contains an epidermal growth factor (EGF) repeat region, and thus known as FReM (Zhang et al. 2008). The mechanism(s) behind the extensive molecular diversification of FREPs has been reported as being generated by point mutation and gene conversion (Zhang and Loker 2004).

Somatic cell RNA diversification has also been shown for several other *B. glabrata* genes. For example, sequence analysis of the multiple transcripts corresponding to galectin showed the existence of multiple variant-sized transcripts corresponding to this lectin (Guillou et al. 2007b; Yoshino et al. 2008). Similar diversity in transcript size has also been shown for transcripts encoding the snail ortholog of human mucin 5 (Ittiprasert et al. 2010). Interestingly, Northern analysis revealed significant differential regulation of these diverse transcripts between resistant and susceptible snails following schistosome infection (Ittiprasert et al. 2010).

From a recent extensive proteomic study reported by Roger et al. (2008), it was suggested that the different isoforms of mucin might serve as a target parasite protein that defines the complex interplay between the parasite and snail host (Roger et al. 2008).

Thus, a significant homology between mucin protein sequences of the snail host and parasite may be a significant determining factor underlying snail host–parasite compatibility. This suggestion has been described as the "match-mismatched hypothesis" and more extensive sequencing of both snail and parasite molecules will have to be completed in order to explain/prove, unequivocally, if indeed the molecular basis of compatibility depends solely on the degree of shared sequences existing between the snail and a given compatible strain of *S. mansoni*.

What molecular mechanisms that help to generate the extensive diversification of the aforementioned proteins has yet to be discovered? Somatic cell recombination remains a strong possibility but key enzymes involved in this process are yet to be characterized. A likely candidate enzyme that might be involved in helping to generate transcript/protein diversification in the snail is cytidine deaminase. The full-length gene encoding this enzyme in the snail has been isolated but its enzymatic mode of action remains to be resolved (Bouchut et al. 2006). Because the enzyme functions in the regulation of immunoglobulin diversification processes, somatic hypermutation, class switch recombination, and gene conversion (Bransteitter et al. 2003), it may be important to bring the full molecular characterization of cytidine deaminase to fruition. Some early studies have shown the differential regulation of this enzyme between parasite-resistant and susceptible snails, with resistant snails showing an induction (up-regulation) of this enzyme after parasite infection (Ittiprasert et al. 2010). The regulation of expression of several genes of *B. glabrata* in response to schistosomes is being intensively studied within the context of snail resistance or susceptibility to schistosomes. Aside from the aforementioned lectins, other genes that are differentially regulated between resistant and susceptible snails have also been identified by a variety of RNA profiling studies that include differential display, subtraction cDNA cloning strategies, and microarray hybridizations (Lockyer et al. 2000; Miller et al. 2001; Schneider and Zelck 2001; Lockyer et al. 2004; Mitta et al. 2005; Hanelt et al. 2008; Lockyer et al. 2008). Table 8.1 shows a list of some of the genes identified as showing differential expression between resistant and susceptible snails in response to parasite infection, and the methodologies that have been utilized toward their identification. Although the direct effect of the protein products encoded by most of these genes remains to be evaluated in relation to either parasite killing or development, it is hoped that gene-silencing techniques, such as RNAi, will one day enable us to assess which genes to target toward the development of parasite transmission blocking strategies in these snails. Although gene silencing by RNAi has not been widely tested in *B. glabrata* (only one study has documented the successful application of RNAi in this snail) (Jiang et al. 2006), the method has, however, been useful in studies investigating the neurologic role of nitric-oxide synthase (NOS) in relation to the feeding activity of another pulmonate gastropod, the pond snail, *Lymneae stagnalis* (Korneev et al. 2002). From these studies, it was revealed that microinjection of NOS dsRNA into *L. stagnalis* resulted in diminished feeding behavior in snails where NOS activity was suppressed. No change in feeding behavior was seen in snails injected with an unrelated dsRNA as control. These results show convincingly that post-transcriptional gene silencing involving dsRNA occurs naturally in snails as with other invertebrates, and underscores the need for

Table 8.1 Transcripts from *B. glabrata* by variety of RNA profiling tools

Gene	Methodology	Differential gene regulation		Reference
		Resistant	Susceptible	
Alpha integrin-like	Hemocyte response to glutamic acid-substituted control peptide (RGDS)	–	↓	Davids and Yoshino (1998)
Cadherin-like	qPCR	–	↑	Bouchut et al. (2006)
Cathepsin B preproprotein	Real time PCR	↑	–	Myers et al. (2008)
Cu/Zn superoxide dismutase	qPCR	↑	–	Goodall et al. (2004); Bender et al. (2007)
Cytidine deaminase	qPCR	↑	↑	Bouchut et al. (2006)[a]; Ittiprasert et al. (2009)
Cytochrome oxidase (CO) subunits	RAPD	⊕	nd	Rollinson et al. (1998)
Cytochrome p450	ddPCR, semi qPCR	nd	↓	Lockyer et al. (2000)
Dermatopontin 1	qPCR	–	↑	Bouchut et al. (2006)[a]
Endo-1,4,-beta-mannase	qPCR	–	⊕	Bouchut et al. (2006)[a]
Fibrinogen-related protein (FREPs) 2 and FREP4	qPCR	↑	↑	Hertel et al. (2005)
Gram-negative bacteria binding protein	Northern blotting	nd	↑	Zhang et al. (2007)
Heat shock protein 70	RT-PCR, qPCR	⊕	↑	Lockyer et al. (2004); Ittiprasert et al. (2009)
Histone H4	qPCR	–	↑	Bouchut et al. (2006)[a]
Low-density lipoprotein receptor (LDLR)	Real time PCR	–	↑	Ittiprasert et al. (2009)
Matrillin	qPCR	–	↑	Bouchut et al. (2006)[a]
Mn superoxide dismutase (MnSOD) and SOD1	qPCR	↑↑	↑	Bender et al. (2007)
Non-LTR retrotransposon nimbus (BgI) reverse transcriptase	Real time PCR	–	↑	Ittiprasert et al. (2009)
Peptidoglycan recognition protein	Northern blotting	nd	↑	Zhang et al. (2007)
Peroxiredoxin; BgPrx	Real time PCR	↑	↓	Knight et al. (2009)
RACK (receptor of protein kinase C)	Real time PCR	↑	–	Ittiprasert et al. (2009)
Reverse transcriptase 52	RT-PCR	↑	⊕	Raghavan et al. (2003)
Schistosomin	qPCR	nd	–	Zhang et al. (2009)
Sel protein/lectin	Semi qPCR, real time PCR	⊕	↑	Guillou et al. (2007b); Ittiprasert et al. (2009)
Type 2 cystatin	qPCR	⊕	nd	

⊕ = exist; nd = no data; ↑ = up-regulation after exposure; ↓ = reduction after exposure; – = no change after exposure
[a]Studied in Echinostome

more intense research efforts to develop and optimize delivery methods (other than microinjection) of either double-stranded (ds) or synthetic interference (si) RNA into the snail. Developing stable transfection-based systems to suppress gene expression in the snail will serve as a useful tool to verify the biological relevance of all the genes thus far identified for roles in parasite development and/or killing.

As shown in Table 8.1, other RNA profiling methods, aside from differential display and microarray analysis, have led to the identification of transcripts that may be important in the *B. glabrata/S. mansoni* relationship. For example, the cDNA cloning strategy of selective cloning by suppression subtractive hybridization has been used extensively in this host–parasite interaction to isolate major genes whose expression is linked to either resistance or susceptibility of this snail to infection. Applying this method to investigate differential gene regulation between resistant and susceptible snails when exposed to the parasite allowed Nowak et al. (2004) to show elevated expression of ferritin heavy chain and alpha crystalline beta chain (hsp) in resistant compared to susceptible snails. Likewise, using the same method, Ittiprasert et al. (2010) showed that mucin, cytidine deaminase (CDA), and low-density lipoprotein receptor (LDLR) are induced by normal but not irradiated attenuated miracidia in juvenile susceptible snails. Additional results showed that normal parasite infection of juvenile resistant snails specifically induced receptor for activated protein kinase C (RACK) and macrophage express gene 1 (Mpeg1). Varying profiles of gene expression (fibrinogen C terminal domain, C-type lectin, theromacin, and Fas [tumor necrosis factor alpha family]) resulting from whether snails were exposed to either normal or attenuated parasites showed, for the first time, that other factors involved with the parasite's penetration of the snail (possibly molecular events related to wound healing) may need to be taken into account when assessing the significance of any gene activity (up- or down-regulation) between resistant and susceptible snails responding to parasite exposure. From this study, it was also discovered that the stress response protein, Heat shock protein (Hsp) 70, was induced in juvenile susceptible snails but not juvenile resistant or non-susceptible snails in response to parasite infection (Ittiprasert et al. 2009). In contrast to these results, Lockyer et al. (2004) showed the specific up-regulation of Hsp70 in adult resistant but not adult susceptible snails, an indication that the age of the snail indeed matters in the regulation of gene expression in the *B. glabrata/* schistosome interaction.

8.4 Stress Response in Snail/Schistosome Interactions

While the induction of Hsp70 was found to be more dramatic in juvenile susceptible snails than in their resistant counterparts in response to parasite exposure, the interesting and unexpected result was that the kinetics of induction of Hsp70 coincided with the up-regulation of the transcript corresponding to the reverse transcriptase (RT) domain of the *B. glabrata* non-LTR retrotansposon, *nimbus* (Figs. 8.1 and 8.2;

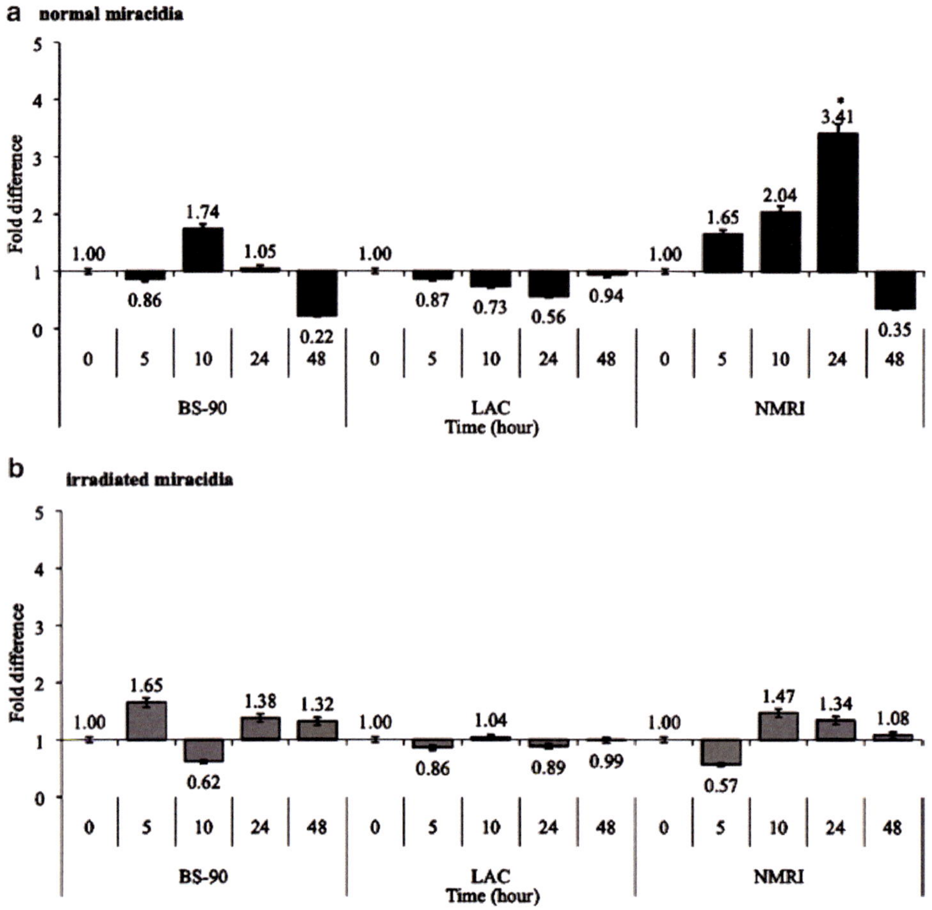

Fig. 8.1 Real-time Q-RT-PCR analysis of the differential gene expression of Hsp70 in snails that are resistant (BS-90), non-susceptible (LAC2), and susceptible (NMRI) following exposure for various time periods (0.5–3 h) to either normal (**a**) or irradiated attenuated miracidia (**b**). Fold difference of gene expression was calculated by comparing expression levels of the transcripts between normal and exposed snails by the comparative Ct method (Livak and Schmittgen 2001). Significant *P*-values of <0.05, <0.01, and <0.005 are indicated by *, **, and ***, respectively, to show the significance of gene expression by using Student's *t*-test. Reproduced from Ittiprasert et al. (2009) with permission from Elsevier

Raghavan et al. 2007). In contrast, results showed that both transcripts, Hsp70 and *nimbus* RT, remained down-regulated when juvenile susceptible snails were exposed to live attenuated miracidia. Because both the normal and the attenuated miracidia were found to penetrate the snail host (the head foot) at the same rate, these results clearly indicated that this differential induction of stress-related genes between the resistant and susceptible juvenile *B. glabrata* snails was caused by a factor(s) released by the developing parasite inside the snail and not due to injury/wound healing during penetration. The discovery that a stress-related mechanism might underlie the snail/schistosome relationship is intriguing and requires more investigation.

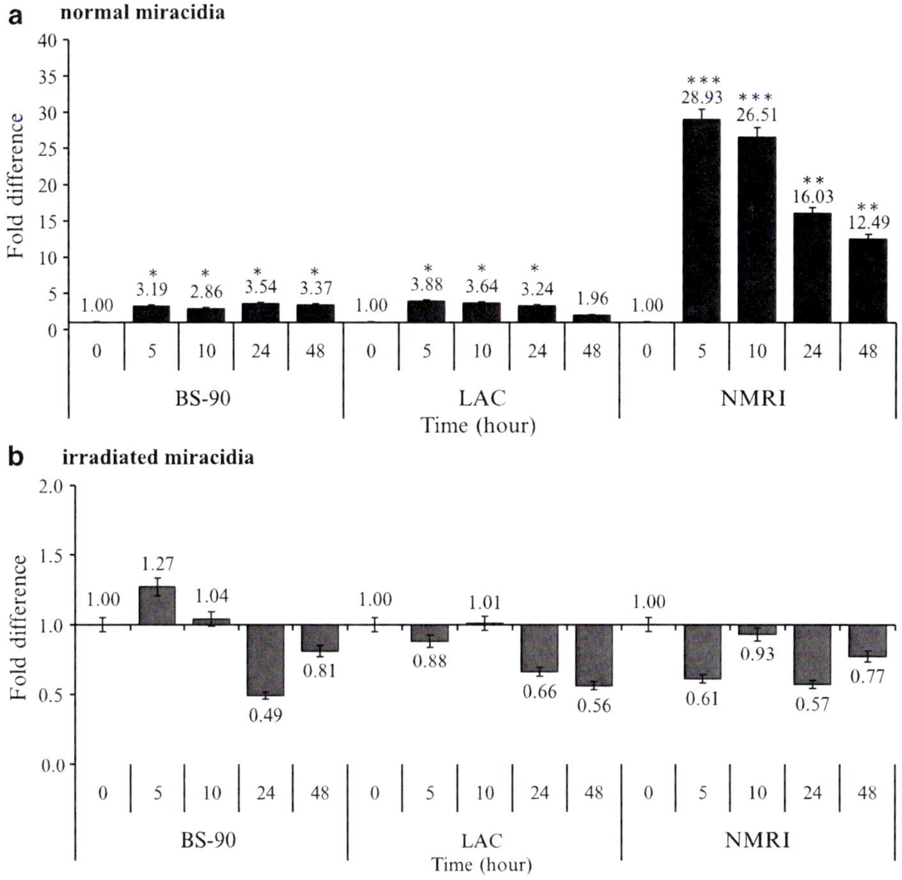

Fig. 8.2 Real-time Q-RT-PCR analysis of the differential gene expression of *nimbus* RT in snails that are resistant (BS-90), non-susceptible (LAC2), and susceptible (NMRI) following exposure for various time periods (0.5–3 h) to either normal (**a**) or irradiated attenuated miracidia (**b**). Fold difference of gene expression was calculated by comparing expression levels of the transcripts between normal and exposed snails by the comparative Ct method (Livak and Schmittgen 2001). Significant *P*-values of <0.05, <0.01, and <0.005 are indicated by *, **, and ***, respectively, to show the significance of gene expression by using Student's *t*-test. Reproduced from Ittiprasert et al. (2009) with permission from Elsevier

With these results now showing that the co-induction of Hsp70 and *nimbus* RT may be associated with the juvenile susceptibility phenotype, it will be possible to use the kinetics of induction of these transcripts as biomarkers for parasite infectivity to identify readily other transcripts (induced within the same time frame) that are relevant in the snail/schistosome interaction (Ittiprasert et al. 2009).

Aside from the above results showing that the expression of stress-related genes in the snail host may be involved in an as yet unknown mechanism(s) to promote and/or sustain the parasite's development, recent gene profiling studies using serial analysis of gene expression [SAGE] (Taft et al. 2009; Williams et al. 2007) of sporocysts (the intramolluscan stage of the parasite) also revealed similar expression of transcripts for Hsp70 in the sporocyst. Collectively, it is possible that these recent data could provide us with a new approach toward uncovering major stress-related genes that play a role in the determination of snail and schistosome compatibility.

8.5 Proteolytic and Redox Enzymes and Snail/Schistosome Interactions

Also identified as playing key roles in the outcome of snail/schistosome interactions are transcripts predominantly encoding proteolytic and redox enzymes (Table 8.1). Although a role for hydrolytic enzymes in the snail host/schistosome relation has long been described, the specific involvement of cysteine proteases has only recently come to light. Thus between resistant and susceptible snail stocks, Myers et al. (2008) showed that a higher basal level of cysteine protease was a characteristic feature of resistant and non-susceptible snails but not susceptible snails. Further molecular analysis revealed that the earlier and higher levels of expression of the transcript encoding the cysteine protease, cathepsin B, occurred in parasite-exposed juvenile resistant and non-susceptible but not susceptible snails. The full-length cDNA of this protease has been sequenced and studies are underway to characterize the cloned transcripts corresponding to other cysteine proteases, namely, cathepsin L and legumain. A role for lysozyme in the *B. glabrata*/schistosome relationship has also been described (Kassim and Richards 1978), and by Northern analysis, we have recently shown that like cysteine proteases, a higher expression of the lysozyme-encoding transcript occurs in the parasite-resistant compared to the susceptible snail (Myers and Knight, unpublished). In addition, we have detected diverse transcripts corresponding to this gene in the snail's hepatopancreas (Fig. 8.3).

Also being determined is the genotypic variation in the corresponding loci of these enzymes by Restriction Fragment Length Polymorphism (RFLP) and RFLP-PCR analysis. It is hoped that the ability to associate a particular variant form to different disease phenotypes will help toward the development of DNA-based tools to distinguish resistant from susceptible snails. The expression of the detoxification enzymes (GST and SOD), protease inhibitor (cystatin), and cysteine protease inhibitor (Table 8.1), including serine protease inhibitor, has also been shown to most likely play a role in the snail's behavior toward the parasite, as elevated levels of the transcripts corresponding to these different inhibitors were discovered in snails exposed to either schistosomes or echinostomes (Darani et al. 1997; Zhang et al. 2007).

Mechanisms of snail host resistance toward trematodes and the strategies that have co-evolved between the two organisms, enabling either survival or evasion, are not completely understood. It is clear, however, that reactive oxygen and nitrogen species, in addition to hydrogen peroxide released from hemocytes, are all damaging to the incoming parasite. Several of the antioxidant enzymes functioning to maintain the reduction–oxidation (redox) balance in the snail host have been described

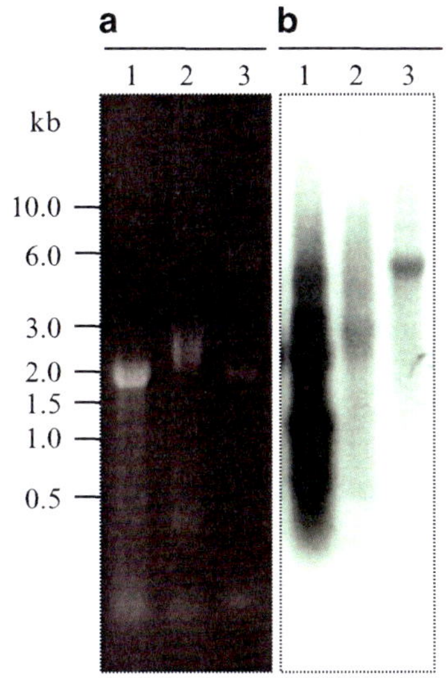

Fig. 8.3 (**a**) The ethidium bromide-stained agarose gel (1%) of resolved total RNA from (1) *B. glabrata* hepatopancreas, (2) adult *S. mansoni*, and (3) mouse liver. The MOPS/formaldehyde method as previously described by Miller et al. (2001) was used to denature the RNA samples, and the gel was blotted onto nitrocellulose paper before hybridization with a probe corresponding to *B. glabrata* lysozyme, shown in *panel B*. (**b**) Northern hybridization of total RNA from (1) *B. glabrata* hepatopancreas, (2) adult *S. mansoni*, and (3) mouse liver with ^{32}P-labeled cDNA insert corresponding to *B. glabrata* lysozyme. Note the presence of multiple transcripts in the tissue ranging in size from approximately 6 to 0.5 kb. Hybridizations and washes were done under stringent conditions as previously described (Miller et al. 2001)

(Zelck and Von Janowsky 2004; Jung et al. 2005; Goodall et al. 2006; Bender et al. 2007; Knight et al. 2009). These include the *B. glabrata* orthologs of copper/zinc superoxide dismutase (SOD1) and peroxiredoxin (BgPrx4). In the case of SOD1, different alleles of the enzyme have been shown to be associated with either resistance or susceptibility phenotypes to parasite infection (Goodall et al. 2006). Furthermore, significantly higher SOD enzymatic activity and transcription were found in hemocytes of resistant snails in accordance with the possession of the allele associated with this phenotype (Davids and Yoshino 1998; Goodall et al. 2004; Bender et al. 2007).

The other antioxidant enzyme that is also highly expressed in resistant snails compared to susceptible ones is peroxiredoxin. The snail enzyme is highly conserved and related (by sequence and phylogenetic analysis) to the human ortholog, PRDX 4. Interestingly, diverse transcripts corresponding to this enzyme were detected between resistant and susceptible snails following exposure to schistosomes (Knight et al. 2009). While various sized transcripts were identified by Northern blot analysis and by the sequencing of two diverse forms (753 bp and 516 bp) of the transcript encoding *Bg*Prx4, Western blot analysis, on the contrary, revealed the occurrence of a single protein product (29–30 kDa) in the hepatopancreas and albumen gland of the snail. The role of Prx and SOD1 in removing harmful free radicals to protect the snail host and parasite has been recognized but less is known about either the natural substrates of these antioxidants, the nature of the soluble parasite molecule [released ES products] that triggers their expression, or the potential role of these antioxidants in cell signaling. The subject of parasite-mediated cell signal transduction in the snail host has been covered extensively in previous reviews on this subject as well as in chapter 5 of this book (see also review by Yoshino et al. 2006).

8.6 Characteristics of the *B. glabrata* Genome

8.6.1 Genome Complexity

Compared to the parasite (*S. mansoni* = 363 Mb), *B. glabrata* possesses a relatively large genome (931 Mb; Gregory 2003) that is organized into 18 haploid chromosomes that are small and monomorphic. Earlier attempts to karyotype these chromosomes designated them into nine groups according to their size, centromere position, and banding (Goldman et al. 1984). Using the only molluscan cell line in existence, the embryonic cell line Bge (Hansen 1976) that was derived in the 1970s by Eder Hansen, gene mapping studies with Fluorecence *In Situ* Hybridization (FISH) were recently undertaken (Odoemelam et al. 2009). Although these studies showed the existence of anomalies within the Bge cell line chromosomes (mostly aneuploidy), single-copy genes were physically mapped for the first time using this cell line. Experiments to map *ex vivo* chromosomes are currently underway and novel techniques for preparing chromosome spreads from the snail's ovotestis are being developed (Bridger and Arican, unpublished).

For physical mapping purposes, probes corresponding to specific Bacterial Artificial Chromosome (BAC) clones are labeled with biotin and used as probes for hybridization onto chromosomes; followed by detection with streptavidin-conjugated cyanine 3, and counterstaining with 4,6-diamidino-2-phenylinodole (DAPI), as described by Odoemelam et al. (2009). An example of metaphase chromosomes of Bge cells hybridized with BAC probes corresponding to *Bg*Prx4 and *piwi* is shown in Fig. 8.4 (Knight et al. 2009). Two BAC libraries for *B. glabrata* are currently available. One of the BAC libraries, with ninefold coverage, was made from a wild-type susceptible snail, BB02, the same snail stock that was chosen for full-genome sequencing. This library is commercially available from the Arizona Genomics Institute (Adema et al. 2006). The other BAC library was made from the resistant BS-90 snail stock (Raghavan et al. 2007). Several individual clones (7 completed and 6 are being finished) from the BAC (BB02) library have been fully sequenced as part of the overall strategy to obtain the genome sequence of this snail (Sandy Clifton, Washington University, personal communication).

Compared to the aforementioned mollusks for which genome sequencing projects are also ongoing, *B. glabrata* alone offers the unique opportunity toward elucidating how the close association of the snail host and it is obligate parasite has helped in

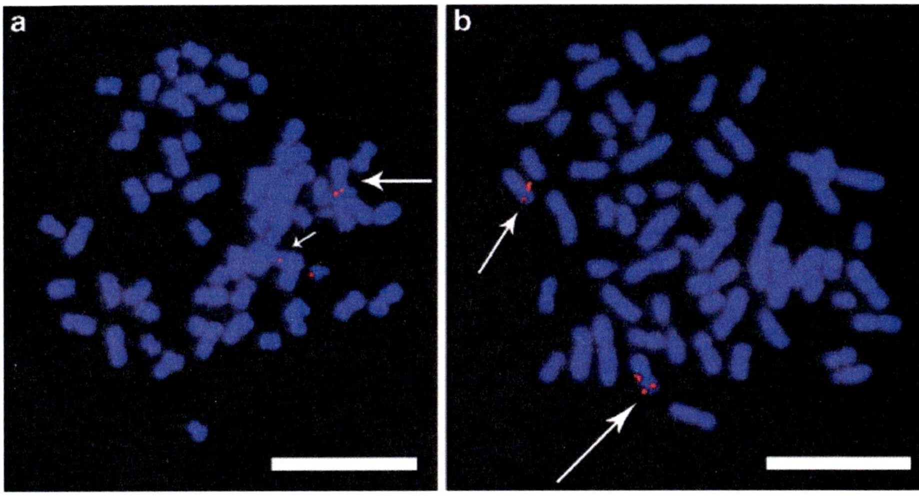

Fig. 8.4 FISH using BAC probes corresponding to (**a**) Prx or (**b**) *piwi* to metaphase spreads from Bge cell chromosomes. "Reprinted from The karyotyping and gene mapping of the *Biomphalaria glabrata* embryonic (Bge) cell line". Reproduced from Odoemelam et al. (2009) with permission from Elsevier

shaping the genomes of both these organisms. It has previously been noted that in the life history traits of schistosomes, the genetic variability of the snail host (rather than the human host) may be a more significant factor in influencing the variability of the parasite (Lewis et al. 2001; Sandland et al. 2007; Zavodna et al. 2008).

It is possible, therefore, that the anticipated forthcoming data from the snail genome project (Raghavan and Knight 2006) will help launch a new field of study involving this snail as a future model organism in our understanding of molecular events in resolving the complex dynamics of host/pathogen interactions.

Of the two organisms (*B. glabrata* and *S. mansoni*), the entire nuclear genome sequence of the parasite has recently been determined (Berriman et al. 2009). Although that of the snail still lags behind, steady progress is being made. The genome-sequencing project for this snail was initiated in 2005 at the Washington University Genome Sequencing (WUGS) Center. Accordingly, the project has produced an exponential increase in the number of *B. glabrata* sequences in GenBank. While most of the sequences correspond to Expressed Sequenced Tags (ESTs) derived from either whole snails or different tissues, other sequences correspond to trace reads from shot-gun sequencing of the genome or from individual BAC clones.

Like its parasite, the *B. glabrata* mitochondrial genome is also AT rich (74.6%) and highly repetitive (DeJong et al. 2004). Presumably, the repetitive nature of the genome could pose a challenge to attain a fully assembled sequence for this organism. This challenge has already been tested in the sequencing of the *S. mansoni* genome, as well as in other projects that involved other parasites with equally repetitive genomes, such as *Plasmodium falciparum*, a causative agent of malaria. The one major drawback is the size of these molluscan genomes. At 931 Mb, the *B. glabrata* genome stands considerably smaller than the 1,800 Mb genome of *A. californica*. It is hoped that once the sequence information from all the ongoing snail-sequencing projects, including *Lottia gigantea* (genome size 359.5 Mbp), become available, we will be able to draw comparisons between these sequences toward fulfilling the goal of obtaining full coverage and quality annotation of all the sequences available. Comparative genome analysis between the snail and human hosts should also help to reveal key molecular targets that may be involved in helping this parasite survive in either host.

8.6.2 Genome Plasticity and Mobile Genetic Elements

The repetitiveness of the *B. glabrata* genome is now considered as a characteristic feature of this organism, but little is known about repeat families/units that make up this genome. As mentioned above, the inherent genetic variation of the snail in relation to schistosomes has long been appreciated but the molecular mechanism(s) that help to generate these variations in the snail genome is (are) unknown. Mobile genetic elements (MGEs) in the genome are responsible for generating mutations within the genomes they occupy (Knight et al. 2009). To date, only one MGE from *B. glabrata* has been fully sequenced and characterized. The element, named *nimbus*, is a non-LTR-retrotransposon and is a close relative of the human *L1*; *Aedes aegypti* element, *mosquI*; *Drosophila melanogaster*, I-factor; *ingi* of trypanosomes; and *CiI* of *Ciona intestinalis* (Raghavan et al. 2007; Knight et al. 2009). *Nimbus* exists as an interspersed middle repetitive element in the snail's genome. Sequence analysis of a BAC clone (65,764 bp, BRIBAC72bg_line5; GenBank Acc No.EF418587) that contained 10 copies of *nimbus* (full-length and truncated forms) among other MGEs (*gypsy*, CR1, and RTE) revealed that the full element was ~5.8 kb, encompassing two open reading frames (ORF1 and ORF 2) that contained hallmark domains typical of all class I non-LTR retrotransposons. ORF1 encodes a nucleic acid-binding protein, and ORF2 encodes the apurinic/apyrimidinic endonuclease (APE), reverse transcriptase (RT), and RNase H (RNH). As mentioned above, the RT domain of the element is co-regulated with Hsp70 as part of a stress-like response operating in susceptible juvenile snails in response to schistosome infection.

An intriguing discovery made while assessing the possible existence of *nimbus*-related sequences in the parasite was that the non-coding 5′ and 3′ regions (5′ and 3′ NCR) of the *nimbus* element is present in the genomic sequences (whole genome shotgun sequences in NCBI) of *S. mansoni*. The presence of homologous 5′ and 3′ NCR sequences of *nimbus* in the parasite may be an indication of possible lateral transfer of host sequences into the parasite. There is as yet no explanation for the presence of *nimbus*-related sequences in the parasite, and more investigations will have to be done to ascertain the biological relevance (if any) of sharing of these sequences between the snail host and parasite.

8.6.3 Inter- and Intra-Snail-Stock Polymorphisms

Studies to determine the degree of polymorphisms between and among snail stocks have mostly been geared toward the development of a stable DNA-based method to identify snails with different parasite susceptibility phenotypes. Earlier studies of inter- and intra-genetic variations in *B. glabrata* snails were done using ribosomal probes to detect polymorphisms within the ribosomal locus. And, since the majority of probes utilized encompassed the non-coding regions of the ribosomal gene cassette (external transcribed spacer and internal transcribed spacer), sequence variations within these regions were readily detected and formed the basis of diagnostics for snail species and strain differentiation (Vidigal et al. 1998; Mavarez et al. 2002). Other studies for snail differentiation were undertaken using Southern blots, followed by hybridizations to either radioactive or non-radioactive probes related to ribosomal genes (Knight et al. 1991; Kane and Rollinson 1994; Vidigal et al. 1994; Stothard et al. 1996). Polymorphisms were revealed by the sequence variations underpinning restriction enzyme recognition sites generating RFLPs associated with variant forms for the different snails in question.

Genotypic variations in different snail isolates were also revealed by PCR-based techniques, such as Random Amplified Polymorphic DNA (RAPD), and variations in Simple Sequence Repeats (SSRs), also known as microsatellites. Using the RAPD method, Vidigal et al. (1994) described the extensive variations in snail populations in Brazil. In the last decade, the RAPD technique, despite its shortfalls, has become the mainstay method for revealing snail and strain genetic variations, and for following heritable DNA markers associated with disease resistance and susceptibility (Knight et al. 1999). Variation in microsatellite loci has also been useful in assessing genetic diversity in snail populations and investigating mating behavior. Because of the successful application of these methods, we are now in a position to develop high-density genetic linkage maps for this snail (Rollinson et al. 1998, 2009; Jones et al. 1999; Campos et al. 2002). Several comprehensive reviews have been written on this subject, and the application of this method in conjunction with the aforementioned physical mapping by FISH will help toward the eventual assembly of the snail genomic sequences once the full genome sequence of this snail becomes available.

8.7 Concluding Remarks

As an important intermediate snail host that is integral to the transmission of a significant human pathogen, it is clear that we will continue to work on the molecular aspects of parasitism using *B. glabrata* as a model organism to understand fully the molecular basis of host/parasite interactions. With the use of the latest technologies, especially high-throughput genome sequencing and proteomics, the next decade should see an exponential increase in molecular information related to this snail and its parasite.

This review has been an attempt to document the progress that has been made in the past two decades in snail genomics and proteomics. Although we have only seen small advances so far, the progress we have made has clearly provided us with important glimpses of some of the molecular events involved in snail–trematode infections. Uncovering the rest of the puzzle must await the anticipated completion of the whole genome-sequencing project for *B. glabrata*. This and the commitment of funding to resolve schistosome/snail-related basic research should provide the impetus to sustain and attract investigators into the field of molecular malacology that will make it possible to fulfill the eventual goal of using transgenic snails to help with the eradication of schistosomiasis.

Acknowledgments We wish to thank Andre Miller for his helpful suggestions with the writing of this manuscript. We also thank Pat Caspar for helping us with the irradiation experiments, and Frances Barnes for her technical support. This work was supported by NIH-NIAID grant no. R01-AI63480.

References

Adema CM, Luo MZ, Hanelt B, Hertel LA, Marshall JJ, Zhang SM, DeJong RJ, Kim HR, Kudrna D, Wing RA, Soderlund C, Knight M, Lewis FA, Caldeira RL, Jannotti-Passos LK, Carvalho Odos S, Loker ES (2006) A bacterial artificial chromosome library for *Biomphalaria glabrata*, intermediate snail host of *Schistosoma mansoni*. Mem Inst Oswaldo Cruz 101:S167–S177

Bayne CJ (2009) Successful parasitism of vector snail *Biomphalaria glabrata* by the human blood fluke (trematode) *Schistosoma mansoni*: a 2009 assessment. Mol Biochem Parasitol 165:8–18

Bender RC, Goodall CP, Blouin MS, Bayne CJ (2007) Variation in expression of *Biomphalaria glabrata* SOD1: a potential controlling factor in susceptibility/resistance to *Schistosoma mansoni*. Dev Comp Immunol 31:874–878

Bergquist R, Johansen MV, Utzinger J (2009) Diagnostic dilemmas in helminthology: what tools to use and when? Trends Parasitol 25:151–156

Berriman M, Haas BJ, Loverde PT, Wilson RA, Dillon GP, Cerqueira GC, Mashiyama ST, Al-Lazikani B, Andrade LF, Ashton PD, Aslett MA, Bartholomeu DC, Blandin G, Caffrey CR, Coghlan A, Coulson R, Day TA, Delcher A, Demarco R, Djikeng A, Eyre T, Gamble JA, Ghedin E, Gu Y, Hertz-Fowler C, Hirai H, Hirai Y, Houston R, Ivens A, Johnston DA, Lacerda D, Macedo CD, Mcveigh P, Ning Z, Oliveira G, Overington JP, Parkhill J, Pertea M, Pierce RJ, Protasio AV, Quail MA, Rajandream MA, Rogers J, Sajid M, Salzberg SL, Stanke M, Tivey AR, White O, Williams DL, Wortman J, Wu W, Zamanian M, Zerlotini A, Fraser-Liggett CM, Barrell BG, El-Sayed NM (2009) The genome of the blood fluke *Schistosoma mansoni*. Nature 460:352–358

Bouchut A, Roger E, Coustau C, Gourbal B, Mitta G (2006) Compatibility in the *Biomphalaria glabrata/Echinostoma caproni* model: potential involvement of adhesion genes. Int J Parasitol 36:175–184

Bransteitter R, Pham P, Scharff MD, Goodman MF (2003) Activation-induced cytidine deaminase deaminates deoxycytidine on single-stranded DNA but requires the action of RNase. Proc Natl Acad Sci USA 100:4102–4107

Campos YR, Carvalho OS, Goveia CO, Romanha AJ (2002) Genetic variability of the main intermediate host of the *Schistosoma mansoni* in Brazil, *Biomphalaria glabrata* (Gastropoda: Planorbidae) assessed by SSR-PCR. Acta Trop 83:19–27

Chitsulo L, Loverde P, Engels D (2004) Schistosomiasis. Nat Rev Microbiol 2:12–13

Cooper LA, Richards CS, Lewis FA, Minchella DJ (1994) *Schistosoma mansoni*: relationship between low fecundity and reduced susceptibility to parasite infection in the snail *Biomphalaria glabrata*. Exp Parasitol 79:21–28.

Darani HY, Curtis RH, McNeice C, Price HP, Sayers JR, Doenhoff MJ (1997) *Schistosoma mansoni*: anomalous immunogenic properties of a 27 kDa larval serine protease associated with protective immunity. Parasitology 115:237–247

Davids BJ, Yoshino TP (1998) Integrin-like RGD-dependent binding mechanism involved in the spreading response of circulating molluscan phagocytes. Dev Comp Immunol 22:39–53

DeJong RJ, Emery AM, Adema CM (2004) The mitochondrial genome of *Biomphalaria glabrata* (Gastropoda: Basommatophora), intermediate host of *Schistosoma mansoni*. J Parasitol 90:991–997

Duclermortier P, Lardans V, Serra E, Trottein F, Dissous C (1999) *Biomphalaria glabrata* embryonic cells express a protein with a domain homologous to the lectin domain of mammalian selectins. Parasitol Res 85:481–486.

Goldman MA, Loverde PT, Chrisman L, Franklin DA (1984) Chromosomal evolution in planorbid snails of the genera *Bulinus* and *Biomphalaria*. Malacologia 25:427–446

Goodall CP, Bender RC, Broderick EJ, Bayne CJ (2004) Constitutive differences in Cu/Zn superoxide dismutase mRNA levels and activity in hemocytes of *Biomphalaria glabrata* (Mollusca) that are either susceptible or resistant to *Schistosoma mansoni* (Trematoda). Mol Biochem Parasitol 137:321–328

Goodall CP, Bender RC, Brooks JK, Bayne CJ (2006) *Biomphalaria glabrata* cytosolic copper/zinc superoxide dismutase (SOD1) gene: association of SOD1 alleles with resistance/susceptibility to *Schistosoma mansoni*. Mol Biochem Parasitol 147:207–210

Gregory TR (2003) Genome size estimates for two important freshwater molluscs, the zebra mussel (Dreissena polymorpha) and the schistosomiasis vector snail (Biomphalaria glabrata). Genome 46:841–844.

Gryseels B, Polman K, Clerinx J, Kestens L (2006) Human schistosomiasis. Lancet 368:1106–1118

Guillou F, Mitta G, Galinier R, Coustau C (2007a) Identification and expression of gene transcripts generated during an anti-parasitic response in *Biomphalaria glabrata*. Dev Comp Immunol 31:657–671

Guillou F, Roger E, Mone Y, Rognon A, Grunau C, Theron A, Mitta G, Coustau C, Gourbal BE (2007b) Excretory-secretory proteome of larval *Schistosoma mansoni* and *Echinostoma caproni*, two parasites of *Biomphalaria glabrata*. Mol Biochem Parasitol 155:45–56

Hahn UK, Bender RC, Bayne CJ (2001) Killing of *Schistosoma mansoni* sporocysts by hemocytes from resistant *Biomphalaria glabrata*: role of reactive oxygen species. J Parasitol 87:292–299

Hanelt B, Lun CM, Adema CM (2008) Comparative ORESTES-sampling of transcriptomes of immune-challenged *Biomphalaria glabrata* snails. J Invertebr Pathol 99:192–203

Hansen EL (1976) A cell line from embryos of *Biomphalaria glabrata* (Pulmonata): Establishment and characteristics. In: Maramorosch K (ed.) In Invertebrate Tissue Culture: Research Applications. New York: Academic Press; pp. 75–97

Hertel LA, Adema CM, Loker ES (2005) Differential expression of FREP genes in two strains of *Biomphalaria glabrata* following exposure to the digenetic trematodes *Schistosoma mansoni* and *Echinostoma paraensei*. Dev Comp Immunol 29:295–303

Hubendick B (1958) A possible method of schistosome-vector control by competition between resistant and susceptible strains. Bull World Health Organ 18:113–116

Ittiprasert W, Miller A, Myers J, Nene V, El-Sayed NM, Knight M (2010) Identification of immediate response genes dominantly expressed in juvenile resistant and susceptible Biomphalaria glabrata snails upon exposure to *Schistosoma mansoni*. Mol Biochem Parasitol 169(1):27–39

Ittiprasert W, Nene R, Miller A, Raghavan N, Lewis F, Hodgson J, Knight M (2009) *Schistosoma mansoni* infection of juvenile *Biomphalaria glabrata* induces a differential stress response between resistant and susceptible snails. Exp Parasitol 123:203–211

Jiang Y, Loker ES, Zhang SM (2006) In vivo and in vitro knockdown of FREP2 gene expression in the snail *Biomphalaria glabrata* using RNA interference. Dev Comp Immunol 30:855–866

Jones CS, Lockyer AE, Rollinson D, Piertney SB, Noble LR (1999) Isolation and characterization of microsatellite loci in the freshwater gastropod, *Biomphalaria glabrata*, an intermediate host for *Schistosoma mansoni*. Mol Ecol 8:2149–2151

Jung Y, Nowak TS, Zhang SM, Hertel LA, Loker ES, Adema CM (2005) Manganese superoxide dismutase from *Biomphalaria glabrata*. J Invertebr Pathol 90:59–63

Kane RA, Rollinson D (1994) Repetitive sequences in the ribosomal DNA internal transcribed spacer of *Schistosoma haematobium, Schistosoma intercalatum* and *Schistosoma mattheei*. Mol Biochem Parasitol 63:153–156

Kassim OO, Richards CS (1978) *Biomphalaria glabrata*: lysozyme activities in the hemolymph, digestive gland, and headfoot of the intermediate host of *Schistosoma mansoni*. Exp Parasitol 46:218–224

Knight M, Brindley PJ, Richards CS, Lewis FA (1991) *Schistosoma mansoni*: use of a cloned ribosomal RNA gene probe to detect restriction fragment length polymorphisms in the intermediate host *Biomphalaria glabrata*. Exp Parasitol 73:285–294

Knight M, Miller AN, Patterson CN, Rowe CG, Michaels G, Carr D, Richards CS, Lewis FA (1999) The identification of markers segregating with resistance to *Schistosoma mansoni* infection in the snail *Biomphalaria glabrata*. Proc Natl Acad Sci USA 96:1510–1515

Knight M, Raghavan N, Goodall C, Cousin C, Ittiprasert W, Sayed A, Miller A, Williams DL, Bayne CJ (2009) *Biomphalaria glabrata* peroxiredoxin: effect of *Schistosoma mansoni* infection on differential gene regulation. Mol Biochem Parasitol 167:20–31

Korneev SA, Kemenes I, Straub V, Staras K, Korneeva EI, Kemenes G, Benjamin PR, O'Shea M (2002) Suppression of nitric oxide (NO)-dependent behavior by double-stranded RNA-mediated silencing of a neuronal NO synthase gene. J Neurosci 22:RC227

Leonard PM, Adema CM, Zhang SM, Loker ES (2001) Structure of two FREP genes that combine IgSF and fibrinogen domains, with comments on diversity of the FREP gene family in the snail *Biomphalaria glabrata*. Gene 269:155–165

Lewis FA, Patterson CN, Grzywacz C (2003) Parasite-susceptibility phenotypes of F1 *Biomphalaria glabrata* progeny derived from interbreeding *Schistosoma mansoni*-resistant and -susceptible snails. Parasitol Res 89:98–101

Lewis FA, Patterson CN, Knight M, Richards CS (2001) The relationship between *Schistosoma mansoni* and *Biomphalaria glabrata*: genetic and molecular approaches. Parasitology 123:S169–S179

Lie KJ, Jeong KH, Heyneman D (1980) Tissue reactions induced by *Schistosoma mansoni* in *Biomphalaria glabrata*. Ann Trop Med Parasitol 74:157–166

Livak KJ, Schmittgen TD (2001) Analysis of relative gene expression data using real-time quantitative PCR and the 2(-Delta Delta C(T)) Method. Methods 25:402–408

Lockyer AE, Jones CS, Noble LR, Rollinson D (2000) Use of differential display to detect changes in gene expression in the intermediate snail host *Biomphalaria glabrata* upon infection with *Schistosoma mansoni*. Parasitology 120:399–407

Lockyer AE, Noble LR, Rollinson D, Jones CS (2004) *Schistosoma mansoni*: resistant specific infection-induced gene expression in *Biomphalaria glabrata* identified by fluorescent-based differential display. Exp Parasitol 107:97–104

Lockyer AE, Spinks J, Kane RA, Hoffmann KF, Fitzpatrick JM, Rollinson D, Noble LR, Jones CS (2008) *Biomphalaria glabrata* transcriptome: cDNA microarray profiling identifies resistant- and susceptible-specific gene expression in haemocytes from snail strains exposed to *Schistosoma mansoni*. BMC Genomics 9:634

Lockyer AE, Spinks J, Noble LR, Rollinson D, Jones CS (2007) Identification of genes involved in interactions between *Biomphalaria glabrata* and *Schistosoma mansoni* by suppression subtractive hybridization. Mol Biochem Parasitol 151:18–27

Loker ES, Bayne CJ, Buckley PM, Kruse KT (1982) Ultrastructure of encapsulation of *Schistosoma mansoni* mother sporocysts by hemocytes of juveniles of the 10-R2 strain of *Biomphalaria glabrata*. J Parasitol 68:84–94

Loker ES, Bayne CJ, Yui MA (1986) *Echinostoma paraensei*: hemocytes of *Biomphalaria glabrata* as targets of echinostome mediated interference with host snail resistance to *Schistosoma mansoni*. Exp Parasitol 62:149–154

Mavarez J, Pointier JP, David P, Delay B, Jarne P (2002) Genetic differentiation, dispersal and mating system in the schistosome-transmitting freshwater snail *Biomphalaria glabrata*. Heredity 89:258–265

Miller AN, Raghavan N, Fitzgerald PC, Lewis FA, Knight M (2001) Differential gene expression in haemocytes of the snail *Biomphalaria glabrata*: effects of *Schistosoma mansoni* infection. Int J Parasitol 31:687–696

Mitta G, Galinier R, Tisseyre P, Allienne JF, Girerd-Chambaz Y, Guillou F, Bouchut A, Coustau C (2005) Gene discovery and expression analysis of immune-relevant genes from *Biomphalaria glabrata* hemocytes. Dev Comp Immunol 29:393–407

Myers J, Ittiprasert W, Raghavan N, Miller A, Knight M (2008) Differences in cysteine protease activity in *Schistosoma mansoni*-resistant and -susceptible *Biomphalaria glabrata* and characterization of the hepatopancreas cathepsin B Full-length cDNA. J Parasitol 94:659–668

Newton WL (1955) The establishment of a strain of *Australorbis glabratus* which combines albinism and high susceptibility to infection with *Schistosoma mansoni*. J Parasitol 41:526–528

Niemann GM, Lewis FA (1990) *Schistosoma mansoni*: influence of *Biomphalaria glabrata* size on susceptibility to infection and resultant cercarial production. Exp Parasitol 70:286–292

Nowak TS, Woodards AC, Jung Y, Adema CM, Loker ES (2004) Identification of transcripts generated during the response of resistant *Biomphalaria glabrata* to *Schistosoma mansoni* infection using suppression subtractive hybridization. J Parasitol 90:1034–1040

Odoemelam E, Raghavan N, Miller A, Bridger JM, Knight M (2009) Revised karyotyping and gene mapping of the *Biomphalaria glabrata* embryonic (Bge) cell line. Int J Parasitol 39:675–681

Paraense W, Correa L (1963) Variations in susceptibility of populations of *Australorbis glabratus* to a strain of *Schistosoma mansoni*. Rev Inst Med Trop Sao Paulo 5:15–22

Raghavan N, Knight M (2006) The snail (*Biomphalaria glabrata*) genome project. Trends Parasitol 22:148–151

Raghavan N, Miller AN, Gardner M, Fitzgerald PC, Kerlavage AR, Johnston DA, Lewis FA, Knight M (2003) Comparative gene analysis of *Biomphalaria glabrata* hemocytes pre- and post-exposure to miracidia of *Schistosoma mansoni*. Mol Biochem Parasitol 126:181–191

Raghavan N, Tettelin H, Miller A, Hostetler J, Tallon L, Knight M (2007) *Nimbus* (BgI): an active non-LTR retrotransposon of the *Schistosoma mansoni* snail host *Biomphalaria glabrata*. Int J Parasitol 37:1307–1318

Richards CS (1973) Susceptibility of adult *Biomphalaria glabrata* to Schistosoma mansoni infection. Am J Trop Med Hyg 22:748–756.

Richards CS (1975a) Genetics of pigmentation in *Biomphalaria straminea*. Am J Trop Med Hyg 24:154–156

Richards CS (1975b) Genetic factors in susceptibility of *Biomphalaria glabrata* for different strains of *Schistosoma mansoni*. Parasitology 70:231–241

Richards CS (1975c) Genetic studies on variation in infectivity of *Schistosoma mansoni*. J Parasitol 61:233–236

Richards CS, Minchella DJ (1987) Transient non-susceptibility to *Schistosoma mansoni* associated with atrial amoebocytic accumulations in the snail host *Biomphalaria glabrata*. Parasitology 95:499–505

Richards CS, Shade PC (1987) The genetic variation of compatibility in *Biomphalaria glabrata* and *Schistosoma mansoni*. J Parasitol 73:1146–1151

Roger E, Gourbal B, Grunau C, Pierce RJ, Galinier R, Mitta G (2008) Expression analysis of highly polymorphic mucin proteins (Sm PoMuc) from the parasite *Schistosoma mansoni*. Mol Biochem Parasitol 157:217–227

Roger E, Grunau C, Pierce RJ, Hirai H, Gourbal B, Galinier R, Emans R, Cesari IM, Cosseau C, Mitta G (2008a) Controlled Chaos of Polymorphic Mucins in a Metazoan Parasite (*Schistosoma mansoni*) Interacting with Its Invertebrate Host (*Biomphalaria glabrata*). PLoS Negl Trop Dis 2:e330.

Roger E, Mitta G, Mone Y, Bouchut A, Rognon A, Grunau C, Boissier J, Theron A, Gourbal BE (2008b) Molecular determinants of compatibility polymorphism in the Biomphalaria glabrata/Schistosoma mansoni model: new candidates identified by a global comparative proteomics approach. Mol Biochem Parasitol 157:205–216.

Rollinson D, Stothard JR, Jones CS, Lockyer AE, De Souza CP, Noble LR (1998) Molecular characterisation of intermediate snail hosts and the search for resistance genes. Mem Inst Oswaldo Cruz 93:111–116

Rollinson D, Webster JP, Webster B, Nyakaana S, Jorgensen A, Stothard JR (2009) Genetic diversity of schistosomes and snails: implications for control. Parasitology 27:1–11

Sandland GJ, Foster AV, Zavodna M, Minchella DJ (2007) Interplay between host genetic variation and parasite transmission in the *Biomphalaria glabrata-Schistosoma mansoni* system. Parasitol Res 101:1083–1089

Schneider O, Zelck UE (2001) Differential display analysis of hemocytes from schistosome-resistant and schistosome-susceptible intermediate hosts. Parasitol Res 87:489–491

Stothard JR, Hughes S, Rollinson D (1996) Variation within the internal transcribed spacer (ITS) of ribosomal DNA genes of intermediate snail hosts within the genus *Bulinus* (Gastropoda: Planorbidae). Acta Trop 61:19–29

Stothard JR, Mgeni AF, Khamis S, Seto E, Ramsan M, Hubbard SJ, Kristensen TK, Rollinson D (2002) New insights into the transmission biology of urinary schistosomiasis in Zanzibar. Trans R Soc Trop Med Hyg 96:470–475

Sullivan JT, Spence JV (1994) Transfer of resistance to *Schistosoma mansoni* in *Biomphalaria glabrata* by allografts of amoebocyte-producing organ. J Parasitol 80:449–453

Sullivan JT, Spence JV, Nunez JK (1995) Killing of *Schistosoma mansoni* sporocysts in *Biomphalaria glabrata* implanted with amoebocyte-producing organ allografts from resistant snails. J Parasitol 81:829–833

Taft AS, Vermeire JJ, Bernier J, Birkeland SR, Cipriano MJ, Papa AR, McArthur AG, Yoshino TP (2009) Transcriptome analysis of *Schistosoma mansoni* larval development using serial analysis of gene expression (SAGE). Parasitology 136:469–485

Vidigal TH, Dias Neto E, Carvalho Odos S, Simpson AJ (1994) *Biomphalaria glabrata*: extensive genetic variation in Brazilian isolates revealed by random amplified polymorphic DNA analysis. Exp Parasitol 79:187–194

Vidigal TH, Dias Neto E, Spatz L, Nunes DN, Pires ER, Simpson AJ, Carvalho OS (1998) Genetic variability and identification of the intermediate snail hosts of *Schistosoma mansoni*. Mem Inst Oswaldo Cruz 93:103–110

Wang LD, Guo JG, Wu XH, Chen HG, Wang TP, Zhu SP, Zhang ZH, Steinmann P, Yang GJ, Wang SP, Wu ZD, Wang LY, Hao Y, Bergquist R, Utzinger J, Zhou XN (2009) China's new strategy to block *Schistosoma japonicum* transmission: experiences and impact beyond schistosomiasis. Trop Med Int Health 14(12):1475–1483

Williams DL, Sayed AA, Bernier J, Birkeland SR, Cipriano MJ, Papa AR, Mcarthur AG, Taft A, Vermeire JJ, Yoshino TP (2007) Profiling *Schistosoma mansoni* development using serial analysis of gene expression (SAGE). Exp Parasitol 117:246–258

Wu W, Niles EG, Hirai HA, LoVerde PT (2007) Evolution of a novel subfamily of nuclear receptors with members that each contain two DNA binding domains. BMC Evol Biol 7:27

Yoshino TP, Dinguirard N, Kunert J, Hokke CH (2008) Molecular and functional characterization of a tandem-repeat galectin from the freshwater snail *Biomphalaria glabrata*, intermediate host of the human blood fluke *Schistosoma mansoni*. Gene 411:46–58

Yoshino TP, Vermeire JJ, Humphreys JE (2006) Signal transduction at the host-parasite interface. In: Maule AG, Mark NJ (eds) Parasitic flatworms molecular biology, biochemistry, immunology and physiology. CAB International, Belfast, pp 210–224

Zavodna M, Sandland GJ, Minchella DJ (2008) Effects of intermediate host genetic background on parasite transmission dynamics: a case study using *Schistosoma mansoni*. Exp Parasitol 120:57–61

Zelck UE, Von Janowsky B (2004) Antioxidant enzymes in intramolluscan *Schistosoma mansoni* and ROS-induced changes in expression. Parasitology 128:493–501

Zhang SM, Leonard PM, Adema CM, Loker ES (2001) Parasite-responsive IgSF members in the snail *Biomphalaria glabrata*: characterization of novel genes with tandemly arranged IgSF domains and a fibrinogen domain. Immunogenetics 53:684–694

Zhang SM, Loker ES (2004) Representation of an immune responsive gene family encoding fibrinogen-related proteins in the freshwater mollusc *Biomphalaria glabrata*, an intermediate host for *Schistosoma mansoni*. Gene 341:255–266

Zhang SM, Nian H, Wang B, Loker ES, Adema CM (2009) Schistosomin from the snail Biomphalaria glabrata: expression studies suggest no involvement in trematode-mediated castration. Mol Biochem Parasitol 165(1):79–86.

Zhang SM, Nian H, Zeng Y, DeJong RJ (2008) Fibrinogen-bearing protein genes in the snail *Biomphalaria glabrata*: characterization of two novel genes and expression studies during ontogenesis and trematode infection. Dev Comp Immunol 32:1119–1130

Zhang SM, Zeng Y, Loker ES (2007) Characterization of immune genes from the schistosome host snail *Biomphalaria glabrata* that encode peptidoglycan recognition proteins and gram-negative bacteria binding protein. Immunogenetics 59:883–898

Chapter 9
The Biological Control of the Snail Hosts of Schistosomes: The Role of Competitor Snails and Biological Invasions

Jean-Pierre Pointier, Patrice David, and Philippe Jarne

Abstract Biological control of the snail hosts of schistosomes has been considered in the last few decades as an alternative to molluscicides. Several groups of organisms have been proposed to control snail hosts, but very few have proven their efficacy in the field. Competitor snails can be considered as the most efficient biological control agents and numerous promising laboratory studies and field experiments have been carried out, mainly in the Caribbean. Two species of competitor caenogastropod snails belonging to the Ampullariidae (*Marisa cornuarietis*) and Thiaridae (*Melanoides tuberculata*) families have succeeded in eliminating or reducing populations of schistosome-transmitting snails, especially *Biomphalaria glabrata* in several different habitats in St Lucia, Martinique, and Guadeloupe. However, their efficiency is context-dependent. Caenogastropods are good competitors in relatively stable habitats only when long-term resource exploitation rather than colonization is the limiting factor. At the same time, unassisted invasions by these species and by other freshwater snails, including numerous pulmonates, were detected in the 1950s, followed by rapid spread in the following decades to most Neotropical areas. These invasions were largely responsible for the general decline of *B. glabrata* in islands, such as Martinique and Guadeloupe, replicating at a larger scale the results of biological control programs. No extinction of local snail species occurred following the invasion by exotic snails, except for *B. glabrata* in Martinique. Thus, biological invasions could qualify as efficient "unintentional biological control" agents. However, the downside of biological invasions is that snail hosts can be invasive and establish new sites of parasite transmission in formerly parasite-free areas. Moreover, the apparent lack of extinctions may mask the ongoing declines of local species leading to future extinctions, especially if new invaders continue to appear on a yearly basis.

J.-P. Pointier (✉)
USR 3278 CNRS-EPHE, CRIOBE, Université de Perpignan,
52 Av. Paul Alduy, 66860 Perpignan, Cedex, France
e-mail: pointier@univ-perp.fr

R. Toledo and B. Fried (eds.), Biomphalaria *Snails and Larval Trematodes*,
DOI 10.1007/978-1-4419-7028-2_9, © Springer Science+Business Media, LLC 2011

9.1 Introduction

Schistosomiasis control has a long history, and our understanding of both local parasite transmission and the array of control techniques has gradually improved. The first control programs mainly focused on populations of snail vectors but the discovery of new diagnostic methods, new effective drugs, and powerful molluscicides led to the development of more integrated approaches implementing several types of interventions simultaneously (Bruun and Aagaard-Hansen 2008). Snail control has remained an essential component of these integrated control programs, but chemical molluscicides are expensive; the organization needed to spray them is complex, and they may negatively impact local faunas (McCullough 1992). Over the last several decades, biological control has been considered as an alternative approach to molluscicides. The logic underpinning biological control is to introduce a new species ("natural enemy") that has the ability to establish self-sustained populations and to eliminate – or to limit the density of – populations of the target species. A scheme for screening and evaluating the cost-efficiency and environmental impact of biological control agents has been proposed by the World Health Organization (1982). This protocol includes five stages of laboratory and field studies before using any biological agent at large scale in integrated disease control operations. However, few biological agents have proven their efficiency against vector snails in the field despite several decades of work. On the other hand, many unassisted biological invasions took place in the whole tropical areas during the same period inducing long-term changes in snail communities.

Four main groups of organisms have been proposed for controlling snail hosts of schistosomes: pathogens, predators, parasites, and competitors. Several micropathogens, mainly bacteria and fungi, and protozoans have been studied, but research remained at the laboratory stage. A long list of micropathogens of freshwater snails is available in the literature (see the review by Madsen 1995). However, very little work has been done for systematically screening snails for micropathogens and more thorough studies on the fitness impact of pathogens on snails are required.

Predators of freshwater snails have been studied more extensively. They include almost all groups of the animal kingdom, including planarians, leeches, insects, and their larvae, crustaceans, fishes, birds and mammals. A lot of empirical observations and laboratory studies are available in the literature on these predators, particularly the sciomyzid flies and cichlid malacolophagous fishes (see the review by Madsen 1995); field experiments remain scarce. An exception is the investigation on the crayfish *Procambarus clarkii* in Kenya (Hofkins et al. 1991; Loker et al. 1993; Mkoji et al. 1995). Moreover, introducing predators is notoriously dangerous, as the lack of specificity of predators may lead to ecological disasters. For example, the introduction of the land snail *Euglandina rosea* into the Society Archipelago, French Polynesia, is responsible for the extinction of a remarkable endemic land snail fauna (Tiller and Clarke 1983).

Parasites, especially larval trematodes, have been also considered as control agents, and three main characteristics of the snail–parasite interactions have attracted the attention of researchers: (1) the antagonism between larvae of different trematode species within the snail host (Lie et al. 1965; Lie 1967; Lie and Heynemann 1972), (2) the parasitic castration of the snail host (Jourdane and Kulo 1981), and (3) the pathogenic or lethal action of larval trematodes on snails (Jourdane and Kulo 1982). Here , field experiments are also scarce and they either failed (Lie et al 1974) or had only ephemeral success (Nassi et al. 1979).

Competitor snails have been considered as the most promising biological control agents; many encouraging laboratory studies have been conducted using several competitor species (Madsen 1995), and trials have proven their efficacy in the field (Pointier and Jourdane 2000). These field experiments were mainly conducted in the Caribbean over the last half-century using ampullariids and thiarid snails. However, several species of the same group accidentally invaded numerous Caribbean islands and countries during the same period, and the number of newcomer freshwater snails steadily increased. These newcomers had substantial influence on those snail species acting as intermediate hosts for schistosomes. Our aim here is to illustrate key aspects of the impact of competitor snails on the snail hosts of schistosomes and on native ecosystems. We consider both intentional introductions of a snail competitor and bioinvasions, drawing heavily from the few situations that have been studied thoroughly, to allow some inferences on biological control and species interactions. These studies were mainly conducted in the Lesser Antilles (Jobin et al. 1977; Prentice 1983; Pointier et al. 1989; Pointier and Guyard 1992; Facon et al. 2005, 2008; Facon and David 2006).

9.2 Field Experiments Using the Ampullariid Snail *Marisa cornuarietis*

One of the classical examples of biological control of schistosome transmitting snails is that of *Biomphalaria glabrata* by the ampullarid snail *Marisa cornuarietis*. This species is native to Venezuela and has been introduced to Florida and several Caribbean islands as an ornamental snail by aquarists (Madsen and Frandsen 1989). Its ability to eliminate *B. glabrata* was reported for the first time in 1952 in natural water bodies on the island of Puerto Rico (Oliver-González et al. 1956). In 1956, a first field trial using *M. cornuarietis* was carried out in 111 irrigation ponds located on the south coast of Puerto Rico. By 1965, *B. glabrata* had been displaced in 89 out of 97 ponds (92%) still in operation (Ruiz-Tibén et al. 1969). Other field experiments confirmed the efficiency of *M. cornuarietis* against *B. glabrata* in other types of Puerto Rican habitats, such as artificial lakes (Jobin et al. 1977) or flowing waters (Jobin and Laracuente 1979). In Tanzania, *M. cornuarietis* was successfully introduced to a dam harboring large populations of *Biomphalaria pfeifferi, Bulinus tropicus* and *Lymnaea natalensis*. Two years after the introduction of the ampullariid snails, all three pulmonate species had been eliminated (Nguma et al. 1981).

A similar field experiment was initiated in 1987 in Grande-Terre of Guadeloupe using *M. cornuarietis* against *B. glabrata* (Pointier and David 2004). Grande-Terre harbors about 2,000 ponds over an area of about 700 km^2 and a few small rivers. The competitor snail was introduced to 15 experimental ponds, while 15 control ponds were left unchanged. Populations of all species of freshwater molluscs as well as representative plant species were surveyed twice a year for the 13 subsequent years. *B. glabrata*, which was initially seen frequently in the studied ponds (and over the entire island), rapidly disappeared from the experimental sites. Concomitantly, a slow and continuous decrease in the number of control ponds harboring *B. glabrata* was observed from 1994 to 2000 (Fig. 9.1). This trend was not restricted to the study sites but rather represented a progressive, general decline in the occurrence of *B. glabrata* over the whole island. In fact, *B. glabrata* occurred in 26.6% of 188 ponds surveyed in 1980–1981 (Pointier et al. 1985), whereas this proportion had decreased to 2% by 2000 (authors' unpublished data; 150 ponds). This decline probably results from both an increase in the rate of local extinction and a decrease in recolonization rates. Although periodic drying of the ponds normally results in a regime of local extinctions during each dry season, additional extinctions have taken place because of competitive interactions with introduced species. The latter include *M. cornuarietis* (in the experimental ponds, see above) and also, in many other ponds, two exotic species, *Biomphalaria kuhniana* (a member of the *Biomphalaria straminea* species complex, formerly identified as *B. straminea* in Pointier et al. 1993a) and *Melanoides tuberculata*. These two species invaded many sites without assistance between 1980 and 2000 (Pointier et al. 1993a; Pointier and Delay 1995; authors' unpublished data). The entire system of ponds can be considered as a single metapopulation of *B. glabrata* and is subject to recurrent extinction–colonization dynamics. The metapopulation structure amplifies

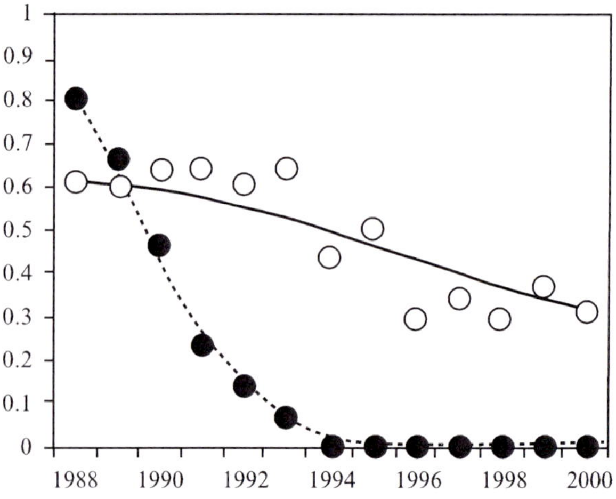

Fig. 9.1 Temporal variation in the frequency of ponds occupied by *Biomphalaria glabrata* in a biological control trial implemented in Grande-Terre (Guadeloupe) using the competitor snail *Marisa cornuarietis*. *Black circles*: experimental ponds (N=15) where *M. cornuarietis* was introduced at the end of 1987; *white circles*: control ponds (N=15). Note a slow, delayed decline of *B. glabrata* in the control ponds from 1994 on

the effects of removing favorable sites through the action of biological control agents. Because sites are sources of colonizers to each other, such a removal can disproportionately decrease the equilibrium occurrence of a species on a metapopulation scale (Pointier and David 2004). According to the hypothesis of metapopulation collapse, and assuming that the presence of *M. cornuarietis* makes a pond unavailable for the establishment of *B. glabrata*, the frequency of the latter should decrease even in sites where *M. cornuarietis* is absent, which is exactly what was observed. Of course, the 15 experimental ponds represent a small fraction of the suitable sites for *B. glabrata*. However, the introduction of *M. cornuarietis* to other ponds (not part of the experiment), as well as the spread of two other alien species may have affected a sufficient number of sites to seriously lower the colonization rate, even in sites devoid of exotic species. This interactive effect is probably the key to the slow decline of *B. glabrata* in Grande-Terre of Guadeloupe in the last few decades.

Importantly, other pulmonate species, such as *Biomphalaria schrammi, Drepanotrema surinamense* or *Drepanotrema depressissimum* were apparently not affected by the presence of *M. cornuarietis*. This observation was interpreted as the consequence of strong differences in life-history traits between these species and *B. glabrata*. More rapid growth and a much shorter life span provide a strong advantage to these species at the beginning of the rainy season (Lévêque and Pointier 1976). They are able to colonize some refuges much more rapidly than *B. glabrata* after the dry season and can resist more effectively competitor introduction. It also appears clearly from this report that biological control might be less straightforward than initially imagined, and may have consequences at scales larger than that at which the control experiment was conducted (Pointier and David 2004).

The efficiency of *M. cornuarietis* against vector snails is probably linked to its feeding habits. It is indeed a voracious snail consuming many types of submerged or floating aquatic plants. Another typical example of biological control mediated through plant destruction can be drawn from Guadeloupe. In the Grand Etang Lake (Basse-Terre) the exclusive habitat of *B. glabrata* was constituted by a belt of the floating plant *Pistia stratiotes*. The introduction of *M. cornuarietis* in 1987 resulted in its complete destruction in the course of 3 years and consequently in the elimination of the only snail host from that lake. This was followed by the eradication of *Schistosoma mansoni* in the animal reservoir *Rattus rattus* (Pointier et al. 1991; Fig. 9.2).

Regarding the 15 experimental ponds of Grande-Terre, the main visible consequence of the introduction of *M. cornuarietis* on the aquatic vegetation was a significant decline of the water lily *Nymphaea ampla* (Pointier and David 2004). Fortunately, this plant is very common in the Caribbean islands and in South America, and its extinction seems very unlikely due to the large number of ponds still free of *M. cornuarietis* and the poor dispersal capacities of this snail (Pointier and David 2004). Caution remains necessary, however, in using it in biological control programs as it can become a pest for cultivated rice, water-cress (*Nasturtium officinale*) or dasheen (*Colocasia esculenta*). The invasion of several South-Eastern Asian countries by ampullariid species from South America, and their impact on agriculture is quite illustrative of this problem (Cowie 2002).

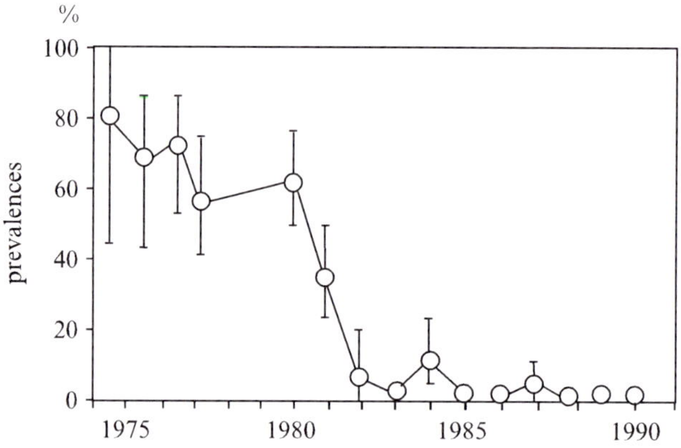

Fig. 9.2 Infection rate of rats, *Rattus rattus*, by *Schistosoma mansoni* at Grand Etang Lake, Basse-Terre of Guadeloupe between 1974 and 1990. The parasite has been eradicated by 1990 following the elimination of the snail host *Biomphalaria glabrata* through biological control

9.3 Field Experiments Using Thiarid Snails

Thiarid snails, native to the Paleotropics, have invaded the Neotropics since the 1930s. They soon attracted the attention of researchers who assessed their efficiency as biological control agents in controlled trials. The history of thiarid introductions to Caribbean countries is well documented (Fig. 9.3). All are unintentional except in the island of St Lucia (see Prentice 1983). Water birds and passive transport by cars or cattle are probably involved in the local spread of invasive species from one watershed to the other, or to nearby islands, yet aquarist trade is probably more efficient as a means of long-distance (transcontinental) introductions, as these species are often associated with aquarium plants (Abbott 1952; Madsen and Frandsen 1989). Two species are mainly concerned, *Tarebia granifera* and *M. tuberculata* (Fig. 9.3).

A first field experiment of biological control using thiarid snails against the snail hosts of schistosomes was conducted in St Lucia in 1978 (Prentice 1983). *M. tuberculata* (erroneously identified as *T. granifera* by Prentice) was introduced to seven sites, including dasheen (*C. esculenta*) and water-cress (*N. officinale*) marshes drained by small streams or springs. Other similar sites, left unchanged, were used as controls. These habitats were very common in St Lucia and highly favorable to *B. glabrata* and to schistosomiasis transmission. Two years later, the competitor snail was widespread in these sites and had eliminated *B. glabrata* (Fig. 9.4). Following this success, *M. tuberculata* was introduced to other sites on the island (A. Calender, pers com). A malacological survey of the whole hydrographic system in 1992 revealed that *M. tuberculata* had become the most common freshwater snail in St Lucia. This survey was undertaken in sites where large populations of *B. glabrata* had been found in the past, and showed (1) the absence of the planorbid snail from seven sites extensively colonized by *M. tuberculata*, (2) the presence of *B. glabrata* at low to very low densities in 17 sites together with *M. tuberculata*, and (3) the presence of *B. glabrata* in large populations in the only two sites where

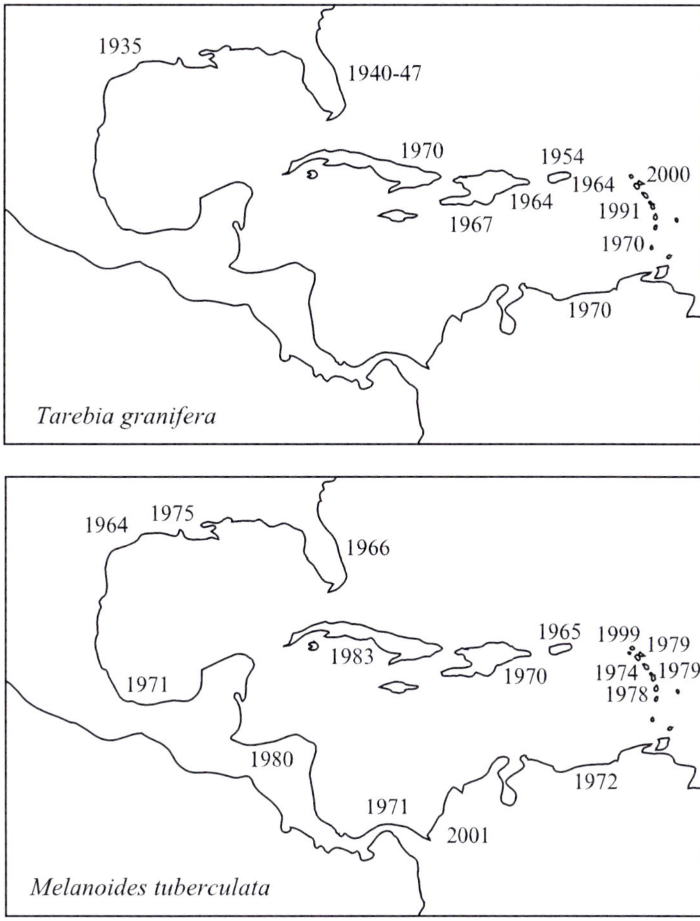

Fig. 9.3 Introduction of thiarid snails to the Caribbean area. *Top map*: *Tarebia granifera*. Sources are Murray and Woopschall (1965) for San Antonio, Texas, USA, Abbott (1952) for Lihia Springs, Florida, USA, Harry and Aldrich (1958) for Puerto Rico, Ferguson (1977) for Vieques and the Dominican Republic, Robart et al. (1979) for Haiti, Ferguson (1977) for Grenada, Chrosciechowski (1973) for Venezuela, Jaume (1972) for Cuba, Pointier et al. (1998) for Martinique and Pointier (2008) for Guadeloupe. *Bottom map*: *Melanoides tuberculata*. Sources are Murray (1964) for Texas, Abbott (1973) for Puerto Rico, Clench (1969) for Florida, Gomez et al. (1986) for Dominican Republic, Abbott (1973) for Mexico and Panama, Chrosciechowski (1973) for Venezuela, Pointier unpublished data and Starmühlner (1984) for Dominica, Dundee and Paine (1977) for Louisiana, Prentice (1983) for St Lucia, Pointier and McCullough (1989); Pointier and Delay (1995) for Martinique and Guadeloupe, Clarke (1987) for Honduras, Perera et al. (1987) for Cuba, Stevens and Waldmann (2001) for Montserrat. Note that the invasion sequence is very difficult to interpret because *M. tuberculata* is in fact a single name for independent strains which were introduced and are still being introduced at different places at different times

M. tuberculata was absent (Pointier 1993). Its scarcity in habitats which were formerly important transmission sites in St Lucia probably means that transmission is presently low or very low if it indeed still occurs. Although other ecological or anthropogenic factors may be involved, there is no doubt that the thiarid snail

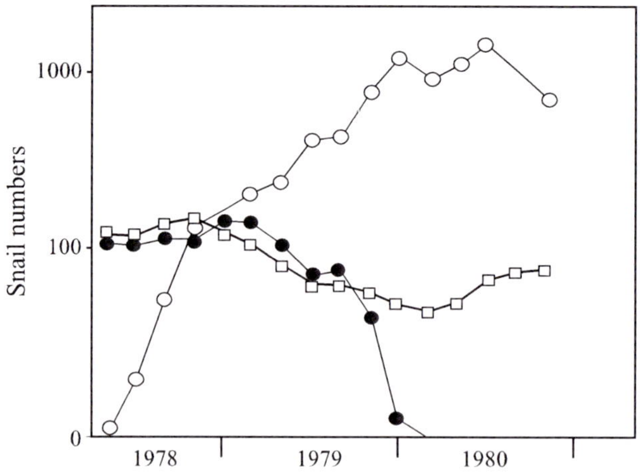

Fig. 9.4 Temporal variation of snail number (log scale) in a dasheen marsh in St Lucia during a biological control trial of *Biomphalaria glabrata* (*black circles*) by *Melanoides tuberculata* (*white circles*). *White squares*: *B. glabrata* in the control site without *M. tuberculata*

M. tuberculata played a major role in the decline of *B. glabrata* populations. Also of importance is the long-term permanence of dense populations of the thiarid snail, maintaining *B. glabrata* at a very low level and preventing the recolonization of sites where the planorbid snail has been eliminated.

Biological control using *M. tuberculata* has also been conducted in Martinique. This species was discovered there for the first time in 1979, and then spread rapidly. In 1982, *B. glabrata* and schistosome transmission were still occurring in several water-cress cultures (Pointier et al. 1984). The biological control program took advantage of previous studies of snail population dynamics conducted in 1982. *M. tuberculata* was introduced in 1983 to a group of water-cress cultures and eliminated the snail hosts *B. glabrata* and *B. kuhniana* (formerly identified as *B. straminea*) in less than 2 years (Pointier et al. 1989; see Fig. 9.5). Following this success, *M. tuberculata* was introduced to other groups of water-cress beds. In 1990, *B. glabrata* and *B. kuhniana* had totally disappeared from eight sites and a few individuals only were recorded from the remaining sites (Pointier and Guyard 1992).

The last example we present is that of Guadeloupe where the marshy forest of *Pterocarpus officinalis* located behind the mangrove swamp is a schistosomiasis focus. The reservoir of adult schistosomes is the black rat *R. rattus* (Théron and Pointier 1995). This focus includes the marshy forest and a wet meadow zone. The area extending from the marshy forest to the herbaceous zone irregularly experiences periods of flooding and drying following the rainfall regime. It is mainly devoted to dasheen culture and harbors large populations of *B. glabrata*. The herbaceous zone used for cattle grazing is also suitable for this snail. Additionally, permanent ponds fed by groundwater can be found in both the marshy forest and herbaceous zones. They are used for watering cattle or for growing water-cress and providing favorable habitats for *B. glabrata*. A biological control trial was initiated in 1984. *M. tuberculata* was introduced to the three main types of habitats of this area,

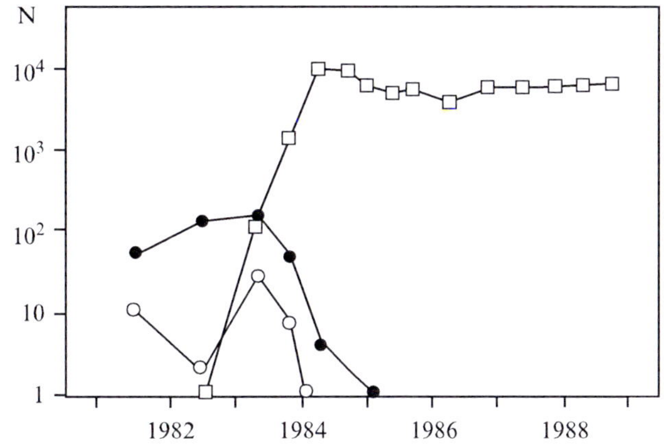

Fig. 9.5 Temporal variation of snail number (log scale) in a water-cress bed of Martinique during a 7-years trial of biological control of *Biomphalaria glabrata* (*black circles*) and *B. kuhniana* (*white circles*) by *Melanoides tuberculata* (*white squares*)

and snail dynamics was followed over the next 6 years (Pointier et al. 1993b). Colonization by *M. tuberculata* was successful in all habitats, but the impact on the snail host differed with habitats. In permanent ponds, *M. tuberculata* rapidly reached high densities 6–10 months after its introduction (see the example in Fig. 9.6, top). Initially low, the densities of *B. glabrata* declined and the snail population disappeared almost completely. In contrast, colonization of the dasheen cultures was more difficult presumably because of irregular periods of desiccation during the dry season. *M. tuberculata* densities remained low and *B. glabrata* densities fluctuated markedly, but there was no apparent correlation between the two species (Fig. 9.6, bottom). These contrasted results can be explained by the quite opposite demographic strategies exhibited by the two snail species. *M. tuberculata* has slow growth, low reproductive rate, and a long life span, whereas *B. glabrata* exhibits rapid growth, high reproductive rates, and a very short life span. Thus, *B. glabrata* has a strong advantage on *M. tuberculata* in unstable and temporary habitats such as dasheen cultures. On the contrary, *M. tuberculata* has enough time to reach and maintain high densities in permanent and stable habitats, such as water-cress cultures, and thus is able to become a serious competitor for food and space (Pointier et al. 1993b).

To summarize, the competitor species *M. cornuarietis* and *M. tuberculata* have been quite successful in eliminating or reducing populations of schistosome-transmitting snails, especially *B. glabrata*. However, their efficiency is context-dependent; these caenogastropods are efficient competitors in sufficiently stable habitats only where long-term resource exploitation rather than colonization ability is the limiting factor. Species with more colonization-oriented life histories, especially other pulmonates, may be more efficient than caenogastropods to limit *B. glabrata* in isolated and unstable habitats, such as small ponds. The combined presence of introduced caenogastropods and introduced pulmonates (the latter being nonassisted invasions) can dramatically limit, if not eliminate, *B. glabrata* populations at the regional scale.

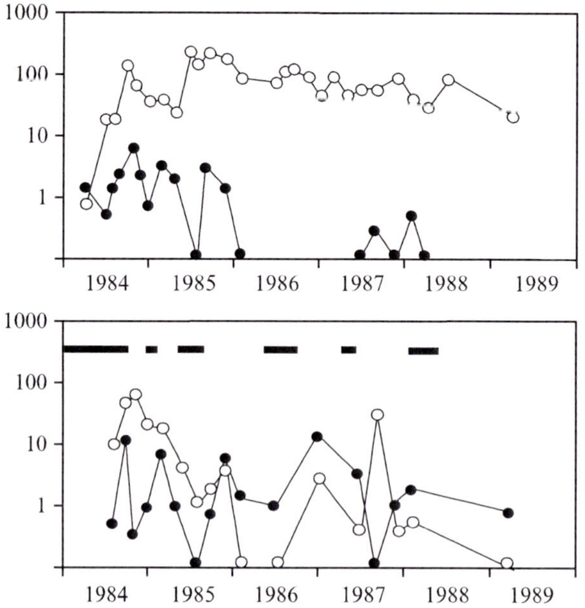

Fig. 9.6 Temporal variation of snail number (log scale) in the marshy forest of Guadeloupe during a 6-years trial of biological control of *Biomphalaria glabrata* (*black circles*) by *Melanoides tuberculata* (*white circles*). Snail dynamics in a permanent pond (*top*) and in a dasheen culture (*bottom*). *Horizontal black bars* indicate periods of site drying

9.4 Snail Invaders and French Antillean Islands: The Decline of *Biomphalaria glabrata*

Unassisted invasions of exotic snails have considerably modified most tropical freshwater faunas in recent decades. Invasive species have often ended up behaving as efficient, if involuntary, control agents against schistosome-transmitting snails, especially *B. glabrata*. In Guadeloupe and Martinique, the initiation of field trials in biological control programs (see above) was paralleled by rapid and unassisted invasions of entire hydrographic systems. They have been well documented by long-term malacological studies initiated in the beginning of the 1970s (Golvan et al. 1981; Pointier and David 2004; Facon and David 2006).

In Martinique (Fig. 9.7), this fauna was first investigated at the end of the nineteenth century by Mazé (1874) and Bordaz (1899) who reported ten species, including *B. glabrata*. Dreyfuss (1953) confirmed the presence of four planorbid species. At that time, *B. glabrata* was apparently widely distributed throughout the island. Starting with the discovery of *B. kuhniana* in 1967 (*Biomphalaria havanensis* in Grétillat (1967)), a continuous influx of exotic species has been detected (Guyard and Pointier 1979; Pointier 2008) so that the number of introduced species now

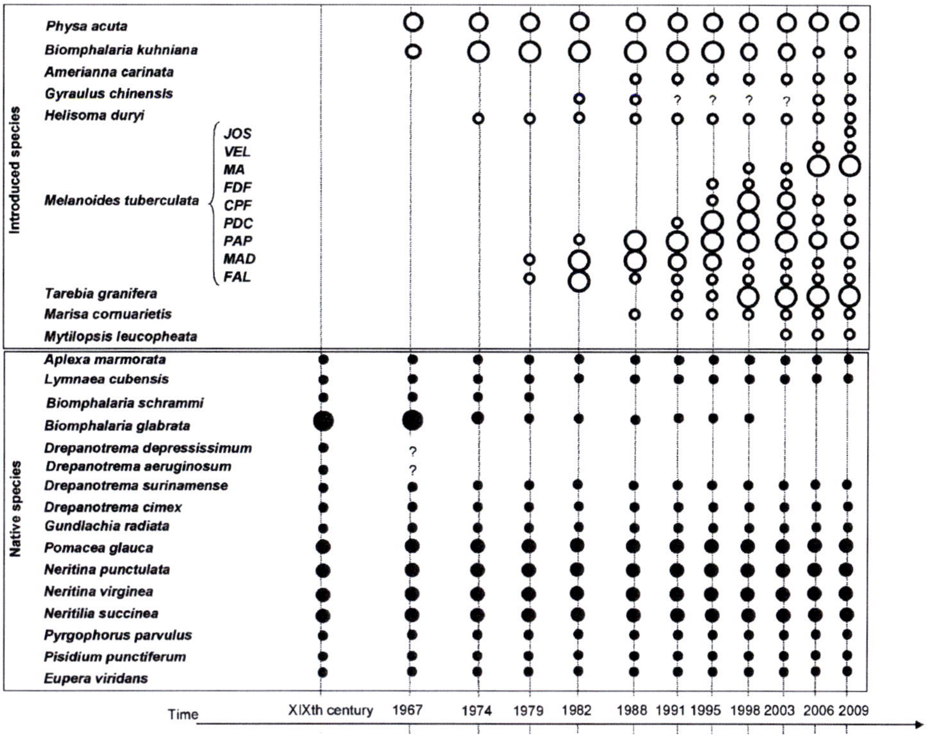

Fig. 9.7 Changes in the freshwater malacofauna of Martinique during the twentieth century distinguishing introduced (*white*) and native (*black*) species. *Circle size* corresponds to taxa abundance; *small circles*: taxon very localized and/or rare; *medium circle*: taxon either found at many sites and low densities, or localized and abundant; *large circle*: dominant taxon, very common and widespread. From Mazé (1874), Bordaz (1899), Dreyfuss (1953), Grétillat (1967), Guyard and Pointier (1979), Pointier et al. (1993c), Pointier (1996, 2008), Facon et al. (2003), Facon and David (2006), and authors (unpublished data)

approximately matches that of native species (Fig. 9.7). The most spectacular invasions are those by members of the Thiaridae family, *M. tuberculata* and *T. granifera*. The former is characterized by a high morphological and genetical diversity; it consists of several well-individualized strains (see Fig. 9.7) perpetuating through clonal reproduction that can ecologically be considered as a different species. These strains can be recognized in the field based on shell sculpture and colors (Fig. 9.8). Those occurring in Martinique, depicted in Fig. 9.7, have been independently introduced from outside, with the exception of two (named CPF and FDF) which are local hybrids produced by rare sexual reproduction events among previously established strains (Facon et al. 2003, 2008).

The freshwater malacological fauna of Guadeloupe includes a larger number of native species than Martinique because of the diversity of habitats offered by the contrasted geomorphological characteristics of the two main islands of the

Fig. 9.8 Thiarid snails that have invaded Martinique over the last 30 years with introduction year. FAL, MAD, PAP, PDC, CPF, FDF, MA, VEL, and JOS are morphs of *Melanoides tuberculata* which have been recognized based on shell and genetical characteristics. TG = *Tarebia granifera*

Guadeloupean archipelago (Fig. 9.9). Basse-Terre is a volcanic mountainous island with numerous rivers, streams, and canals, whereas Grande-Terre is flat, calcareous, offering numerous lentic habitats, such as natural or artificial ponds and marshes. The introduction history is summarized in Fig. 9.9 and shows marked similarities with that of Martinique. Many species have invaded the two islands at approximately the same time (*T. granifera* and *M. tuberculata*) or at different times (*B. kuhniana*). However, other invasions have reached a single island, some of which concern taxa that are now both abundant and widespread (e.g., the PDC and MA strains of *M. tuberculata* in Martinique, the GOS strain of *M. tuberculata* and the lymnaeid *Pseudosuccinea columella* in Guadeloupe). To some extent, the two islands can, therefore, be considered as independent natural experiments. The current status of introduced species is very diverse and

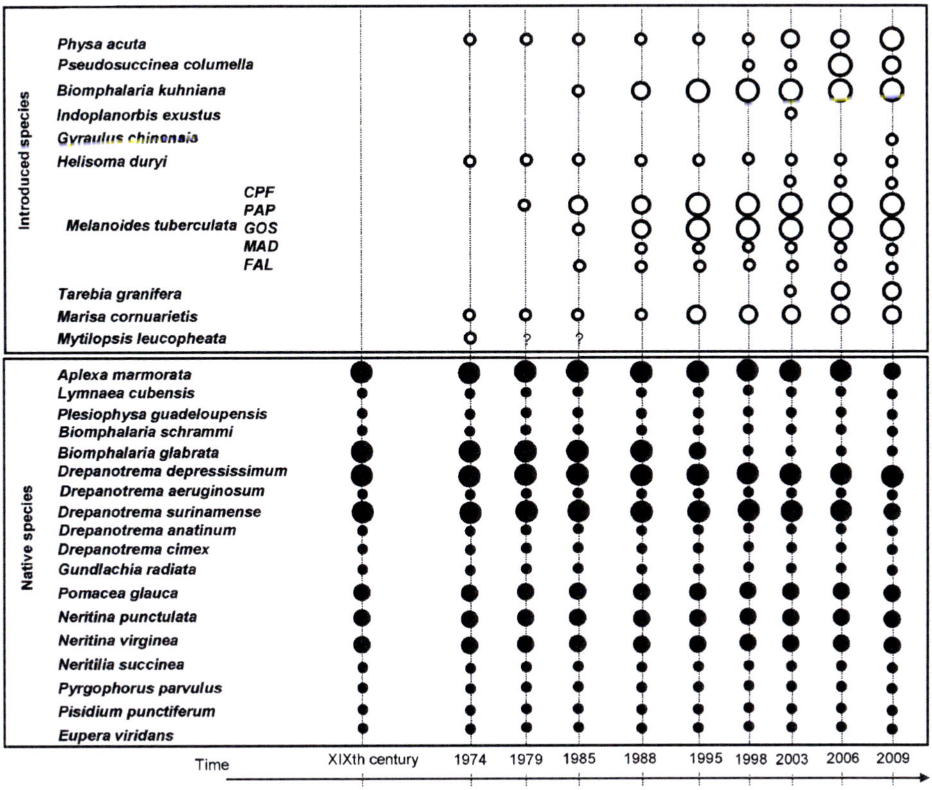

Fig. 9.9 Changes in the freshwater malacofauna of Guadeloupe during the twentieth century. *Circle size* corresponds to taxa abundance, as in Fig. 9.7. From Schramm (1869), Mazé (1890), Floch (1969), Pointier (1974), Pointier et al. (1993a), Pointier and Delay (1995), Pointier (2008), and authors (unpublished data)

a large range of situations may be encountered (see Table 9.1). Some taxa are now widely distributed throughout the islands and may even occur in huge populations (e.g., the PAP and MA strains of *M. tuberculata* and *T. granifera* in Martinique) while others are declining (the FAL morph of *M. tuberculata* and *B. kuhniana* in Martinique) or are restricted to a particular type of aquatic environment (e.g., *Amerianna carinata* and *Gyraulus chinensis* to oxbow lakes, *Helisoma duryi* to artificial habitats). Others even failed to persist for more than 2 or 3 years after their introduction (e.g., *Indoplanorbis exustus* in Guadeloupe).

As already pointed out, although ecological and anthropogenic factors may also be involved, a clear correlation has been observed in the Caribbean area in the last 30 years between the invasion by exotic snails and the decline of *Biomphalaria* species. In Martinique, for example, *B. kuhniana* (which was itself a previously established invader) has strongly regressed in the last decades, while thiarid snails were on the rise, and *B. glabrata* can be considered as eradicated since 2005

Table 9.1 Introduced species of freshwater snails in Martinique and Guadeloupe (in chronological order), and current status

Introduction year	Alien species	Present status
Martinique		
?	*Physa acuta*	Widely distributed mainly in anthropized sites
1953–1967	*Biomphalaria kuhniana*	Strongly regressed between 1979 and 2009
1972–1978	*Helisoma duryi*	Restricted to a very few sites (artificial tanks and ponds)
1979	*Melanoides tuberculata* FAL and MAD	FAL morph widely distributed in 1986, now strongly regressed. MAD now widely distributed
1981	*Melanoides tuberculata* PAP	Widely distributed
1982	*Gyraulus chinensis*	Restricted to some oxbow lakes
1987	*Marisa cornuarietis*	Restricted to two marshes
1987	*Amerianna carinata*	Restricted to some oxbow lakes
1990	*Melanoides tuberculata* PDC	Widely distributed
1991	*Tarebia granifera*	Widely distributed
1993	*Melanoides tuberculata* CPF	Uncommon
1994	*Melanoides tuberculata* FDF	Appeared at a single site in 1994. Now disappeared
1997	*Melanoides tuberculata* MA	Widely distributed. Probably the most competitive taxon
2002	*Melanoides tuberculata* VEL	Restricted to a few sites
2005	*Melanoides tuberculata* JOS	Restricted to a very few sites
Guadeloupe		
1969	*Helisoma duryi*	Restricted to a very few sites (artificial tanks and Grand Etang Lake)
Around 1970	*Physa acuta*	Widely distributed (invasion in progress?)
1973	*Marisa cornuarietis*	Widely distributed in ponds of Grande-Terre. Invasion in progress in marshes
1979	*Melanoides tuberculata* PAP	Widely distributed
1984	*Melanoides tuberculata* GOS	Widely distributed
1984	*Melanoides tuberculata* FAL	Restricted to a few sites
1985	*Biomphalaria kuhniana*	Widely distributed in Grande-Terre. Restricted to a few rivers in Basse-Terre
1986	*Melanoides tuberculata* MAD	Restricted to a few sites
1997	*Pseudosuccinea columella*	Widely distributed in Grande-Terre. Invasion in progress in Basse-Terre
2000	*Tarebia granifera*	Restricted to few ponds of Grande-Terre. Invasion in progress in Basse-Terre since 2007

(continued)

Table 9.1 (continued)

Introduction year	Alien species	Present status
2002	*Indoplanorbis exustus*	Introduced to a single pond. Species not collected after 2004. Example of an invasion failure
2008	*Gyraulus chinensis*	Restricted to some oxbow lakes in Basse-Terre

Morphs of *Melanoides tuberculata* have names based on a three-letter code. CPF is an hybrid between FAL and PAP, and FDF between FAL and PDC (see Samadi et al. 1999)

(Pointier and Théron 2006; see Fig. 9.10). In Guadeloupe, the situation is quite different. *B. kuhniana* has invaded numerous ponds of Grande-Terre from 1985 on and is now a major component of malacological communities in this area. Thiarids reach insufficient abundance in most temporary ponds to eliminate *B. kuhniana*. In contrast, *B. kuhniana* seems to have much more difficulty in invading streams and rivers of the Basse-Terre Island, probably because these habitats were already widely occupied by thiarids. In the 1970s, this island harbored the largest foci of intestinal schistosomiasis and huge populations of the snail host *B. glabrata* (Théron and Pointier 1995). In the last several decades, *B. glabrata* has strongly regressed from Basse-Terre following the invasion by thiarids and the elimination of numerous canals that had become useless. It is currently restricted to a few sites such as Vieux Fort pool in the Northern part of the island (authors' unpublished data). In Grande-Terre, the situation depends on the type of aquatic environment. As already pointed out, a phenomenon of metapopulation collapse was observed for *B. glabrata* in ponds (see the first paragraph on field experiments using *M. cornuarietis*). The situation is quite different in the marshy forest zone located behind mangrove swamps (see above). The impact of the introduction of alien species, such as *M. tuberculata* or *P. columella* upon *B. glabrata* populations, was apparently limited and this aquatic environment may be considered as a refuge for the planorbid snail. A consequence is the persistence in this area of an active focus of schistosomiasis transmission. This situation might change with the recent arrival of *M. cornuarietis* in some of these sites (authors' unpublished data). To summarize, unassisted snail species introductions in Martinique and Guadeloupe have reproduced at a larger scale the results of field experiments: introduced species have been instrumental in reducing populations of schistosome-transmitting snails (*Biomphalaria spp.*) in the habitats that were massively invaded, such as streams and rivers in which thiarids snails literally thrive. This resulted in the regional collapse of *Biomphalaria spp.* populations in Martinique and Basse-Terre where these habitats are dominant. However, the impact of alien species has been more limited in some specific habitats (marshy forest in Guadeloupe) where invasive snails did not outcompete local ones, and where they now coexist with *B. glabrata*.

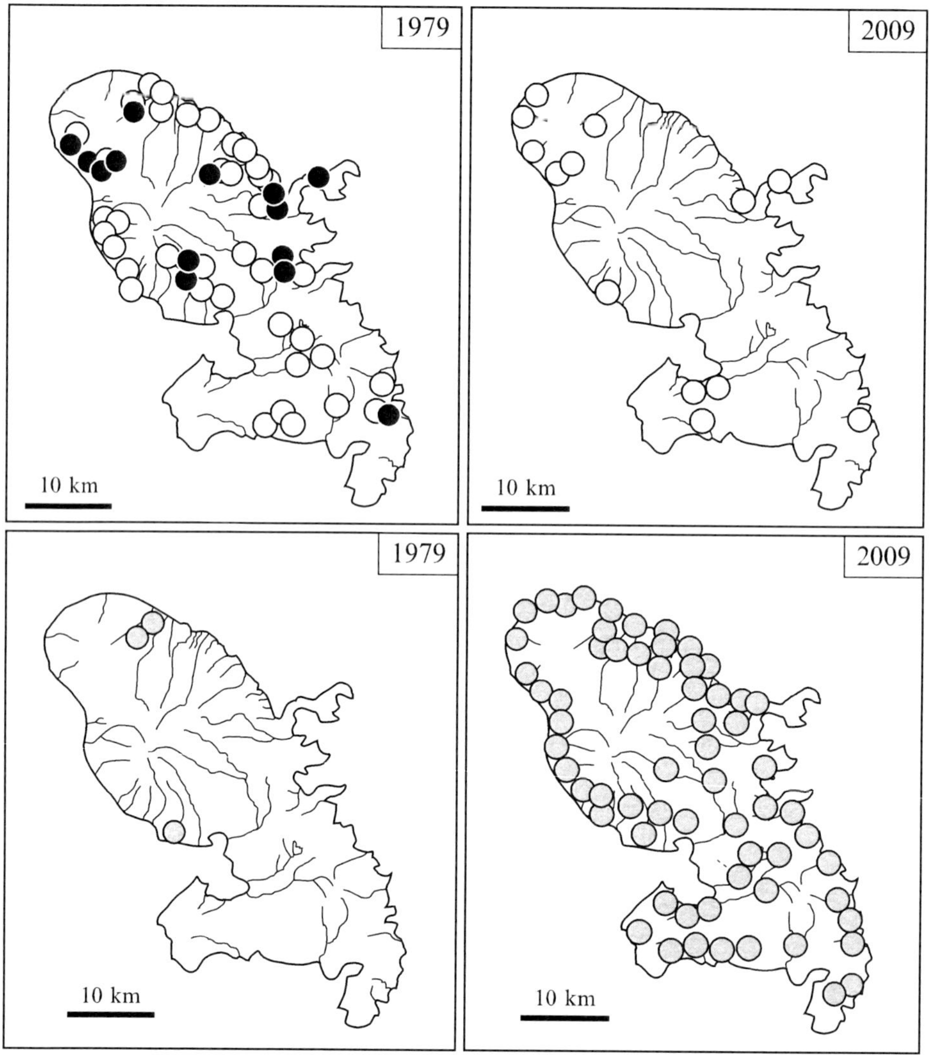

Fig. 9.10 *Top*: distribution of *Biomphalaria glabrata* (*black circles*) and *Biomphalaria kuhniana* (*white circles*) in 1979 and 2009 in Martinique. *Bottom*: distribution of thiarid snails (*gray circles*) in 1979 (introduction time) and 2009

9.5 Freshwater Snail Invasions: Threats for Environment and Biodiversity?

We have so far described the effects of exotic snail introductions (whether unassisted or done on purpose) on schistosome-transmitting snails, and found that introduced competitors have been the most powerful cause of decline of the latter, especially *B. glabrata*. However, assessing the overall cost-benefit balance of snail introductions requires considering the threats that introduced snails pose to biodiversity and environment. Two types of threats can be considered (we do not consider herein the economic costs entailed by the introduction of apple snails, especially in Asia, since they are unrelated to *Biomphalaria* species; see Cowie 2002 for a review): the first is that introduced species may carry new parasites, especially trematodes parasitizing humans or domestic animals; the second is that they may alter competitive or predator–prey equilibria in such a way that extinctions of local species occur, and species diversity or genetic diversity decrease in native communities and taxa.

On the first aspect, several species belonging to the genus *Biomphalaria* can be considered as invasive in different parts of the world, and this is of particular importance since they are involved in the transmission of *S. mansoni*. An example is the accidental introduction to Egypt of *B. glabrata*, the best snail host for this parasite, originally restricted to the Neotropical area (around 1981; Pflüger 1982). During the following years, *B. glabrata* invaded the irrigation and drainage systems of the Nile Delta area (Yousif et al. 1996) and hybridized with the local *B. alexandrina* (Kristensen et al. 1999). This hybrid was also found naturally infected with *S. mansoni* suggesting that it played a role in the parasite transmission (Yousif et al. 1998). A recent extensive survey failed to detect any *B. glabrata* or hybrids (Lofty et al. 2005). The application of molluscicides in most of the putative *B. glabrata* localities by the Egyptian Snail Control Section may explain this absence.

B. straminea is also of Neotropical origin, but it has increased its distribution to such an extent over the last few decades that it can be considered, together with *Physa acuta*, as the most invasive mollusc species. It has invaded several Brazilian States as well as new habitats in Paraguay, Argentina and Uruguay (Paraense and Corrêa 1989; Paraense 2001; Teles et al. 2003). Despite its low susceptibility to *S. mansoni, B. straminea* is presently heavily involved in the transmission of schistosomiasis in numerous sites from North-Eastern Brazil (Carvalho 1992). In the Caribbean area, the introduction of *B. kuhniana* (a member of the *B. straminea* complex) was reported in Colombia in 1966 (Barbosa 1968), in Costa Rica in 1976 (Paraense et al. 1981), as well as in several islands of the Lesser Antilles, including Martinique around 1950 (Grétillat 1967), Grenada in 1970 (Ferguson and Buckmire 1974), Guadeloupe in 1985 (Pointier et al. 1993a) and St Lucia in 1992 (Pointier 1993). Intestinal schistosomiasis was already occurring in most of these islands with *B. glabrata* as intermediate host, and the role of *B. kuhniana* in parasite transmission has never been clearly established except in Martinique (WL Paraense, pers. comm.). Outside the Neotropics, *B. straminea* was introduced to Hong Kong in 1973 (Meier-Brook 1974) and then began to colonize the adjacent territories (Yipp 1990). The study of Tang (1983) did not reveal the presence of trematodes infecting *B. straminea* in the invaded habitats. The Neotropical *Biomphalaria tenagophila* has a more restricted distribution area than *B. glabrata* and *B. straminea*. However, this snail was recently introduced to Africa where it colonized the Kinshasa area, Democratic Republic of Congo and was responsible for the creation of a new focus of intestinal schistosomiasis (Pointier et al. 2005a).

Another trematode parasite, the liver fluke *Fasciola hepatica*, was introduced to the New World over the last 400 years through cattle importation. Recent studies have demonstrated that the snail host *Lymnaea truncatula* was also introduced to several South American countries and was responsible for parasite transmission to humans (Jabbour-Zahab et al. 1997; Pointier et al. 2009). Another lymnaeid species, *P. columella* also has a successful history of introductions and is now one of the most widespread freshwater snails worldwide (in tropical as well as in temperate countries). It occurs in a large fraction of ponds in Guadeloupe (Pointier 2008). This species can also serve as intermediate host for the liver fluke. Fascioliasis has never been reported in Guadeloupe, but the current presence of *P. columella* must be considered as a serious threat.

M. tuberculata may also serve as intermediate host for numerous trematodes infecting wild animals. It is also the snail host of *Centrocestus formosanus*, the parasite responsible for a food-borne intestinal infection in Asia. The source of infection for humans is freshwater fish (when eaten raw or improperly cooked). This parasite was introduced to Oaxaca State, Mexico (Amaya-Huerta and Almeyda-Artigas 1994). More recently, *C. formosanus* was discovered in Medellín, Colombia with a prevalence of 73% in the snail host *M. tuberculata* (Velásquez et al. 2006).

From these examples, we see that the increased rate of biological invasion over the last few decades has both positive and negative effects on parasite transmission to humans: while host snails have been outcompeted by invasive species in some regions (such as Guadeloupe and Martinique), other regions have been invaded by susceptible snails. This resulted in new foci of parasitic transmission, sometimes with dramatic sanitary consequences (e.g., liver fluke transmission in the Altiplano region).

The second threat deriving from the introduction of exotic snails is their potential impact on native biodiversity. Generally, this impact appears to have been very modest. Strikingly, we are aware of no example of local snail species that became extinct at a regional scale following invasions by exotic freshwater snails, to the exception of *B. glabrata* in Martinique and probably in St Lucia ("welcome" extinctions, considering that the relictual populations of *B. glabrata* were associated with the last foci of schistosome transmission). Because of the absence of extinctions, the two aquatic snail communities that have been followed in most detail (Guadeloupe and Martinique) experienced a steady increase in snail biodiversity with the accumulation of exotic species (see Figs. 9.7 and 9.9). This is because of the diversity of habitats that offer gradients from temporary to permanent, from lotic to lentic, and from easily accessible to completely isolated habitats. As already mentioned, local pulmonate species are adapted to more or less rapidly colonize the habitats created by occasional changes in water level. Different species will occupy different positions in this spatial and temporal mosaic, and it appears that these habitats are not collectively saturated in terms of pulmonate diversity, at least in islands such as Guadeloupe and Martinique. Moreover, these islands were initially mainly devoid of benthic caenogastropods (the only important species were the ampullariid *P. glauca* and the neritids *Neritina punctulata* and *Neritina virginea*). This group had no obvious effect on either *P. glauca,* or *Neritina* spp. Thiarids can occur in both rivers and permanent ponds where they might potentially compete with virtually all resident species. Several pulmonate species (including *B. glabrata* and *B. kuhniana*) have decreased following the invasion of thiarids, but have been able to escape extinction whenever they were able to occupy unstable habitats unfavorable to thiarids. The rarity of such habitats in Martinique may explain why *B. glabrata* did go extinct there. In fact, the most intense competitive interactions occur among introduced species (or strains of *M. tuberculata*) with similar ecologies. For example, the invasion of *T. granifera* in Martinican rivers, and later that of the MA strain of *M. tuberculata*, drastically impacted the other strains of *M. tuberculata*, leading one of them (FDF) to extinction, and several others (FAL and CPF) became rare. It is possible that species diversity might be adversely affected in countries where the diversity of indigenous thiarids and other benthic caenogastropods would not be as low as in Guadeloupe or Martinique. For example, Genner et al. (2004) showed that the diversity of endemic thiarids in the Great African Lakes could be threatened by the invasion of Asian strains of *M. tuberculata*. Similar threats are expected wherever thiarid diversity exists, such as in the Greater Antilles or continental South America where numerous endemic *Hemisinus, Aylacostoma* and *Pachychilus* occur (Pointier et al. 2005b; Simone 2006). Studying such situations in details through temporal surveys would certainly bring clearer answers.

Another lesson from the history of aquatic snail invasions in the Lesser Antilles is that invasive populations should not be regarded as genetically depauperate. Owing to many successive introductions, the *M. tuberculata* populations in Martinique and Guadeloupe have indeed accumulated high levels of genetic diversity, in terms of both genetic variation and ecologically important traits (Facon et al. 2008). Remarkably, rare sexual events between introduced lines give rise to natural hybrids (FDF and CPF) which to our knowledge do not occur anywhere else in the world. These hybrids strikingly differ from their parents in both morphology and life history, and have locally outperformed them (Facon et al. 2005). This story illustrates that invasions can sometimes be evolutionarily creative. Yet the long-term balance of invasions is difficult to establish, especially in the Lesser Antilles. The apparent absence of extinctions may mask ongoing declines of local species leading to future extinctions, especially if new invaders continue to appear year after year.

9.6 Concluding Remarks

There is little doubt that both voluntary introductions within biological control programs and unassisted invasions are strongly associated with the general decline of populations of schistosome-transmitting snails (*B. glabrata*) in several places, especially the Caribbean area, over the last several decades. However, very different results have been reported at the local level depending on the species involved, certainly resulting from a wide diversity of ecological situations. In this respect, the *B. glabrata* model in the Caribbean area seems to be exemplary, as it encompasses a variety of freshwater habitats and eco-epidemiological situations from small-scale insular systems (Lesser Antilles) to large-scale continental systems (Venezuela, Brazil). In general, the impact of exotic snail introductions on freshwater biodiversity has been moderate until now, although this does not mean that it will remain so in the decades to come.

Species introductions, whether voluntary or unassisted, come with another potential risk, namely, the establishment of new transmission foci, for example, schistosomiasis and fasciolosis. Several examples have been provided above calling for scrutiny whenever a new potential host is detected somewhere.

References

Abbott RT (1952) A study of an intermediate snail host (*Thiara granifera*) of the oriental lung fluke (*Paragonimus*). Proc US Nat Mus 102:71–116

Abbott RT (1973) Spread of *Melanoides tuberculata*. Nautilus 87:29

Amaya-Huerta D, Almeyda-Artigas RJ (1994) Confirmation of *Centrocestus formosanus* (Nishigori, 1924) Price, 1932 (Trematoda: Heterophyidae) in Mexico. Res Rev Parasitol 54:99–103

Barbosa FS (1968) *Biomphalaria straminea* (Dunker) en Colombia. Antioquia Med 18:753–758

Bordaz G (1899) Liste des coquilles recueillies à la Martinique. Bulletin de la Société d'Histoire Naturelle d'Autun 12:5–24

Bruun B, Aagaard-Hansen J (2008) The social context of schistosomiais and its control. An introduction and annotated bibliography. Special programme for research and training in tropical diseases, World Health Orgainzation, Switzerland

Carvalho OS (1992) Intermediate hosts of *Schistosoma mansoni* in Brazil. Mem Inst Oswaldo Cruz 87:307–309

Chrosciechowski P (1973) Un caracol en busca de nueva residencia. Lago 39:813–814

Clarke AH (1987) Apparent competitive exclusion of *Biomphalaria havanensis* and *Pachychilus largilierti* by *Melanoides tuberculata* in a large Honduran lake. First International Congress on Medical and Applied Malacology, Monterrey

Clench WJ (1969) *Melanoides tuberculata* (Müller) in Florida. Nautilus 83:72

Cowie RH (2002) Apple snails (Ampullariidae) as agricultural pests: their biology, impacts and management. In: Barker GM (ed) Molluscs as Crop Pests, CABI Publishing, Wallingford, pp 145–192

Dreyfuss R (1953) Les planorbes de la Martinique. Bulletin de la Société Française d'Histoire Naturelle des Antilles 2:41–45

Dundee DS, Paine A (1977) Ecology of the snail, *Melanoides tuberculata* (Müller) intermediate host of the human liver fluke (*Opisthorchis sinensis*) in New Orleans, Louisiana. Nautilus 91:17–20

Facon B, David P (2006) Metapopulation dynamics and biological invasions: spatially explicit model applied to a freshwater snail. Am Nat 168:769–783

Facon B, Pointier JP, Glaubrecht M, Poux C, Jarne P, David P (2003) A molecular phylogeography approach to biological invasions of the new World by parthenogenetic thiarid snails. Mol Ecol 12:3027–3039

Facon B, Jarne P, Pointier JP, David P (2005) Hybridization and invasiveness in the freshwater snail *Melanoides tuberculata:* hybrid vigour is more important than increase in genetic variance. J Evol Biol 18:524–535

Facon B, Pointier JP, Jarne P, Sarda V, David P (2008) High genetic variance in life-history strategies within invasive populations by way of multiple introductions. Curr Biol 18:363–367

Ferguson FF (1977) The role of biological agents in the control of schistosome-bearing snails. US Department of Health and Education Welfare Public Health Services, 1–107

Ferguson FF, Buckmire KW (1974) Notes on the freshwater molluscs of Grenada, British West Indies. Carib J Sci 14:147–148

Floch HA (1969) Campagne prophylactique contre la bilharziose intestinale en Guadeloupe. Données et conditions de réalisation. Archives de l'Institut Pasteur de la Guadeloupe, 86–91

Genner MJ, Michel E, Erpenbeck D, De Voogd N, Witte F, Pointier JP (2004) Camouflaged invasion of Lake malawi by an Oriental gastropod. Mol Ecol 13:2135–2141

Golvan YJ, Combes C, Euzéby J, Salvat B (1981) Enquête d'épidémiologie écologique sur la schistosomose à *Schistosoma mansoni* en Guadeloupe (Antilles françaises). Mémoires du Museum National d'Histoire Naturelle 119:1–229

Gomez JD, Vargas M, Malek EA (1986) Moluscos de agua dulce de Republica Dominicana, vol 535. Universidad Autonoma de Santo Domingo, Editora Universitaria, Santo Domingo, pp 1–135

Grétillat S (1967) Prospection malacologique aux Antilles françaises. Observations sur l'écologie et l'élevage au laboratoire de *Lymnaea cubensis*, vol 2. Revue d'Elevage et de Médecine Vétérinaire des pays Tropicaux, pp 279–289

Guyard A, Pointier JP (1979) Faune malacologique dulçaquicole et vecteurs de la schistosomose intestinale en Martinique (Antilles françaises). Ann Parasitol 54:193–205

Harry HW, Aldrich DV (1958) The ecology of *Australorbis glabratus* in Puerto rico. Bull World Health Organ 18:829–832

Hofkins BV, Mkoji GM, Koech DK, Loker ES (1991) Control of schistosome-transmitting snails in Kenya by the North American crayfish *Procambarus clarkii*. Am J Trop Med Hyg 45:339–344

Jabbour-Zahab R, Pointier JP, Jourdane J, Jarne P, Oviedo J, Bargues MD, Mas-Coma S, Angles R, Perera G, Balzan C, Khallayoune K, Renaud F (1997) Phylogeography and genetic divergence of some lymnaeid snails, intermediate hosts of human and animal fascioliasis with special reference to lymnaeids from the Bolivian Altiplano. Acta Trop 64:191–203

Jaume ML (1972) Nuevo género y especie de molusco fluvial para la fauna malacológica Cubana (Gastropoda: Prosobranchia: Thiaridae). Circulares del Museo y Biblioteca de Zoología de la Habana, pp 1523–1525

Jobin WR, Brown RA, Veléz SP, Ferguson FF (1977) Biological control of *Biomphalaria glabrata* in major reservoirs in Puerto Rico. Am J Trop Med Hyg 26:1018–1024

Jobin WR, Laracuente A (1979) Biological control of schistosome transmission in flowing water habitats. Am J Trop Med Hyg 28:916–917

Jourdane J, Kulo SD (1981) Etude expérimentale du cycle biologique de *Echinostoma togoensis* n. sp., parasite à l'état larvaire de *Biomphalaria pfeifferi* au Togo. Ann Parasitol 56:477–488

Jourdane J, Kulo SD (1982) Perspectives d'utilisation de *Echinostoma togoensis* Jourdane and Kulo, 1981 dans le contrôle biologique de la bilharziose intestinale en Afrique. Ann Parasitol 57:443–451

Kristensen TK, Yousif F, Raahauge P (1999) Molecular characterization of *Biomphalaria* spp. in Egypt. J Mollus Stud 65:133–136

Lévèque C, Pointier JP (1976) Study of the growth of *Biomphalaria glabrata* (Say) and other Planorbidae in Guadeloupe (West Indies). Ann Trop Med Parasitol 70:199–204

Lie KJ (1967) Antagonism of *Paryphostomum segregatum* rediae and *Schistosoma mansoni* sporocysts in the snail *Biomphalaria glabrata*. J Parasitol 53:969–976

Lie KJ, Basch PF, Umathevy T (1965) Antagonism between two species of larval trematodes in the same snail. Nature 206:422–423

Lie KJ, Heynemann D (1972) Intramolluscan inter-trematode antagonism: a review of factors influencing the host-parasite system and its possible role in biological control. Adv Parasitol 10:192–268

Lie KJ, Schneider CR, Sornmani S, Lanza GR, Impand P (1974) Biological control by trematode antagonism. II. Failure to control Schistosoma spindale in a field trial in Northeast Thailand. South East Asian J Trop Med Public Health 5:60–64

Lofty WM, DeJong RJ, Abdel Kader A, Loker ES (2005) A molecular survey of *Biomphalaria* in Egypt: is *B. glabrata* present ? Am J Trop Med Hyg 73:131–139

Loker ES, Hofkins BV, Mkoji GM, Mungai B, Kihara J, Koech DK (1993) Distributions of freshwater snals in Southen Kenya with implications for the biological control of schistosomiasis and othersnail-mediated parasites. J Med Appl Malacol 5:1–20

Madsen H (1995) Methods for biological control of schistosome intermediate hosts, an update. In: Proceedings from a status of research on medical malacology in relation to schistosomiasis in Africa, pp 347–376

Madsen H, Frandsen F (1989) The spread of freshwater snails including those of medical and veterinary importance. Acta Trop 46:139–146

Mazé H (1874) Catalogue des coquilles terrestres et fluviatiles recueillies à la Martinique en 1873. J Conchyliol 22:158–173
Mazé H (1890) Supplément au catalogue révisé des mollusques terrestres et fluviatiles de la Guadeloupe et de ses dépendances. J Conchyliol 38:19–34
McCullough FS (1992) The role of mullusciciding in schistosomiasis control. World Health Organization/Schisto/92.107
Meier-Brook C (1974) A snail intermediate host of *Schistosoma mansoni* introduced into Hong Kong. Bull World Health Organ 51:661
Mkoji GM, Hofkins BV, Ouma JH, Kuris AM, Koech DK, Loker ES (1995) Research on biological control of schistosome-transmitting snails in Kenya. In: Proceedings from a status of research on medical malacology in relation to schistosomiasis in Africa, 265–274
Murray HD (1964) *Tarebia granifera* and *Melanoides tuberculata* in Texas. American Malacological Union Incompany, 15–16
Murray HD, Woopschall LJ (1965) Ecology of *Melanoides tuberuclata* (Müller) and *Tarebia granifera* (Lamarck) in South Texas. American Malacological Union Incompany, pp 25–26
Nassi H, Pointier JP, Golvan YJ (1979) Bilan d'un essai de contrôle de *Biomphalaria glabrata* en Guadeloupe à l'aide d'un trématode stérilisant. Ann Parasitol 54:185–192
Nguma JFM, MlcCullough FS, Lasha E (1981) Elimination of *Biomphalaria pfeifferi, Bulinus tropicus* and *Lymnaea natalensis* by the ampullariid snail *Marisa cornuarietis* in a man-made dam in Northern Tanzania. Acta Trop 39:85–90
Oliver-González J, Bauman PM, Benenson AS (1956) Effect of the snail *Marisa cornuarietis* on *Australorbis glabratus* in natural bodies of water in Puerto Rico. Am J Trop Med Hyg 5:290–296
Paraense WL (2001) The schistosome vectors in the Americas. Mem Inst Oswaldo Cruz 96:7–16
Paraense WL, Corrêa LR (1989) A potential vector of *Schistosoma mansoni* in Uruguay. Mem Inst Oswaldo Cruz 84:281–288
Paraense WL, Zeledon R, Rojas G (1981) *Biomphalaria straminea* and other planorbid molluscs in Costa Rica. J Parasitol 67:282–283
Perera G, Yong M, Sanchez R (1987) First record and ecological studies of *Melanoides tuberculata* (Müller). Walkerana 2:165–171
Pflüger W (1982) Introduction of *Biomphalaria glabrata* to Egypt and other African countries. Trans R Soc Trop Med Hyg 90:340–346
Pointier JP (1974) Faune malacologique dulçaquicole de l'île de la Guadeloupe (Antilles françaises). Bull Mus Natl Hist Nat 235:905–933
Pointier JP (1993) The introduction of *Melanoides tuberculata* (Mollusca: Thiaridae) to the island of Saint-Lucia (West Indies) and its role in the decline of *Biomphalaria glabrata*, the snail intermediate host of *Schistosoma mansoni*. Acta Trop 54:13–18
Pointier JP (1996) An Australian planorbid freshwater snail recently introduced to the island of Martinique: *Amerianna carinata* (H. Adams, 1861). Haliotis 25:25–31
Pointier JP (2008) Guide to the frehswater molluscs of the Lesser Antilles. Conchbooks, Hackenheim
Pointier JP, David P (2004) Biological control of *Biomphalaria glabrata*, the intermediate host of schistosomes, by *Marisa cornuarietis* in ponds of Guadeloupe: long-term impact on the local snail fauna and aquatic flora. Biol Control 29:81–89
Pointier JP, Delay B (1995) Spread of the introduced freswater snail *Melanoides tuberculata* (Müller, 1774) on the island of Guadeloupe, French West indies (Prosobranchia, Thiaridae). Haliotis 24:109–116
Pointier JP, Guyard A (1992) Biological control of the snail intermediate host of *Schistosoma mansoni* in Martinique, French West Indies. Trop Med Parasitol 43:98–101
Pointier JP, Jourdane J (2000) Biological control of the snail hosts of schistosomiasis in areas of low transmission: the example of the Caribbean area. Acta Trop 77:53–60
Pointier JP, McCullough F (1989) Biological control of the snail hosts of *Schistosoma mansoni* in the Caribbean area using *Thiara* spp. Acta Trop 46:147–155
Pointier JP, Théron A (2006) Transmission de la bilharziose intestinale aux Antilles Guyane. Bull d'Alerte Surveill Antilles Guyane 1:1–4

Pointier JP, Guyard A, Théron A, Dumoutier A (1984) Le fonctionnement d'un site de transmission de *Schistosoma mansoni* en Martinique (Antilles françaises). Ann Parasitol 59:589–595

Pointier JP, Théron A, Vathaire F (1985) Corrélations entre la présence du mollusque *Biomphalaria glabrata* et quelques paramètres du milieu dans les mares de la Grande Terre de Guadeloupe (Antilles françaises). Malacol Rev 18:37–49

Pointier JP, Guyard A, Mosser A (1989) Biological control of *Biomphalaria glabrata* and *B. straminea* by the competitor snail *Thiara tuberculata* in a transmission site of schistosomiasis in Martinique, French West Indies. Ann Trop Med Parasitol 83:263–269

Pointier JP, Théron A, Imbert-Establet D, Borel G (1991) Eradication of a sylvatic focus of *Schistosoma mansoni* using biological control by competitor snails. Biol Control 1:244–247

Pointier JP, Paraense WL, Mazille V (1993a) Introduction and spreading of *Biomphalaria straminea* (Dunker, 1848) (Mollusca: Pulmonata: Planorbidae) in Guadeloupe, French West indies. Mem Inst Oswaldo Cruz 88:449–455

Pointier JP, Théron A, Borel G (1993b) Ecology of the introduced snail *Melanoides tuberculata* (Gastropoda: Thiaridae) in relation to *Biomphalaria glabrata* in the marshy forest zone of Guadeloupe, French West Indies. J Mollus Stud 59:421–428

Pointier JP, Thaler L, Delay B, Pernot AF (1993c) Invasion of the Martinique island by the parthenogenetic snail *Melanoides tuberculata* and succcession of morphs. Acta Oecol 14:33–42

Pointier JP, Samadi S, Jarne P, Delay B (1998) Introduction and spread of *Thiara granifera* (Lamarck, 1822) in Martinique, French West Indies. Biodiv Conserv 7:1277–1290

Pointier JP, DeJong RJ, Tchuem-Tchuenté LA, Kristensen T, Loker ES (2005a) A neotropical snail hosts of *Schistosoma mansoni* introduced into Africa and consequences for the schistosomiasis transmission: *Biomphalaria tenagophila* in Kinshasa (Democratic Republic of Congo). Acta Trop 93:191–199

Pointier JP, Yong M, Gutiérrez A (2005b) Guide to the freshwater Molluscs of Cuba. Conchbooks, Hackenheim

Pointier JP, Noya O, Alarcón de Noya B, Théron A (2009) Distribution of Lymnaeidae (Mollusca, Pulmonata), intermediate snail hosts of *Fasciola hepatica* in Venezuela. Mem Inst Oswaldo Cruz 104:790–796

Prentice MA (1983) Displacement of *Biomphalaria glabrata* by the snail *Thiara granifera* in field habitats in St Lucia, West Indies. Ann Trop Med Parasitol 77:51–59

Robart G, Mandahl-Barth G, Ripert C (1979) Inventaire, répartition géographique et écologie des mollusques dulçaquicoles d'Haiti (Caraibes). Haliotis 8:159–172

Ruiz-Tibén E, Plamer JR, Ferguson FF (1969) Biological control of *Biomphalaria glabrata* by *Marisa cornuarietis* in irrigation ponds in Puerto Rico. Bull World Health Organ 41:329–333

Samadi S, Mavárez J, Pointier JP, Delay B, Jarne P (1999) Microsatellite and morphological analysis of population structure in the parthenogenetic freshwater snail *Melanoides tuberculata*: insights into the creation of clonal variability. Mol Ecol 8:1141–1153

Schramm, MA (1869) Catalogue des coquilles et des crustacés de la Guadeloupe envoyés à l'exposition universelle de 1867 par l'administration de la colonie. Collection Caillet et Desbonne. Basse-Terre: Imprimerie du gouvernement

Simone LRL (2006) Land and freshwater Molluscs of Brazil. EGB, Fapesp, São Paulo

Starmühlner F (1984) Occurrence, longitudinal distribution and geographical range of the fresh and brackish water molluscs of the Lesser Antillean islands (Guadeloupe, Dominica and Martinica). Soosiana 12:83–102

Stevens M, Waldmann G (2001) Animal biodiversity of the Lesser Antillean island of Montserrat. (British West Indies): an annotated checklist of terrestrial and freshwater animals. Arch Zoologischer Publikationen 6:140

Tang CT (1983) A survey of *Biomphalaria straminea* (Dunker, 1848) (Planorbidae) for trematode infection, with a report on larval flukes from other Gastropoda in Hong Kong. In: Proceedings of the second international workshop in malacology in Honk Kong, pp 393–408

Teles HM, Hiramoto RM, Oliveira JC, Aveiro JA (2003) Occurrence of Schistosomiasis mansoni vectors on the northern coast of the State of São Paulo. Cadratico Saude Publica 19:1887–1891

Théron A, Pointier JP (1995) Ecology, dynamics, genetics and divergence of trematode populations in heterogeneous environments: the model of *Schistosoma mansoni* in the insular focus of Guadeloupe. Res Rev Parasitol 55:49–64
Tiller S, Clarke BC (1983) Lutte biologique et destruction du patrimoine génétique: le cas des mollusques gastéropodes pulmonés dans les territoires français du Pacifique. Genet Sel Evol 15:559–566
Velásquez LE, Bedoya JC, Areiza A, Vélez I (2006) Primer registro de *Centrocestus formosanus* (Digenea: Heterophyidae) en Colombia. Rev Mex Biodivers 77:119–121
World Health Organization (1982) The role of biological agents in integrated vector control and the formulation of protocols for field testing of biological agents. TDR/VEC-SWG (6)/82/.3
Yipp MW (1990) Distribution of the schistosome vector snail, *Biomphalaria straminea* (Pulmonata: Planorbidae) in Hong Kong. J Mollus Stud 56:47–55
Yousif F, Haroun N, Ibrahim A, El-Bardicy S (1996) *Biomphalaria glabrata:* a new threat for schistosomiasis transmission in Egypt. J Egypt Soc Parasitol 26:191–205
Yousif F, Ibrahim A, Abdel-Kader A, El-Bardicy S (1998) Invasion of the Nile Valley in Egypt by a hybrid of *Biomphalaria glabrata* and *Biomphalaria alexandrina*, snail vectors of *Schistosoma mansoni*. J Egypt Soc Parasitol 28:569–582

Index